FUTURE CROPS

—Volume 2—

The Editor

Prof. K.V. Peter, Director World Noni Research Foundation, Chennai is the Former Vice-Chancellor, Kerala Agricultural University (KAU); Director, Indian Institute of Spices Research, Calicut; Director of Research KAU and Professor of Horticulture from 1979. He is an acknowledged teacher and science manager. A post graduate from G.B. Pant University of Agriculture and Technology Pantnagar (1969-1975), he is associated with development of biotic stress resistant varieties in Chilli (Pant-C1, Pant-C2), tomato (Sakthi) and brinjal (Surya) which are grown through out the country. He is recipient of several National Awards like Rafi Ahmad Kidwai Award; Dr. M.H. Marigowda National Award; HSI-Shivasakthi Award for Life Time Contributions to vegetable research; Recognition award and Dr. K. Ramiah Memorial Award of National Academy of Agricultural Sciences, New Delhi; PNASF Gold Medal and Gold medal from Indian Society of Vegetable Science, Varanasi. He is Fellow of NAAS New Delhi, NASI, Allahabad and NABS Chennai. National Book Trust, New Delhi published his authored book "*Plantation Crops*" in three languages-English, Hindi and Malayalam- and "*Tuber Crops*" in English. Woodhead Publishing Company UK published the three volume series titled "*Handbook of Herbs and Spices*" now revised into two volumes (2012). ICAR New Delhi published the book "*Genetics and Breeding of Vegetable Crops*". New India Publishing Agency, New Delhi published a 12 volume series on "*Horticultural Science*"; five volume series on "*Underutilized and Underexploited Horticultural Crops*" and two volumes of "*Science of Horticulture*". Daya Publishing Company (Astral International Pvt. Ltd.) published four volumes of "*Biodiversity of Horticultural Crops*" and two volumes of "*Future Crops*". He co-edited the book "*Handbook of Industrial Crops*" with Prof. V.L. Chopra and published by Taylor and Francis, USA. The monograph on "*Jackfruit*" co-edited by him and published by Studium Press, LLC is informative and widely read. He edited the much referred monograph on Noni (*Morinda citrifolia* L.) authored by Prof. P.I. Peter, The Divine Servant of the People and Chairman, Noni Biotech, Chennai. His immediate family consists of Vimala; two sons Anvar and Ajay and daughter-in-law Cynara. He comes from the famous tourist village Kumbalanghy, Cochin, Kerala and family Kuruppacharil.

FUTURE CROPS

— Volume 2 —

Editor
Professor K.V. Peter

2014
Daya Publishing House®
A Division of
Astral International Pvt. Ltd.
New Delhi – 110 002

ISBN 9789351301226

Published by : **Daya Publishing House**®
A Division of
Astral International Pvt. Ltd.
– ISO 9001:2008 Certified Company –
4760-61/23, Ansari Road, Darya Ganj
New Delhi-110 002
Ph. 011-43549197, 23278134
E-mail: info@astralint.com
Website: www.astralint.com

Laser Typesetting : **Classic Computer Services**, Delhi - 110 035

Printed at : **Thomson Press India Limited**

PRINTED IN INDIA

Devotion

FUTURE CROPS Vol. 2 is devoted to the memory of Dr. N.K. Anant Rao (Dean Rao), Dr. R.K. Arora, Dr. B. Chaudhury, Dr. B.P. Ghildyal, Dr. S. Jayaraj, Dr. S. Kannaiyan, Dr. B.R. Murthy, Dr. R.M. Pandey, Dr. J.S. Pruthi, Dr. P.S. Rao, Dr. G.M. Reddy, Dr. Joseph Thomas, Dr. J.S.P. Yadav, Dr. Sant Ram, Dr. R.L. Paliwal and others who moulded me from the days at Pantnagar and whom I miss a lot. All these men contributed to physical, spiritual and scientific knowledge which made our stay in planet earth worth living. I believe all above are in safe hands of God Almighty.

Acknowledgement

Prof. P.I. Peter Chairman NoniBiotech Chennai renders me unstinct support and physical facilities to make my academic dreams to reality. No words are sufficient to thank him from bottom of my heart. Dr. Kirti Singh, Chairperson World Noni Research Foundation Chennai is always supportive and encouraging. I am grateful to Dr. N. Krishna Kumar, Deputy Director General (Horticulture), KAB-II, ICAR, New Delhi for the FOREWORD and Dr. C. Devakumar, Assistant Director General (Education) ICAR, New Delhi for the PREFACE. I acknowledge the silent support and unfathomable sacrifice of my dear wife Vimala Peter, sons Anvar and Ajay and daughter-in –law Cynara which enabled me to be at WNRF Chennai and compile this volume. I acknowledge all the contributors who supported me with information with patience and good spirits. The Volume 2 reflects their scientific acumen and commitment to science. I also acknowledge Mr Anil Mittal, Astral International Pvt. Ltd., New Delhi for attractive and error free printing.

डा. एन.के. कृष्णा कुमार
उप महानिदेशक (बागवानी)
Dr. N.K. Krishna Kumar
DEPUTY DIRECTOR GENERAL (Horticulture)

भारतीय कृषि अनुसंधान परिषद
कृषि अनुसंधान भवन-II,
पूसा, नई दिल्ली 110 012
INDIAN COUNCIL OF AGRICULTURAL RESEARCH
KRISHI ANUSANDHAN BHAVAN-II,
PUSA, NEW DELHI 110 012

Foreword

Agri-horticulture and Agro-forestry are in critical turning points resulting from pressure on land, water, energy, human labour, urbanization and new life styles. Deforestation is rampant and shifting cultivation has done irreparable damage to eco-systems. Plundering of natural resources especially plants of medicinal values and large scale deforestation are making many plants extinct. Climate change and globalization are constraining food, feed and fuel production. Sustainability of many crops like rice, sugarcane, sugarbeet and forages are being questioned due to high water requirement.

Migration and urbanization for better employment opportunities are creating new life styles and new dietary requirements. Habitat pollution is throwing new challenges in the form of diseases like cancer, diabetes, hypertension, cardiovascular diseases and above all disturbed mind leading to unhealthy body. Customer convenient fast foods, frozen and ready to cook items; bottled water and ready to serve drinks are infusing chemical preservatives to the long term detriment of people. By 2060 world population is expected to touch 9 billion with India surpassing China. Food and Nutrition requirement of 2060 demand doubling the present production and that too from a drastically reduced land-area, water and energy. The crops are to be tailored to meet the emerging requirements. Aquatic vegetables are getting popular in South East Asian countries and the aquatic weeds are projected as future foods.

There are a large number of underexploited and underutilized fruits in North East India especially Mizoram with its congenial climate and virgin lands. Further, there are several underutilized spices demanding lesser natural resources for growth and development. Fruits like cactus, pear and citron are getting attention for suitability to areas with limited rainfall. Medicinal plants like camelina, embelia, salada and saraca are dealt as promising plants against new life style diseases. Lasoda and mappia are arid crops demanding research attention. Water chestnut is getting attention as a economic crop in areas with natural ponds and water reserves. Palmyrah palm is emerging as a source of health food and timber for houses.

FUTURE CROPS Vol. 2 is a continuation of the series edited by Dr. K.V. Peter, Director World Noni Research Foundation, Chennai. Twenty chapters contributed by 50 well known scientists from 18 Research Institutes constitute the informative book and deserve appreciation.

N.K. Krishna Kumar

Preface

Fruits and vegetables are the main source of many essential vitamins and minerals in our diet. They also provide much needed fibre and are a veritable source of anti-toxicants. According to the National Sample Survey 66th Round (July 2009-June 2010), fruits and nuts accounted for only 2.4 per cent of consumer expenditure in rural India and 3.2 per cent in urban India. Banana, apple, mango, grapes, papaya and oranges together accounted for around 84 per cent of the value of fresh fruits consumed in rural areas and 86 per cent of the value in urban areas. In case of each fruit and nut, per capita urban consumption outstripped rural consumption not only in value but also in quantity. Thus, there is enormous scope for fruits in the coming years.

According to Darrow and Yerks (1937), "All of our present cultivated plants, it must be remembered, have been derived from wild plants." Evolution of edible plants from wild relative is a continuous process mediated by environmental, ecological, geo-physical and anthropogenic factors. Some vegetables like tomato were not considered edible till 15th century. Cruciferous vegetables like cabbages, cauliflower, Brussels sprout and sprouting broccoli were not grown in tropical regions of India because of sensitivity to higher temperature and higher rainfall. Now with introgression of thermo-resistant genes from Asiatic cauliflowers into temperate cauliflower, the vegetable is now grown in tropics.

The basket of economic plants is at present small as numerous plants remain either underutilized or underexploited. Maize, wheat, rice, potato, cassava, soybean, sweet potato, sorghum, yam and banana are the ten top crops in decreasing order that feed the world. Fortunately, India is rich in biodiversity with varied agro-ecological zones. India being a sub-continent can grow fruits of temperate, arid and sub-tropical grades. It is in this context, the series *FUTURE CROPS* edited by Dr K V Peter, Director, World Noni Research Foundation, Chennai is relevant and timely. The Vol. 2 of this series contains 20 chapters authored by 50 scientists from 18 research institutes and provides elaborate details on aquatic vegetables, medicinal plants, oil seeds, fruits and energy plants of future.

The editor of this volume is a well-known author of several excellent treatises such as *Underutilized and Underexploited Horticultural Crops, Genetics and Breeding* of *Vegetables, Handbook* of *Herbs and Spices,*

Biodiversity in Horticultural Crops, Science of *Horticulture* etc. I congratulate Prof. K V Peter for this excellent publication which, I am sure, will immensely benefit teachers, students, researchers and all engaged in enriching the fruit basket of India. I commend M/s Astral International Pvt. Ltd. New Delhi for this publication.

C. Devakumar

Assistant Director General Education
Planning and Development,
ICAR, New Delhi

Contents

List of Contributors

Agarwal, Ankur
Defence Institute of Bio-Energy Research (DIBER), DRDO, Ministry of Defence, Field Station, Pithoragarh, Uttarakhand – 262 501
E-mail: darl_ankur@rediffmail.com

Ahmed, Zakwan
Defence Institute of Bio-Energy Research (DIBER), DRDO, Ministry of Defence, Field Station, Pithorgarh, Uttarakhand – 262 501

Anuradha, T. Sane
Indian Institute of Horticultural Research, Bangalore – 56 0089

Arya, M.C.
Defence Institute of Bio-Energy Research (DIBER), DRDO, Ministry of Defence, Field Station, Pithorgarh, Uttarakhand – 262 501

Bandopadhyay, Rajib
Department of Biotechnology, Birla Institute of Technology, Mesra, Ranchi – 835 215, Jharkhand
E-mail: rajib_bandopadhyay@bitmesra.ac.in

Bhattacharjee, P.
Department of Food Technology and Biochemical Engineering, Jadavpur University, Kolkata – 700 032

Chaturvedi, A.K.
Indian Institute of Vegetable Research, Varanasi

Chaudhary, Bratati
Department of Biotechnology, Birla Institute of Technology, Mesra, Ranchi – 835 215
E-mail: bratatichaudhary@yahoo.co.in

Chaudhary, R.K.
Department of Botany, University of Calcutta, Kolkata – 700 019

Chauhan, Nitesh
Central Institute for Arid Horticulture, Beechwal, Bikaner – 334 006, Rajasthan

Choudhary, M.K.
Krishi Vigyan Kendra, Central Arid Zone Research Institute (CAZRI), Pali-Marwar – 306 401, Rajasthan

Datta, S.
Department of Spices and Plantation Crops, B.C.K.V.V., Mohanpur– 741 252, Nadia, West Bengal

Dayal, Hari
Krishi Vigyan Kendra, Central Arid Zone Research Institute (CAZRI), Pali-Marwar – 306 401, Rajasthan

Dhatt, K.K.
Department of Floriculture and Landscaping, Punjab Agricultural University, Ludhiana – 141 004, Punjab
E-mail: dhattkiran@gmail.com

George, Jiji
Research Scientist-II, Manipal Centre for Virus Research, (Regional Reference Laboratory for Influenza Viruses, and ICMR Virology Network Laboratory-Grade-I) Manipal University, Manipal – 576 104, Karnataka

Ghosh, P.K.
Department of Food Technology and Biochemical Engineering, Jadavpur University, Kolkata – 700 032

Hazarika, T.K.
Department of Horticulture, Aromatic and Medicinal Plants, School of Earth Sciences and Natural Resources Management, Mizoram University, Aizawl, Mizoram
E-mail: tridip28@gmail.com

Jain, Renuka
Avesthagen Pvt. Ltd., Bangalore

Jayasankar, P.
Central Institute of Freshwater Aquaculture (ICAR), Kausalyaganga, Bhubaneswar – 751 002, Orissa
E-mail: pjayasankar@yahoo.com

Joglekar, A.
Department of Biodiversity, Abasaheb Garware College, Deccan Gymkhana, Karve Road, Pune – 411 004, Maharashtra
Research and Action in Natural Wealth Administration (RANWA), 16, Swastishree Society, Ganesh Nagar, Kothrud, Pune – 411 052, Maharashtra

Karun, Anitha
Principal Scientist, Biotechnology Section, Central Plantation Crops Research Institute, Kasaragod – 671 124, Kerala
E-mail: anithakarun2008@gmail.com

Krishna, Hare
Central Institute for Arid Horticulture, Beechwal, Bikaner – 334 006, Rajasthan

Kumar, S. Shashi
Chromus Seeds, Bangalore

Kumar, Vasantha
Indian Institute of Horticultural Research, Bangalore – 560 089
E-mail: thuppil@iihr.ernet.in

Kumari, Anuja
Defence Institute of Bio-Energy Research (DIBER), DRDO, Ministry of Defence, Field Station, Pithorgarh, Uttarakhand – 262 501

Lavanya, C.
Directorate of Oilseeds Research Rajendranagar, Hyderabad – 500 030

Maheshwari, S.K.
Central Institute for Arid Horticulture, Beechwal, Bikaner – 334 006, Rajasthan

Mandal, R.N.
Central Institute of Freshwater Aquaculture (ICAR), Kausalyaganga, Bhubaneswar – 751 002, Orissa

Meena, M.L.
Krishi Vigyan Kendra, Central Arid Zone Research Institute (CAZRI), Pali-Marwar – 306 401, Rajasthan

Mhaskar, M.
Department of Biodiversity, Abasaheb Garware College, Deccan Gymkhana, Karve Road, Pune – 411 004, Maharashtra
Research and Action in Natural Wealth Administration (RANWA), 16, Swastishree Society, Ganesh Nagar, Kothrud, Pune – 411 052, Maharashtra

Mukhopadhyay, Kunal
Department of Biotechnology, Birla Institute of Technology, Mesra, Ranchi – 835 215

Pandey, A.K.
Indian Institute of Vegetable Research, Varanasi
E-mail: pandey.ajai1@gmail.com

Parashar, A.
Central Institute for Arid Horticulture, Beechwal, Bikaner – 334 006, Rajasthan

Patwardhan, A.
Department of Biodiversity, Abasaheb Garware College, Deccan Gymkhana, Karve Road, Pune – 411 004, Maharashtra
Research and Action in Natural Wealth Administration (RANWA), 16, Swastishree Society, Ganesh Nagar, Kothrud, Pune – 411 052, Maharashtra
E-mail: ankurpatwardhan@gmail.com

Rai, Mathura
Indian Institute of Vegetable Research, Varanasi, U.P.

Rajasekharan, P.E.
Indian Institute of Horticultural Research, Bangalore – 56 0089

Ram, Lallan
Processing Laboratory, National Research Centre for Citrus, Amravati Road, Nagpur – 440 010, Maharashtra
E-mail: lallanram09@gmail.com

Rao, V.K.
Indian Institute of Horticultural Research, Bangalore – 56 0089

Sane, Anuradha
Indian Institute of Horticultural Research, Bangalore – 560 089

Sharangi, A.B.
Department of Spices and Plantation Crops, B.C.K.V.V., Mohanpur– 741 252, Nadia, West Bengal
E-mail: dr_absharangi@yahoo.co.in

Sharma, S.K.
Central Institute for Arid Horticulture, Beechwal, Bikaner – 334 006, Rajasthan
E-mail: ciah@nic.in

Sharma, Sunil Kumar
IARI Regional Station, Agricultural College Estate, Shivajinagar, Pune – 411 005
E-mail: sunilksharma@yahoo.co.in

Singh, B.
Indian Institute of Vegetable Research, Varanasi, U.P.

Singh, Dheeraj
Krishi Vigyan Kendra, Central Arid Zone Research Institute (CAZRI), Pali-Marwar – 306 401, Rajasthan
E-mail: dheerajthakurala@yahoo.com

Singh, P.M.
Indian Institute of Vegetable Research, Varanasi, U.P.

Singhal, R.S.
Institute of Chemical Technology, Nathalal Parekh Marg, Matunga, Mumbai – 400 019
E-mail: rsinghal7@ediffmail.com

Sinha, Swati
Department of Biotechnology, Birla Institute of Technology, Mesra, Ranchi – 835 215, Jharkhand

Tadwalkar, M.
Department of Biodiversity, Abasaheb Garware College, Deccan Gymkhana, Karve Road, Pune – 411 004, Maharashtra
Research and Action in Natural Wealth Administration (RANWA), 16, Swastishree Society, Ganesh Nagar, Kothrud, Pune – 411 052, Maharashtra

Vasudeva, R.
Department of Forest Biology and Tree Improvement, College of Forestry, University of Agricultural Sciences, Sirsi Campus – 581 401, Dharwad, Karnataka

Wagh, R.
Department of Biodiversity, Abasaheb Garware College, Deccan Gymkhana, Karve Road, Pune – 411 004, Maharashtra
Research and Action in Natural Wealth Administration (RANWA), 16, Swastishree Society, Ganesh Nagar, Kothrud, Pune – 411 052, Maharashtra

Introduction

In an editorial in the journal Science titled 'Combating Hunger', Prof. M S Swaminathan highlights prioritization of an innovative science-based marriage of nutrition and agriculture."At the close of the recent London Olympics, The UK announced a programme designed to rescue 25 million children worldwide from malnutrition by the time of the next Olympics in Brazil in 2016. The problem of hunger is not simply a lack of sufficient quantities of food. The chronic hunger caused by protein and calorie undernutrition (the hidden hunger caused by the deficiency of micronutrients, which include iron, iodine, zinc, Vit. A and Vit. B12) and sometimes by human diseases which disable the body's ability to absorb what micronutrients it receives"-Prof. Swaminathan. Prof. Carlos Perez del Castillo of CGIAR in his Millemmium Lecture delivered on November 21st 2012 elaborated challenges of rural poverty and world food security. Green revolution of 1970's and 1980's happened in irrigated areas and the rainfed areas need technologies and new crops to increase food (rice, wheat, millet and pulses) to feed 9.0 billion people by 2050. Water scarcity affects more than 1.2 billion people calling for water saving crops and technologies. Dryland systems, humid tropics and aquatic agriculture systems are the less tapped for food and nutrition security. CGIAR has three centres-Bioversity, IWMI and Worldfish to deal with aquatic agriculture systems. There are underutilized potential in root, tuber, fruit, vegetable, spice and plantation crops to improve food security, nutrition and livelihoods.

Aquatic vegetables are possible future crops for food, nutrition, energy and raw materials for industry. Nurturing them needs protection, propagation and proliferation. Velvet leaf, Indian blue water lily, water lettuce, marsilea, pygmy water lily, Manchurian rice, common water plantain, water mimosa, parrot's feather, duck lettuce, aponogeton, keysoor, achyranthe, hydrolea, water pepper, water fern, giant chickweed, eel grass, water shield, frog bit, para cress, golden club, papyrus, Chinese arrowhead, cattail, Indian red water lily, fragrant water lily, swamp taro, sonneratia, water spinach, lotus, water cheshnut, buffalo spinach, gorgon nut, water dropwort, monochoria, water cress, Chinese water cheat nut and water sprite are a few aquatic plants edible and nutritious. Water spinach is now a common leaf vegetable. Rhizomes of lotus are edible and dehydrated rhizomes are raw materials for delicious chips and fries. Water hyacinths were used as human food in the Philippines in war time. Water hyacinth protein shows deficient levels in only two of the essential amino acids-valine and methionine-. Medicinal uses of aquatic plants like water spinach, cattail, eel grass, duck lettuce, common water plantain, water shield, papyrus, dwarf waterclover, wasambi, fragrant water lily, sonneratia, Chinese arrowhead, paracress, Indian blue lily, keysoor, hydrolea and water pepper are reported. Aquatic plants are life line to wet land ecosystem, one of the most threatened habitats of the world due to human encroachments, dumping of wastes and above all pollutants from industries. Role of aquatic plants in phyto-remediation has been well understood. Future threats to aquatic vegetables will come from contamination, accumulation of nitrate and oxalates, heavy metals and pathogenic organisms. Depletion of oxygen in wet land water is another issue demanding attention.

Aquatic weeds as potential future foods are debated intensively. Fresh, brackish and saline water bodies host different plants as weeds as of now. Food, feed, fodder and vegetables are four groups of aquatic plants based on usage. Makhana, water cheshnut, lotus, water lily and colocasia are used as food. Wild macrophytes, azolla, duckweed, water spinach and water cheshnut for feed and fodder and wild plants for vegetables are other notable aquatic weeds now."If one fifth area of the total fallow wetlands are utilized for the purpose of harnessing potential of nutrient substances of selective aquatic plants, it can surely bring about a remarkable change in the scenario of future crops"-reports Dr Mandal and Dr Jayasankar(2013).

Underutilized fruit crops are getting evolved to meet the requirements of nutrients, energy and diet of a developing and growing populace. Bael, seethaphal, jackfruit, bamboo fruit, chironji, karonda, lasoda, timru, gab/tumbi, aonla, loquot, baincha, banyan treefigs, papaltree, gullar, anjeer, kaith, phalsa, mahua, rayan/khirni, molsiri, mulberry, senjana, neemphal, cape gooseberry, dates/khajur, pomegranate, jamun, tamarind and ber are a few underexploited fruits of Indian arid zone. Many are sources of food, nutrients, vitamins, antioxidants, pectins, celluloses, lipids and pigments. Many value added products are made from above underexploited fruits. Tamarind pulp, toffees, juice, juice concentrate, pulp powder, pickle, jam, syrup, candy and seed based products are available. The American pharmaceutical industry processes 100 tones of tamarind pulp annually and it is a common ingredient in cardiac and blood sugar reducing medicines. Mizoram, one among the seven states of N. E. India is an abode of several minor fruits of considerable value. Many have outstanding adaptation capabilities to drought tolerance, salt tolerance and are resistant to major pests and diseases.

As per document 676 of International Standards Organization(ISO) there are 109 spices out of which 52 are grown commonly in India, the major ones being black pepper, ginger, cardamom, onion, garlic, cinnamon, clove, nutmeg, allspice, coriander, fennel, fenugreek, cumin and celery. Spices impart colour, flavor and fragrance to food preparations and have several medicinal uses for health and wellness. Many of the above spices face stresses from biotic and abiotic agents. The future spices are the present herbs and minor spices like ajmund, ajowain, aniseed, asafetida, black cumin, curry leaf, horse radish, Japanese mint, kokam, leek, mango ginger, opium poppy, paprika, parsley, rosemary, sage, sweet flag, sweetmarjoram, tamarind, tarragon, tejpat and thyme. Each of above plants has its own flavor and fragrant characteristics.

With desertification looming large and water resources dwindling, attention is focused on cuticle thick and lenticels free plants. Opentia and cactus are water saving crops of interest. Cacti are succulent plants which live in dry(xeric) environment and survive long period of drought. Low water requirements and a high water use efficiency ratio make cacti suitable for arid regions. Cactus pear or Prickly pear is an edible fruit from the plant *Opentia ficus-indica.* At present Mexico grows 79.6 per cent of the world production. In view of the changes in global climate, cactus pear is suggested as a promising crop by the FAO. It can act as a boon for the underdeveloped countries providing nutrition at an affordable costs and is projected as future fruit.

Oil seed crops like soybean, ground nut, canola-mustard, sunflower, sesame, safflower, niger, oil palm and coconut are under threat of several diseases and pests. Water requirements of above crops are also higher. Aftermath of climate change tells up on oil yield and quality of above crops. GM Soybean and GM canola are facing opposition in Europe. Castor bean is emerging as a future oil seed crop because of its adaptability, lesser water requirement and resistance to drought and salinity. Use of castor oil is known since 2000 B. C. in Ayurvedic and Unani medicines. Presence of ricin, a toxic phytochemical, is the limiting factor for use of castor oil as edible oil. Attempts are made to locate ricin free castor varieties.

Camelina belonging to cabbage family is another oil yielding annual crop with potential as an industrial oil crop. Linolenic fatty acid in camelina places it exclusively in industrial use only now. Once the unsaturated fatty acid content is brought down to levels prescribed for edible oil, there is scope for this crop to be a future crop. At present the crop is grown in Europe and was an important crop during Iron and Bronze ages. Camelina is also known as German sesame, Siberian oilseed, False flax, Dutch flax, Linseed dodder and Gold of pleasure. Camelina oil being rich in omega-3 fatty acids also offers opportunity to meet the growing demands for high quality fatty acids after removal of non-edible fractions.

Lime, lemon and oranges are sub-tropical fruits of considerable value both health and processing industry wise. Being the richest source of ascorbic acid(Vit. C), lime and lemon are taken fresh and processed forms. Rapid decline due to viral and nematode infected soils is reducing area under these crops. Among processed fruit juices, oranges occupy top most position. The Nagpur based National Research Centre for Citrus conducts research on citrus and despite all out efforts oranges are getting affected by several pests and diseases. Citron(*Citrus medica* L.) is emerging as a future citrus plant with all the qualities of oranges. In addition, several medicinal uses are attributed to the fruits. Citron seeds are also ingredients of medicines administered in S E Asia as vermifuge.

Ornamental plants decorate men and women, habitats, ceremonies and add natural bliss to living environment. Gardens-Mughal, Japanese, Chinese, European, etc-have characteristics of their own. Flowers and foliages are used in religious functions-birth, marriage, departure-, other functions-house swarming, opening of offices and factories, inaugural occasion-. Possession of a home garden is a dream of many in urban areas. There are flowering seasonals, annuals, biannuals and perennials and like wise foliage plants. There are ornamental trees providing shade and housing to birds and squirrels. Features of a garden like hedges, edges, arches and pargolas, lawns, pandals, water spots for lotus and birds like ducks and peacons are made and decorated by plants of varied colours, textures and heights. Landscape planning and architecture are unique arts and science marrying knowledge on plants with engineering skills. Floriculture industry deals with extraction of pigments, flavourants and volatiles. Novelty, uniqueness and rarety are characteristics of a vibrating floriculture industry. New forms of orchids, cacti, ferns, succulents, palms and foliage plants are coming out using technologies of molecular biology and biotechnology. Many weeds of yesterdays are now much demanded ornamentals. Salvenia, Eupatorium and Aquatic lilies are now evolved into much demanded plants. Delphiniums are emerging as much demanded ornamentals. Tropicalisation of delphinium is in progress.

Medicinal and aromatic plants are receiving attention for their pro-nature and non-residual properties. Every plant in the planet earth has one or other medicinal properties. Traditional wisdom, tribal practices and writings in vedic books, Caraka Samhita, several flora –Flora of Madras etc- by Dutch and British army men and missionaries, Hortus Malabaricus by Van Rheedes and after independence by the Botanical Survey of India, Wealth of India by CSIR, publications of Central

Institute of Medicinal and Aromatic Plants, Lucknow;Directorate of Medicinal and Aromatic Plants, Boriavi Anand;All India Co-Ordinated Research Project on Medicinal and Aromatic Plants under ICAR;Tropical Garden and Research Institute, Palode Trivandrum; Departments of Medicinal and Aromatic Plants in several State Agricultural Universities conduct research on medicinal plants. Ayurveda Universities maintain live collections of these plants. There are threats to these plants because of neglect in regeneration and loss of habitats. Red Data Book carries list of plants under threat of extinction. *Embelia ribes* is one of such plants facing extinction and is listed"endemic". It is an Indo-Malaysian species, commonly seen in places upto a height of 1666 m amsl. Roots, fruits and leaves are used. The plant is used in about 75 formulations of Ayurveda. Pharmacological activities include diabetic dyslipidemia, antioxidant activity, wound healing activity and anticancer activity. *Embelia ribes* being an endemic species, attempts are made to standardize propagation and cultivation techniques. Propagated mainly through seeds, it can be multiplied through tissue culture also. Leaf explants are taken for direct shoot organogenesis. Rooted plants are established in soil with 70 per cent success.

Ber(Jujube) is an emerging fruit tree from a state of oblivion due to its suitability for scanty rain fall areas. There are several varieties ranging from bush to shrub to tree plant types. Two distinct species-Indian jujube(*Ziziphus mauritiana*) and Chinese jujube(*Z. jujube*)-are available. Other minor cultivated species are *Z. spina-christi* and *Z. lotus*. Ber provides nutritious fruit, honey, forage and environmental protection. Jujube is popular as fresh crispy fruits and processed products-dehydrated seedless fruits, dehydrated bits, jujube in brine, jujube sweet and sour pickles and jujube mixed fruit preserves-. Being tap and deep rooted the shrub/tree withstands drought. Grafts of jujube on wild *Zizyphus sp* are now available which start fruiting moderately 3 years after planting. Ber is projected as a fruit of future. Jamun (*Zyzygium cumini*) is another tropical and subtropical fruit tree providing needed shade along highways and yielding fruits of medicinal value especially against diabetes and obeisity. Matured fruits are highly perishable and need to be preserved into products like juice concentrates, preserves, jams, jellies, vinegar and puddings and ready to serve drinks. Jamun is a rich source of anthocyanin imparting pink, red, magenta, purple and dark blue colours. Anthocyanins impart protective effects to human and animal health by antagonistic activities to a few bacteria, virus and fungi. It protects food from microbial spoilage. The Hindus regard jamun as "fruit of Gods" as Lord Ram survived on the fruits in the forest during his exile from Ayodhya. Edible pulp is rich in Vitamins C, A and B complex(riboflavin, choline, folic acid, malaic acid) and several amino acids. Seed contains the glycoside jamboline with hypoglycaemic properties. Jamun has become a much wanted fruit.

There is desertification in progress due to manmade and extraneous reasons. There are limited number of economic crops which can survive the rigours of deserts. One such fruit tree is lasoda(*Cordia myxa*) or cherry of the desert. It is underutilized and underexploited. A multipurpose tree spread over arid and semi-arid regions, folk names include lehsua, goonda/goondi, lasura, assyrian plum, sebestem, Indian cherry and naruvilli. All parts of lasoda have uses-medicinal, nutritional, feed and fodder, wood to make toys, ornamental furniture, utensils and handling sour recipes-. Through crop improvement studies, a few plus trees are identified at Central Horticulture Experiment Station, Bhubaneshwar; NBPGR, Jodhpur and CIAH, Bikaner.

Mappia (*Nothapodytes foetida*) is a rare and endemic medicinal plant in the Western Ghats. It has use in treating cancer, HIV and malaria-camptothecin, an alkaloid to treat cancer and HIV-I is extracted. The current demand for woodchips of mappia is 500-700 tones, the supply being 50 per cent of demand putting pressure on collection from forests. Antifungal and anti bacterial activities of organic solvent extracts of mappia wood chips are established. Indiscriminate cutting and rapid deforestation cause severe shortage of wild and cultivated mappia. Attention is focused on rapid multiplication of mappia through cuttages, tissue culture and other micropropagation methods. Developing a cultivation strategy combined with optimizing the growing conditions to increase the biological yield and campothecin would result in consistant supply of high quality wood chips to industries.

Palmyrah palm is projected as a future palm yielding leaves for shelter, truck for carpentry, nuts for culinary drink, kernel for value added products and root masses for soil conservation and the whole palm as wind break. Being grown in rain shadow and dry areas, palmyrah has advantages of growing in drought prone areas. It requires limited quantity of irrigation water, soil depth and organic matter. It is a multipurpose tree of great utility. Sweet sap(neera) is a primary product from the tapped inflorescence from which palm jaggery, palm honey, palm candy and toddy are made. Palmyrah fruit pulp from endosperm(nungu) is relished for nutritive value, flavor and taste. Attention on palmyrah is getting lesser due to non availability of climbers and conversion of lands for other purposes.

Salacia oblonga, Saraca asoca and *Trapa bispinosa* are future crops of economic value, the first two for medicinal and ornamental purposes and the third(water cheshnut) for vegetable purpose. *Salacia oblonga* has uses in cardiovascular diseases and diabetes. Ashoka *(Saraca asoca)* is a tree of cultural and religious significance. Ashoka stem bark contains tannin, catechol, haematoxylin, a saponin and organic iron compounds. Two formulations based on the bark are Asokarista and Asokaghrita. Antimicrobial properties of *Saraca asoca* are established. Water cheshnut(Singhara) is an aquatic vegetable grown for its starch rich fruits. Singhara is now a poor mans fruit vegetable getting attention for its value as an economic water loving herb.

FUTURE CROPS Vol. 2 deal with fruits, vegetables, spices, ornamentals, medicinal and aromatic plants, oil seeds and crops for timber, fuel and fodder.

K.V. Peter

2014, Future Crops Vol. 2 *Pages 1–15*
Editor: Professor K.V. Peter
Published by: DAYA PUBLISHING HOUSE, NEW DELHI

Chapter 1
Aquatic Vegetables

A.K. Pandey, P.M. Singh, B. Singh,*
A.K. Chaturvedi and Mathura Rai
Indian Institute of Vegetable Research, Varanasi
**E-mail: pandey.ajai1@gmail.com*

Aquatic vegetables are predominantly in vogue where they are naturally available in plenty. Areas rich in water bodies like lakes, lagoons, ponds, ditches, marshy wet places are natural abode of most of the aquatic vegetables. South-East Asia is considered one of the suitable niches and most of the countries lying in the boundaries of SE Asia have got unique gift of nature of unprecedented high rain fall and unique topography to harvest and preserve the nature's drop, resulting in round the year availability of green lustrous aquatic edible greens, carbohydrate rich rhizomes and nutritionally packed flowers and fruits suitable for various vegetable uses. Communities dominated in the wetland areas of the globe do not only get their requirements of vegetables completed, it is an indispensable part of their life. Gorgon fruit of eastern India, lotus and water chest nut are some of the few examples which have been in use at various religious occasions since time immemorial. In Thailand, nutritional contribution of naturally occurring vegetables (NOV) were evaluated within the overall diet of 101 and 110 women in the Mekong Delta (flood period and rainy season, respectively) and of 103 women in the Central Highlands (rainy season). The concentrations of Ca, Fe and Zn were analysed in 28 species, and presence of tannins and phytic acid in 19 species. Provitamin carotenoids were determined in 11 species. Over 90 per cent of the women consumed NOV, and they contributed 43, 72 and 75 per cent (Central Highlands, Mekong rainy and Mekong flood period, respectively) of the total quantities of vegetables eaten. Aquatic species such as *Ipomoea aquatica* and *Limnocharis flava*, and terrestrial vegetables like *Schismattoglottis calyptrata*, were among the most commonly consumed vegetables. They were eaten in mean daily quantities exceeding 50 g and had high concentrations of Ca, Fe and β-carotene. Other major aquatic vegetables included the stems of *Eleocharis* sp., *Nelumbo nucifera* and *Nymphea lotus*, all of which were used as salad vegetables and had relatively low nutrient density (Ogle *et al.*, 2001).

Aquatic Vegetables: A Major Component of Aquaculture

Aquaculture can be broadly defined as the farming or culture of aquatic organisms, including fish, molluscs, crustaceans and aquatic plants where according to FAO (1995) 'farming implies some form of intervention in the rearing process to enhance production. Aquatic vegetable production in semi-intensive and intensive systems is widespread and commercially significant around many cities in South -East Asia. According to Phuong and Tuan (2005) in Hanoi, water spinach (*Ipomoea aquatica*) is produced throughout the year, whilst water mimosa (*Neptunia oleracea*) is cultivated only in the summer (April to August) and water dropwort (*Oenanthe stolonifera*) and water cress (*Rorippa nasturtium-aquaticum*) are produced in the winter (September to March). Most production occur in flooded fields, some of which were converted from rice production to generate a higher income; water spinach is also cultivated floating on canals within the city. Water mimosa and water spinach production are reported from peri-urban provinces around Bangkok (Yoonpundh *et al.*, 2005). Around Ho Chi Minh City, Vietnam, many farmers in Binh Chanh District have combined water mimosa cultivation with fish production in separate ponds; mimosa providing a daily income whilst the fish consume the duckweed that grows alongside the mimosa (Hung and Huy, 2005). A list aquatic vegetables and their common name, scientific name, family and edible parts are given in Table 1.1.

Table 1.1: Aquatic vegetables, their scientific name, family and edible parts.

Common Name	*Scientific Name*	*Family*	*Edible Parts*
Velvet leaf	*Limnocharis flava* (L.) Buch	Limnocharitaceae	Leaf vegetable
Indian blue water Lily	*Nymphaea stellata* Willd.	Nymphaeaceae	Rhizome
Water lettuce	*Pistia stratiotes* L.	Araceae	Leaf vegetable
Marsilea	*Marsilea minuta*	Marsileaceae	Pot herb
Pygmy water lily	*Nymphaea tetragona* Georgi	Nymphaeaceae	leaf buds and seeds
Manchurian Rice	*Zizania latifolia*	Poaceae	*Infected culms*
Common Water plantain	*Alisma plantago-aquatica* L.	Alismataceae	Roots, leaves and petioles
Water mimosa	*Neptunia oleracea* Loureiro	Fabaceae	Young shoots
Parrot's Feather	*Myriophyllum aquaticum* (J. M. da ConceicaoVellozo) Verdc.	Haloragaceae	Edible leaves
Duck lettuce	*Ottelia alismoides*	Hydrocharitaceae	Petioles and tender leaves
Aponogeton	*Aponogeton crispum*	Aponogetonaceae	Starchy tuberous rhizome
Keysoor	*Scirpus grossus* Linn. f. Tikiu	Cyperaceae	Starchy root
Achyranthe	*Achyranthes aquatica*	Amaranthaceae	Tender leaves and twigs
Hydrolea	*Hydrolea zeylanica* (L.) Vahl	Hydroleaceae	Young shoots
Water pepper	*Polygonum hydropiper* L.	Polygonaceae	Leaf vegetable
Water fern	*Dentella repens* J. R. Forst. and G. Forst.	Rubiaceae	Leaf
Giant chickweed	*Stellaria aquatica* L. Scop.	Caryophyllaceae	Young leaves and stem
Eel grass	*Vallisneria spiralis* L.	Hydrocharitaceae	Leaf vegetable
Water shield	*Brasenia schreberi* J. F. Gmel	Cabombaceae	Leaf vegetable
Frog bit	*Hydrocharis dubia* (Blume) Backer	Hydrocharitaceae	Mucilaginous leaves and inflorescence

Contd...

Table 1.1–Contd...

Common Name	*Scientific Name*	*Family*	*Edible Parts*
Para cress	*Spilanthes acmella* AH Moore	Asteraceae	Leaf vegetable, salad
Goldenclub	*Orontium aquaticum* L.	Araceae	Edible root
Papyrus	*Cyperus papyrus* L.	Cyperaceae	Pith of young shoots
Chinese Arrowhead	*Sagitaria trifolia*	Alismataceae	Tuberous root
Cattail	*Typha angustifolia* L.	Typhaceae	Rhizome. petiole
Indian Red Water Lily	*Nymphaea nouchali*	Nymphaeaceae	Rhizome
Fragrant Water Lily	*Nymphaea odorata*	Nymphaeaceae	Rhizome
Swamp taro	*Cyrtosperma hamissonis* (Schott) Merr.	Araceae	Rhizome
Sonneratia	*Sonneratia caseolaris* (L.) Engler	Sonneratiaceae	Immature fruit
Water spinach	*Ipomoea aquatica* Forsskal	Convolvulaceae	Leaf vegetable
Lotus	*Nelumbo nucifera*	Nymphaeaceae	Rhizomes, young leaves and lower buds
Water chestnut	*Trapa bispinosa*	Trapaceae	Kernel
Buffalo spinach	*Enydra fluctuans* Loureiro	Asteraceae	Leaf vegetable
Gorgon nut	*Euryale ferox*	Nymphaeaceae	Nut
Water dropwort	*Oenanthe stolonifera* DC.	Apiaceae	Pot herb
Monochoria	*Monochoria hastata*	Pontederiaceae	Leaf vegetable
Water cress	*Rorippa nasturtium-aquaticum*	Cruciferae	Leaf vegetable
Chinese water chest nut	*Eleocharis* sp.	Cyperaceae	Rhizome
Water sprite	(*Ceratopteris thalictroides* (L.) Brongn	Parkeriaceae	Edible fern

Nutritional Importance of Aquatic Vegetables

A number of aquatic vegetables *viz.* water chestnut (*Trapa* spp), lotus (*Nelumbo nucifera*) and watercress (*Nasturtium officinale*) are common. *Ipomoea aquatica* is grown as a green vegetable. Earlier mentions indicate that the Egyptians were fond of to consume rhizomes of *N. lotus* which when cooked resembled egg-yolks. The culms of *Cyperus papyrus* were also chewed raw, boiled or roasted. Oriental *Nelumbo nucifera* has been widely cultivated and its fruit and rhizomes used in a variety of cooked and fresh dishes for many centuries, as also have the rhizomes of *Sagittaria trifolia*, the floury seeds of the lily *Euryale ferox*, and the starch- and fat-laden horned fruits of *Trapa* spp. The fruits of *T. bicornis*, *T. natans* and *T. incisa* form a staple food in continental Asia, Malaysia and India. Canadian wild rice (*Zizania aquatica*) is eaten by North American Indians for over 300 years and is now grown commercially in the U. S. A. The swollen tubers of *Aponogeton* species and *Cyperus esculentus*, and the rhizomes of *Typha latifolia* and *T. angustifolia* and the rhizomes of several species of *Sagittaria*, and the rhizomes and seeds of *Nelumbo lutea* and Orontium aquaticum have been an important carbohydrate source for the North American Indians. The dried and powdered rhizomes of *Butomus umbellatus*, *Calla palustris* and *Menyanthes trifoliate* have been extensively used to make bread by the Eskimoes and other people of northern Eurasia. In India young leaves, stems and roots of *Ipomoea aquatica* are eaten, and both there and in Malaysia, the foliage of *Ottelia alismoides*, and all the plant (except roots) of *Monochoria hastata* and *M. vaginalis* are eaten. In Java juvenile plants of *Limnocharis flava* are a common vegetable. Several species of *Ceratopteris* are cultivated as a green salad crop in parts of Africa and tropical Asia. In temperate regions watercress, (*Rorippa* spp.) is cultivated.

What is needed is knowledge as to how many other aquatic vegetables, especially those which grow in profusion and thus are described as weeds, may be usefully converted to human food, because this is obviously more efficient and rewarding than having first to feed the plant material to an animal for conversion into edible products. Alligator weed (*Alternanthera philoxeroides*) has so much mineral content that it should be introduced for human food, as salad or garnish. It tastes better than some of the common salad greens and is much improved with salt and salad dressing. A major scope exists for protein extraction from aquatic plants for human food. Analyzing the amino acid content of some aquatic plants, Boyd (1968) reported that the samples were somewhat low in lysine and methionine when compared to meat proteins. Phenylalanine was present in very small quantities in *Alternanthera philoxeroides* and *Sagittaria latifolia*. Otherwise amino acid levels were similar to those for meat proteins and crop plant leaf protein isolates; so leaf protein from aquatic plants was of sufficiently high amino acid quality to be useful as a dietary supplement. Boyd (1968) emphasized that for natural vegetation to be useful as food; large stands containing sufficient quantities to justify harvest must be present. Stands should also be mono -specific so that the quality of the product is somewhat predictable. Many underutilized aquatic vegetables fulfill these conditions. Utilization of aquatic vegetables as food could probably alleviate protein shortages in local populaces of many developing nations. Concerted efforts are needed to assess the food value of the native aquatic flora for their exploitation at commercial scale.

Water hyacinths have been used as human food in the Philippines in war time. The soft white bud of the plant was eaten either raw, or as a salad, or as an ingredient in vegetable dishes. It was called 'repollo' (cauliflower). Under conditions of relative food abundance it is considered unlikely that it would still be used as food. However, protein extracted from hyacinth could be used as fodder for animals or even for human food. Leaf protein judged by its amino acid content appears to be nutritionally better than the seed proteins and as good as animal proteins other than the exceptional proteins of egg and milk. Amino acids of water hyacinth could serve as a supplemental source to improve the nutritional quality of grain proteins. Lysine and tryptophan are the first two limiting amino acids in grains, therefore interest centres on the concentration of these two amino acids in water hyacinth. Since any tryptophan present would be destroyed by the acid hydrolysis, only lysine can be considered at present. Comparison of amino acids obtained from water hyacinth and other sources are given in the Table 1.2.

Table 1.2: Comparison of amino acid content in water hyacinth (per cent).

Amino Acid	*Water Hyacinth*	*FAO Ref. Pattern*	*Cow's Milk*	*Corn Grits*	*Leaf Protein Conc.*
Methionine	0.7	2.2	2.4	2.5	2.1
Cystine	11.6	1.9	–	1.1	0.7
Phenylanaline	4.7	2.8	4.9	6.4	6.0
Tyrosine	3.0	2.8	5.1	6.7	4.2
Threonine	4.3	2.8	4.6	4.1	5.2
Lysine	5.3	4.2	7.8	0.8	6.3
Isoleucine	4.3	4.2	6.4	6.4	5.3
Valine	0.3	4.2	6.9	5.3	6.3
Leucine	7.2	4.8	9.9	15.0	9.8

Water hyacinth protein shows deficient levels in only two of the essential amino acids, valine and methionine, as compared to the FAO reference pattern. A diet containing an adequate protein level will be balanced in lysine if it contains 4.2 per cent of that amino acid in its protein content. Corn is deficient in lysine, containing only 0.8 per cent. However water hyacinth contains 5.3 per cent lysine, while milk contains 7.8 per cent. It is evident that water hyacinth could serve to improve the lysine content of a corn diet. Leaf protein which contains 6.3 per cent lysine has been reported to be an effective supplement for barley in pig rations. Present evidence indicates that water hyacinth could be used to improve the lysine content of grains. Tryptophan content of water hyacinth reveals that this amino acid occurs in substantial amounts, the protein of hyacinth may be developed into useful dietary supplements for grain diets, especially in the underdeveloped countries. If a number of uses are found for other constituents of water hyacinth so as to make processing economically feasible, the protein could be extracted. Such a protein concentrate could be used as an effective supplement to grains for animal or human consumption."

Medicinal Importance of Aquatic Vegetables

A number of aquatic vegetables possess immense potential of several health related principles. They are medicinally utilized in different parts of the world and local people are well acquainted for their healing properties. A list of the aquatic vegetables and medical attributes has been highlighted in Table 1.3.

Maintaining the Genetic Diversity of Aquatic Vegetables in Adversity

In India, lakes, rivers and other freshwaters support a large diversity of biota representing almost all taxonomic groups. Algae in open waters represent the floristic diversity and macrophytes dominate the wetlands. It is difficult to analyze the algal diversity in India with reference to different habitats, endemicity to India, as well as the changes which occur due to anthropogenic disturbances. From an ecological point of view, the diversity of species present in the wetlands is an indication of the relative importance of the aquatic biodiversity issue as a whole. The total numbers of aquatic plant species exceed 1200 and aquatic vegetation is a valuable source of food. In the winter, migratory waterfowl search the sediment for nutritious seeds, roots and tubers. Resident waterfowl may feed on different species of aquatic vegetation year-round. China is home to the world's largest gene bank for aquatic vegetables. The repository, which holds more than 1,700 wild and cultivated specimens from 13 kinds of aquatic vegetables, is based in the city of Wuhan and run by the Wuhan Institute of Vegetable Science. Thirteen vegetables - which include lotus, water bamboo, taro, water celery, water chestnut and watercress - are the most frequently planted aquatic vegetables in the world and out of these11 come from China. Aquatic vegetables are favourite foods in southern parts of China. In Wuhan, where more of the vegetables are produced than anywhere else in the country, 500,000 hectares of them are planted each year, yielding 500,000 tons of produce worth 1 billion yuan (US$123.46 million). The Wuhan Institute of Vegetable Science has been developing the gene bank for 15 years. Most of the vegetables are conserved as plants, replanted annually in open fields. Reproducing them asexually allows the researchers to ensure the genes of the different plants do not mix, preserving their genetic identity. Aquatic vegetables are "environmentally friendly": they suffer from a few diseases and pests and can be grown without chemical fertilizers. Concerted efforts are being made to unravel the genes lie behind some of these attractive traits, such as resistance to pests. The information could be used to improve other crops. Efforts will be made to map the genetic sequences of some of the species.

Table 1.3: Aquatic vegetables and their medical attributes.

Aquatic Vegetables	*Family*	*Medicinal Properties*	*Ref.*
Water spinach (*Ipomoea aquatica*)	Convolvulaceae	Emetic, purgative; in Assam, the plant given for nervous and general disability. It is also used for piles. In Cambodia, the herb is applied as poultice in febrile delirium.	
Cattail (*Typha angustifolia* and *T. Latipholia*)	Typhaceae	The leaves are diuretic. The pollen is astringent, desiccant, diuretic, haemostatic and vulnerary and, used in the treatment of nose bleeds, haematemesis, haematuria, uterine bleeding, dysmenorrhoea, post-partum abdominal pain and gastralgia, scrofula and abscesses. The seeds are haemostatic. The rootstock is astringent and diuretic.	Duke and Ayensu, 1985; Yeung, 1985.
		In China, "Puhuang" is the traditional drug obtained from *Typha* plants. Plant is useful in angina, cystitis, diarrhea, urethritis, menstrual cramps, burns, insect bites, rash and wounds. Pollens are dusted on bleeding wounds to prevent the flow of blood.	Chopra *et al.,* 1986; Gao and Liao, 1998.
Eel grass (*Vallisneria spiralis*)	Hydrocharitaceae	The plant is aperitif, demulcent, refrigerant, stomachic and is also used in the treatment of women's complaints. It is also used in the treatment of leucorrhoea and is made into a tea with sesame (*Sesamum indicum*) to improve the appetite.	Chopra *et al.,* 1986 Duke and Ayensu, 1985
Duck-lettuce (*Ottelia alismoides*)	Hydrocharitaceae	Leaves are used to cure hemorrhoids. It is claimed that this plant has rubeficient properties	
Common water plantain (*Alisma plantago-aquatica* L.)	Alismataceae	The powdered seed is an astringent, used in cases of bleeding. The seed is also said to promote sterility. The root contains an essential oil and has a wide range of medicinal uses. It is antibacterial, anticholesterolemic, diuretic and hypotensive. It lowers blood pressure, cholesterol and blood sugar levels. The root is used in the treatment of oliguria, oedema, nephritis, acute diarrhoea, cholesterolaemia and fatty liver.	Grieve, 1985; Duke and Ayensu, 1985; Yeung, 1985. Bown, 1995. Yeung, 1985.
Water shield (*Brasenia schreberi* J.F. Gmel)	Cabombaceae	Leaves contain Gallic acid and traces of vitamin B12	Facciola,1990.
Papyrus (*Cyperus papyrus* L)	Cyperaceae	The pith is recommended for widening and drying of fistula. Its ash is useful in certain eye diseases; its ash checks malignant ulcers from spreading in the mouth or elsewhere.	
Dwarf waterclover (*Marsilea minuta*)	Marsileaceae	Leaf juice is used to stop nose bleeding. Indigestion is treated by eating the pounded leaves cooked with rice, and swelling of the gums is reduced by applying leaves which have been boiled rolled in a leaf of *Shorea robusta* P. Gaertn	

Contd...

Table 1.3–Contd...

Aquatic Vegetables	*Family*	*Medicinal Properties*	*Ref.*
Wasabi *(Wasabia japonica* (Miq.) Matsum)	Brassicaceae	It suppresses the growth of stomach cancer cells. It is a powerful digestive stimulant and exhibits strong antioxidant properties with fifteen powerful hydroxyl radical scavenging agents. It also possesses anti - diarrhoeal property. All parts of plant possess bactericidal activities against *Helicobacter pylori* strain NCTC 11637, YS 27 and YS 50	Nakayama *et al.,* 1998. Shin *et al.,* 2004.
Fragrant Water Lily (*Nymphaea odorata* Ait.)	Nymphaeaceae	The rhizomes are cooling, sweet, bitter and tonic and useful in diarrhoea, dysentery, dipsia and general debility. The flowers are astringent and cardio tonic. The seeds are sweet, cooling, constipating, aphrodisiac, stomachic and restorative	
Sonneratia (*Sonneratia caseolaris* (L.) Engler)	Sonneratiaceae	In Myanmar, fruits are used for poultices. Malayans use old fruit walls for worms, half-ripe fruits for coughs, and pounded leaves for hematuria and smallpox. Plant possesses antinociceptive and anti-diarrhoeal activities	Perry, 1980.
Chinese arrowhead (*Sagittaria trifolia*)	Alismataceae	Four new biologically active diterpenes, trifoliones A, B, C, and D, have been isolated from the tuber together with three new glycosides, sagittariosides a and b and arabinothalictoside. Trifoliones A, B, C, and D possess inhibitory effects on the histamine release from rat mast cells induced by compound 48/80 or calcium ionophore A-23187.	Yoshikawa *et al.,* 1993.
Paracress (*Spilanthes acmella* L)	Asteraceae	The flower heads are chewed to relieve the toothache and other mouth related troubles. Leaves are used externally in treatment of skin diseases. Root decoction is used as purgative. Leaf decoction is used as diuretic and lithotriptic. Whole plant *has* been well documented for its uses as spices, antiseptic, anti-bacterial, anti-fungal, anti-malarial and as remedy for toothache, flu, cough, rabies diseases and tuberculosis.	Burkill, 1966; Di Stasi *et al.,* 1994; Akah and Ekekwe, 1995; Singh, 1995; Storey and Salem, 1997; Ramsewak *et al.,* 1999.
Indian blue water Lily (*Nymphaea stellata* Willd.)	Nymphaeaceae	The powdered rhizome is given in dyspepsia, diarrhoea and piles. An infusion of rhizome and stem is considered emollient and diuretic. It is used for blennorrrhagia and diseases of the urinary tract. Plant possesses hypoglycemic activity.	
Keysoor (*Scirpus grossus*)	Cyperaceae	Keysoor root has astringent properties, and is given to check diarrhoea and vomiting. The root possesses as cooling and laxative properties and acts as a tonic to the liver. It is effective against infection and poison, diuretic, and useful in burning sensations, fevers, and gonorrhoea.	
Hydrolea (*Hydrolea zeylanica* (L.) Vahl)	Hydroleaceae	Leaves are used as a poultice for ulcers and in traditional Cambodia medicine; the leaves are prescribed for intestinal disorders.	
Water pepper (*Polygonum hydropiper*)	Polygonaceae – Buckwheat family	The leaves are anti-inflammatory, astringent, carminative, diaphoretic, diuretic, emmenagogue, stimulant, stomachic, styptic. They contain rutin, which helps strengthen fragile capillaries and thus helps prevent bleeding. The seed is carminative, diuretic and stimulant.	Grieve,1984; Launert, 981; Foster and Duke,1990; Duke and Yuensu,1985l Coon, 1975.

Aquatic Vegetable: Lifeline of Wetland Ecosystem

India, with its annual rainfall of over 130 cm, varied topography and climatic regimes support and sustain diverse and unique wetland habitats. Natural wetlands in India consist of the high altitude Himalayan lakes, followed by wetlands situated in the flood plains of the major river systems, saline and temporary wetlands of the arid and semi-arid regions, coastal wetlands such as lagoons, backwaters and estuaries; mangrove swamps; coral reefs and marine wetlands, and so on. In fact, with the exception of bogs, fens and typical salt marshes, Indian wetlands over the whole range of the ecosystem types are found. In addition to the various types of natural wetlands, a large number of man-made wetlands also contribute to the faunal and floral diversity. These man-made wetlands, which have resulted from the needs of irrigation, water supply, electricity, fisheries and flood control, are substantial in number. The various reservoirs, shallow ponds and numerous tanks support wetland biodiversity and add to the countries wetland wealth. It is estimated that freshwater wetlands alone support 20 per cent of the known range of biodiversity in India (Deepa and Ramachandra 1999). Wetlands in India occupy 58.2 million hectares, including areas under wet paddy cultivation (Directory of Indian Wetlands). Majority of the inland wetlands are directly or indirectly dependent on the major rivers like, Ganga, Bhramaputra, Narmada, Godavari, Krishna, Kaveri, Tapti. They occur in the hot arid regions of Gujarat and Rajasthan, the deltaic regions of the east and west coasts, highlands of central India, wet humid zones of south peninsular India and the Andaman and Nicobar and Lakshwadeep islands.

Wetlands provide many services and commodities to humanity. Regional wetlands are integral parts of larger landscapes, their functions and values to the people in these landscapes; depend on both their extent and their location. Each wetland thus is ecologically unique. Wetlands perform numerous valuable functions such as recycle nutrients, purify water, attenuate floods, maintain stream flow, recharge ground water, and also serve in providing drinking water, fish, fodder, fuel, wildlife habitat, control rate of runoff in urban area, buffer shorelines against erosion and recreation to the society. The interaction of man with wetlands during the last few decades has been of concern largely due to the rapid population growth - accompanied by intensified industrial, commercial and residential development further leading to pollution of wetlands by domestic, industrial sewage, and agricultural run-offs as fertilizers, insecticides and feedlot wastes.

Wetland Losses: A Threat to Ecological Balance

Wetlands are one of the most threatened habitats of the world. Wetlands in India, as elsewhere are increasingly facing several anthropogenic pressures. Thus, the rapidly expanding human population, large scale changes in land use/landcover, burgeoning development projects and improper use of watersheds have all caused a substantial decline of wetland resources of the country. Significant losses have resulted from its conversion threats from industrial, agricultural and various urban developments. These have led to hydrological perturbations, pollution and their effects. Unsustainable levels of grazing and fishing activities have also resulted in degradation of wetlands. The current loss rates in India can lead to serious consequences, where 74 per cent of the human population is rural (Anon. 1994) and many of these people are resource dependent. Healthy wetlands are essential in India for sustainable food production and potable water availability for humans and livestock. They are also necessary for the continued existence of India's diverse populations of wildlife and plant species; a large number of endemic species are wetland dependent. Most problems pertaining to India's wetlands are related to human population. India contains 16 per cent of the world's population, and yet constitutes only 2.42 per cent of the earth's surface. Indian landscape has contained a fewer natural wetlands over time. Restoration of these converted wetlands is quite difficult once these sites are occupied for non wetland uses. The demand for wetland products (*e. g.*, water, fish, wood, fiber, medicinal plants etc.) will increase with increase in population. Wetland loss refers to physical loss in the spatial extent or loss in the wetland function. The loss of one km^2 of wetlands in India will have much greater impacts than the loss of one km^2 of wetlands in low population areas of abundant wetlands (Foote Lee *et al*,. 1996). The National wetland committee of the Ministry of Environment and Forests (MoEnF), Govt. of India recommended a number of proactive steps. Accordingly, twenty-two sites were declared tentatively as wetlands of National and international significance for long term conservation. The committee recommended a nation wide inventory of wetlands to be undertaken under the guidance of National wetland committee with the support of the Standing Committee on Bioresources (SC-B) of the National Natural Resource Management System (NNRMS) and MoEF. The Space Application Center (SAC) Ahmedabad, of the Indian Space Research Organization (ISRO) had undertaken, in collaboration with various state remote sensing application centers, a nation wide mapping of Inland wetlands.

Table 1.4: Extent of inland wetlands. State wetland area (ha).

Andhra Pradesh	366609 (1: 250,000 Scale)
Bihar	177683
Gujarat	209206
Jammu and Kashmir	406780
Karnataka	254015
Kerala	34200
Madhya Pradesh	294118
Maharashtra	284942
Orissa	162774
Rajasthan	344964
Tamil Nadu	161521
Uttar Pradesh	328690
Pondicherry	59
Arunachal Pradesh	56325 (1: 50,0000 Scale)
Assam	101232
Delhi	4717
Goa	2145
Haryana	27057
Himachal Pradesh	54766
Manipur	52959
Meghalaya	2222
Mizoram	152
Nagaland	918
Punjab and Chandigarh	71879
Sikkim	1985
Tripura	9896
West Bengal	143859
Andaman and Nicobar Islands	3204
Daman Diu & Nagar Haveli	38
Lakshadweep	918
Total	**3558915**

Past research on wet land conservation in the country has shown conclusively that micro-wetlands or satellite wetlands around a bigger wetland act as a constellation of habitat mosaic for resident and migratory waterfowl (Vijayan, 1991). This is of special importance for inland wetland habitats in the flyways of migratory birds in the Indo-Gangetic plains and in Deccan peninsula. Often, the size of these micro wetlands is much smaller than 50 ha. Therefore, there is a great need to map wetlands of size smaller than 50 ha. Thus, the objective is to develop an inland wetland inventory for the entire country. This will be carried out by means of available data and by also fresh data generation using modern spatial technologies. Thus by using digital remote sensing data for wetland mapping and analysis, information at any scale of all Wetlands will be available according to the management and conservation requirements. Realizing the importance of wetlands, the Ramsar convention in 1971 has urged the member countries to designate noted wetlands as Ramsar sites or wetlands of International significance. A wetland conservation strategy should therefore have an extensive bias of participatory process (Choudhury 2000). A hierarchical watershed based approach will have a positive impact in not only reversing the chronic cases of wetland resource depletion but also help design a network of wetland conservation preserves. These preserves would strive to not only conserve precious aquatic biodiversity but also help serve as refuge for important economically useful wild plants and animal genetic resources.

Reuse of Wastewater for Growing Aquatic Vegetables

Wastewater reuse for aquaculture has been practiced in many countries for a considerable period of time. It has the potential of wider application in the cultivation of fish and aquatic vegetables such as water spinach, lotus, chestnut etc. Human excreta and wastewater contain useful materials. These are water, organic carbon and nutrients and should be regarded as a resource. In their natural cycles they are broken down by micro-organisms and become useful to plants and animals, thus sustaining natural ecosystems.

Role of Aquatic Vegetables in Phyto-Remediation

The leaf vegetables *viz. Amaranthus spinosus, Alternanthera philoxeroides* and *A. sessiles* growing on the sewage sludge of Musi river located in greater Hyderabad City (close to 17°26' N latitude and 78°27' E longitude), Andhra Pradesh, India was investigated for metal accumulation. Metal content Cd (non-essential), Zn and Fe (essential) in plant parts of these selected species indicate their ability to bioconcentrate in their tissues. The concentration of these metals was invariably high in leaf tissue (Bañuelos and Meek, 1989; Prasad, 2001b). Thus it is possible to use these species to restore the biosolid and sewage sludge contaminated sites, while exercising caution on human consumption. *Alternanthera philoxeroides* was used for removal of lead and mercury from polluted waters. It is also possible to supplement the dietary requirement of human food with Zn and Fe as these being essential nutrients and the plant species are edible. However, there is a need to monitor the metal transfer factor through food chain (Bañuelos and Meek, 1989; Bañuelos *et al.*, 1993a; Bañuelos *et al.*, 1993b). Glutathione and organic acids metabolism plays a key role in metal tolerance in plants (Arisi *et al.*, 1997; Huang *et al.*, 1998; Schäfer *et al.*, 1998; Zhu *et al.*, 1999a; Arisi *et al.*, 2000; Ma *et al.*, 2001a). Glutathione is ubiquitous component cells from bacteria to plants and animals. In plants, it is the major low molecular mass thiol compound (28). Glutathione occurs in plants mainly as reduced GSH (95 per cent). Its synthesis is mediated by the enzymes glutamylcysteine synthetase (EC 6.3.2.2) and glutathione synthetase (EC 6.3.2.3). Glutathione metabolism is also connected with cysteine and sulphur metabolism in plants. Cysteine concentration limits gluthatione biosynthesis. Low-molecular thiol peptides phytochelatins (PCs) often called class III metalothioneins are synthetized in plants from glutathione induced by heavy metal ions (Mejare and Bulow, 2001; Prasad and Strzalka, 2002).

These peptides are synthetized from glutathione by means of α-glutamylcysteine transferase enzyme (EC 2.3.2.15), which is also called phytochelatin synthase (PCS) catalyzing transfer reaction of (α-Glu-Cys) group from a glutathione donor molecule to glutathione, an acceptor molecule. PCS is a cytosolic, constitutive enzyme and is activated by metal ions *viz.*, Cd^{2+}, Pb^{2+}, Ag^{1+}, Bi^{3+}, Zn^{2+}, Cu^{2+}, Hg^{2+}, and Au^{2+}. PCs thus, synthesized chelate heavy metals and form complexes and these complexes are transported through cytosol in an ATP–dependent manner through tonoplast into vacuole. Thus the toxic metals are swept away from cytosol. Some high-molecular weight complexes (HMW) with S-2 can also be formed from these LMW complexes in vacuole (Clemens *et al.*, 1999; Clemens, 2001).

Transgenic plants with modified genes of PCS and genes of glutathione synthesis enzymes: α-GCS and GS, and enzymes connected with sulphur metabolism, *i.e.* serine acetyl transferase, need special attention to achieve success (Liang *et al.*, 1999; Pilon-Smits *et al.*, 1999b). Plants under heavy metals stress produce free radicals and reactive oxygen species and have to withstand the oxidative stress before aquiring tolerance to toxic metals. Glutathione is then used for the synthesis of PCs as well as for dithiol (GSSG) production. The ascorbate-glutathione pathway is involved in plant defence against oxidative stress. Organic acids play a major role in metal tolerance (Ma *et al.*, 2001a) by forming complexes with metals, a proces of metal detoxification. Chelation of metals with excluded organic acids in the rhizosphere and rhizosphereic processes indeed forms an important aspect of investigation for remediation. These metabolic pathways underscore the physiological, biochemical and molecular basis for heavy metal tolerance (Prasad and Strzalka, 2002).

Future Threats

Contamination

A number of authors have described potential health hazards associated with urban aquaculture, and in particular those activities where wastewater reuse is practised (Mara and Cairncross, 1989; Strauss, 1991; Edwards, 1992; Edwards, 2001; Howgate *et al.*, 2002). Although several of these reviews make hazards associated with aquaculture explicit, it is much harder to quantify the associated level of risk. For example, the risks associated with products grown using waste resources vary, depending on characteristics of the waste resource, the degree of treatment prior to use, the design and operation of the culture system, husbandry and processing practices, subsequent handling and preparation and susceptibility of the consumer. Reviewing health hazards associated with aquaculture employing wastewater reuse, Mara and Cairncross (1989) identified four groups of people at risk: field workers, crop handlers, local residents and consumers. Bunting (2004) provides a more detailed review of the hazards faced by these different groups, describes factors that influence the degree of risk and outlines potential strategies for mitigation. An emerging threat that requires greater attention is the possible relationship between aquaculture operations, where poultry waste is used as a fertiliser, and the transmission of bird flu to humans. When waste water is improperly disposed these substances can cause pollution, because the organic materials exert oxygen demand, and the nutrients promote algal growth in lakes, rivers and near-shore marine environments. Human excreta and wastewater also contain pathogens. Reuse of the wastes must ensure that public health is maintained. Planned reuse is the key to wastewater reuse. Planning for reuse ensures that public health and protection of the environment are taken into account. Reuse of treated wastewater for irrigation of crops, for example, will need to meet (i) standards for indicator pathogens, and (ii) plant requirement for water, nitrogen and phosphorus. WHO and others have developed standards for reuse of wastewater for various purposes. For restricted irrigation, WHO recommends the treated wastewater should contain no more than one human intestinal nematode egg per litre. These intestinal nematodes are: *Ascaris lumbricoides*, the human roundworm, *Trichuris trichiura*, the human whipworm, and *Ancylostoma duodenale* and *Necator americanus*, the human hookworms. For unrestricted irrigation, *i.e.* including salad crops and vegetables eaten raw, WHO recommends the same helminth egg value, and additionally no more than 1000 faecal coliform bacteria per 100 ml of treated wastewater.

Accumulation of Nitrate and Oxalate Content

The calcium oxalate content of *Alternanthera philoxeroides* (alligator weed) was estimated to be a maximum of 4.4 per cent of the dry weight of nitrogen-fertilized terrestrial shoots. This is well below the amount of soluble oxalate (about 20 per cent) found in poisonous plants such as *Halogeton*. Nevertheless the level is high enough to be of interest to nutritionists and medical personnel in those countries, *e.g.* Myanmar, where alligator weed is being used as human food. About 4 g of oxalic acid is the minimum fatal dose for an average sized man. Thus 100 g of fertilized terrestrial alligator weed (on a dry weight basis) could potentially provide a fatal dose if the toxic principle were not solubilized and removed by boiling before eating. Diets with high oxalate-containing vegetables and low calcium have sometimes been implicated in the incidence of kidney stones.

Accumulation of Heavy Metals

Most of the water bodies being used for the cultivation of edible aquatic plants (*Trapa natans* and *Ipomoea aquatica*) were contaminated with a variety of toxic metals (Fe, Cu, Cr, Mn and Pb). The concentration of metals Cr, Pb and Fe in water was much higher than recommended permissible limits of WHO (1995). The edible parts of these plants had bioconcentrated metals from their surrounding water significantly. Despite varying levels of metals found in various fruit parts of *T. natans*, the metal accumulation in kernel was alarming. However, metal content decreased significantly in various parts after boiling the fruit. Similarly, *I. aquatica* also accumulated significantly higher amounts of these metals in leaves, however the metal accumulating potential varied considerably depending upon level of metal contamination in the water body in which they were growing (Rai and Sinha, 2001).

Conclusion

Although aquatic vegetables may be important to only local communities and contributing to very tiny part of food security, however, its multifaceted benefits in terms of maintaining urban bodies in flood control, amenity uses, wild life and broader environmental benefits may be considered in very holistic manner. In the most parts of the world, there are swamp lands and shallow ponds totally unsuited to fish culture or agriculture. The present policy for such areas is to drain and "develop" them for uses not in accord with their nature. There is a need to find techniques that will allow us to use wetlands as wetlands. Cultivating aquatic vegetables is one possible way. The paper highlights a few species that, with research, could become important aquatic vegetables. Some of these plants are pernicious weeds. They should never be introduced to new locations or encouraged to spread in old ones. They are included for the benefit of those areas that already have them and would like to see them used productively.

References

Akah, P. A. and. Ekekwe, R. K 1995. Etnopharmacology of some Asteraceae family used in Nigerian traditional medicine. Fitoterapia, 66: 351-355.

Anonymous, 1994. World Development Report. World Bank Development Report.

Arisi, A. C. M.; Mocquot, B.; Lagriffoul, A.; Mench, M.; Foyer, C. H. and Jouanin, L. 2000. Responses to cadmium in leaves of transformed poplars overexpressing g-glutamylcysteine synthetase. Physiologia Plantarum, 109(2): 143-149.

Arisi, A. C. M.; Noctor, G.; Foyer, C. H. and Jouanin, L. 1997. Modification of thiol contents in poplars (*Populus tremula* x *P. alba*) overexpressing enzymes involved in glutathione synthesis. Planta, 203(3): 362-372.

Bañuelos, G. S. and Meek, D. W. 1989. Selenium accumulation in selected vegetables. Journal of Plant Nutrition, 12(10): 1255-1272.

Bañuelos, G. S.; Cardon, G.; Mackey, B.; Ben-Asher, J.; Wu, L. P.; Beuselinck, P., Akohoue, S. and Zambrzuski, S. 1993a. Boron and selenium removal in B-laden soils by four sprinkler irrigated plant species. Journal of Environmental Quality, 22: 4786-4797.

Bañuelos, G. S.; Mead, R. R. and Hoffman, G. J.1993b. Accumulation of selenium in wild mustard irrigated with agricultural effluent. Agriculture, Ecosystems and Environment, 43(2): 119-126.

Bown. D. 1995. Encyclopaedia of Herbs and their Uses. Dorling Kindersley, London.

Boyd, C. E., 1968. Fresh water plants: a potential source of protein. Econ. Bot. 22: 359–68

Bunting, S. W.2001. Appropriation of environmental goods and services by aquaculture: a reassessment employing the ecological footprint methodology and implications for horizontal integration. Aquaculture Research, 32, 605-609.

Bunting, S. W. 2004. Wastewater aquaculture: perpetuating vulnerability or opportunity to enhance poor livelihoods? Aquatic Resources, Culture and Development 1, 51-75.

Burkill, I. H. 1966. A dictionary of the economic products of the Malay Peninsula. Vol II. Governments of Malaysia and Singapore by the Ministry of Agriculture and Co-operatives, Kuala Lumpur.

Chopra. R. N., Nayar. S. L. and Chopra. I. C 1986. Glossary of Indian Medicinal Plants. Council of Scientific and Industrial Research, New Delhi.

Choudhury, B. C. 2000. Conserving wetlands: emerging scenario. pp. 131-138. In: A. Bhardwaj, R. Badola andB. M. Rathore (Eds.) Proceedings of the workshop on the 'Conserving Biodiversity in the 21st Century, through Integrated Conservation and Development Planning on a Regional Scale'. LBSNA Mussorie and WII Dehradun.

Clemens, S. 2001. Molecular mechanisms of plant metal tolerance and homeostasis. Planta, 212,(4): 475-486.

Clemens, S.; Kim, E. J.; Neumann, D. and Schroeder, J. I. 1999. Tolerance to toxic metals by a gene family of phytochelatin synthases from plants and yeast. EMBO J.18(12): 3325-3333.

Coche, A. G. 1982. Cage culture of tilapias. (In) Pullin, R. S. V. and Lowe-McConnell, R. H. (Eds.), *Biology and Culture of Tilapias*, Metro Manila, Philippines: International Centre for Living Aquatic Resource Management. p.205-246

Coon, N. 1975. The Dictionary of Useful Plants. Rodale Press

Deepa, R. S., and Ramachandra, T. V.1999. Impact of Urbanization in the Interconnectivity of Wetlands. Paper presented at the National Symposium on Remote Sensing Applications for Natural Resources:

Di Stasi, L. C., C. A. Hiruma, E. M Guimarães and C. M. Santos, 1994. Medicinal plants popularly used in Brazilian Amazon. Fitoterapia, 65: 529-540.

Duke. J. A. and Ayensu. E. S.1985. Medicinal Plants of China Reference Publications, Inc.

Edwards, P. 1992. Reuse of Human Waste in Aquaculture, a Technical Review. Washington, USA: UNDP-World Bank Water and Sanitation Program.

Edwards, P. 2001. Public health issues of wastewater-fed aquaculture. Urban Agriculture 3, 20-22.

Facciola. S. 1990. *Cornucopia - A Source Book of Edible Plants.* Kampong Publications.

FAO. 1995. *Aquaculture production statistics 1984-1993*. FAO Fisheries Circular 815, Rev. 7, Fishery Information, Data and Statistics Service, FAO, Rome, p.186.

Foster, S. and Duke, J. A. 1990. A Field Guide to Medicinal Plants. Eastern and Central N. America. Houghton Mifflin Co.

Foote, Lee, Pandey S., and. Krogman, N. T. 1996. Processes of wetland loss in India. Environmental Conservation, 23: 45-54.

Gao, G. and Liao, M. 1998. Determination of flavonoids and quality evaluation of Chinese traditional drug "Puhuang". Yaoxue Xuebao 33(4): 300-303. Inst. Med. Plant, Chinese Acad. Med. Sci. and Peking Union Med. Coll., Beijing 100094, China.

Grieve. 1984. A Modern Herbal. Penguin 1984 ISBN 0-14-046-440-9

Howgate, P., Bunting, S. Beveridge M. and Reilly. A. 2002. Aquaculture associated public, animal and environmental health issues in non-industrialized countries. In: Jahncke, M., Garrett, S., Martin, R., Cole, E., Reilly, A. (Eds.), Public, Animal and Environmental Aquaculture Health, Wiley.

Huang, J. W.; Blaylock, M. J.; Kapulnik, Y. and Ensley, B. D. 1998. Phytoremediation of Uranium-Contaminated Soils: role of organic acids in triggering uranium hyperaccumulation in plants. Environmental Science and Technology, 32(13): 2004-2008.

Hung, L. T. and Huy, H. P. V. 2005. Production and marketing systems of aquatic products in Ho Chi Minh City. *Urban Agriculture* 14: 16-19.

Launert. E.1981. Edible and Medicinal Plants. Hamlyn.

Liang Zhu, Y.; Pilon-Smits, E. A.; Jouanin, L. and Terry, N.1999. Overexpression of glutathione synthetase in Indian mustard enhances cadmium accumulation and tolerance. *Plant Physiology*, 119(1) : 73-80.

Ma, J. F.; Ryan, P. and Delhaize, E.2001a. Aluminium tolerance in plants and the complexing role of organic acids. Trends in Plant Science,. 6(6): 273-278.

Mara, D. and Cairncross. S. 1989. *Guidelines for the Safe Use of Wastewater and Excreta in Agriculture and Aquaculture, Measures for Public Health Protection*. Geneva, Switzerland: World Health Organisation.

Mejare, M. and Blow, L. 2001. Metal binding proteins and peptides in bioremediation and phytoremediation of heavy metals. Trends in Biotechnology, 19(2): 67-75.

Nakayama, H. Suzuki, T. and Suzuki, Y. 1998. Effect of wasabi on ion secretion in guinea pig colon. Nippon Nogeikagaku Kaishi 72(4) : 499-507

Ogle, B. M., Dao, H. T. A.; Mulokozi, G. and Hambraeus, L. 2001. Nutritional importance of gathered vegetables in Vietnam. International Journal of Food Sciences and Nutrition, 52(6) : 485-499

Perry, L. M. 1980. Medicinal plants of east and southeast Asia. MIT Press, Cambridge.

Phuong, N. T. D. and Tuan, P. A. 2005. Current status of periurban aquatic production in Hanoi. Urban Agriculture 14: 10-12.

Pilon-Smits, E. A. H.; Hwang, S.; Lytle, C. M.; Zhu, Y. L.; Tay, J. C.; Bravo, R. C.; Chen, Y.; Leustek, T. and Terry, N. 1999b. Over expression of ATP sulfurylase in Indian mustard leads to increased selenate uptake, reduction, and tolerance. Plant Physiology, 119(1): 123-132.

Prasad, M. N. V. and Strzalka, K. 2002. Physiology and biochemistry of metal toxicity and tolerance in plants. Dordrecht, Kluwer Academic Publishers, 2002, 432 p.

Prasad, M.N.V. 2001. Bioremediation Potential of Amaranthaceae. (In) *Proceedings of the 6th International In Situ and On-Site Bioremediation Symposium*. (4th - 7th June, 2001, San Diego, California, USA

Rai, U. N. and Sinha, S. 2001. Distribution of Metals in Aquatic Edible Plants: *Trapa natans* (Roxb.) Makino and *Ipomoea aquatica* Forsk. Environmental Monitoring and Assessment, 70(3): 241-252.

Ramsewak, R. S., Erickson A. J. and Nair, M. G. 1999. Bioactive Nisobutylamides from the flower buds of *Spilanthes acmella*. Phytochemistry, 51: 729-732.

Schäfer, H. J.; Haag-Kerwer, A. and Rausch, T. 1998. cDNA cloning and expression analysis of genes encoding GSH synthesis in roots of the heavy-metal accumulator *Brassica juncea* L.: evidence for Cd-induction of a putative mitochondrial γ-glutamyl cysteine synthetase isoform. Plant Molecular Biology, 37: 87-97.

Shin, I. S., Masuda, H., and Naohide, K. 2004. Bactericidal activity of wasabi (*Wasabia japonica*) against *Helicobacter pylori*. International Journal of Food Microbiology. 2004; 94(3): 255-261.

Singh, V. 1995. Herbal folk remedies of Morni hills (Haryana), India. Fitoterapia, 66: 425-430.

Storey, C. and Salem J. I., 1997 Lay use of Amazonian plants for the treatment of tuberculosis. Acta Amazonica, 27: 175-182.

Strauss, M. 1991. Human waste use: health protection practices and scheme monitoring. *Water Science and Technology* 24(9): 67-79.

Vijayan, V. S. 1991. Keoladeo National Park Ecology Study. Final Report. Bombay Natural History Society.

Yeung., Him-Che 1985. Handbook of Chinese Herbs and Formulas. Institute of Chinese Medicine, Los Angeles.

Yoonpundh, R., Dulyapurk, V. and. Srithong, C. 2005. Aquatic food production systems in Bangkok. Urban Agriculture, 14, 8-9.

Yoshikawa M, Yamaguchi S, Murakami T, Matsuda H, Yamahara J. and Murakami N.1993. Absolute stereostructures of trifoliones A, B, C, and D new biologically active diterpenes from the tuber of *Sagitarria trifolia* L. Chem Pharm Bull.41(9): 1677-1679.

Zhu, Y. L.; Pilon-Smits, E. A.; Tarun, A. S.; Weber, S. U.; Jouanin, L. and Terry, N.1999 Cadmium tolerance and accumulation in Indian mustard are enhanced by over expressing gamma-glutamylcysteine synthetase. *Plant Physiology*, 121: 1169-1178.

2014, Future Crops Vol. 2 *Pages 17–50*
Editor: Professor K.V. Peter
Published by: DAYA PUBLISHING HOUSE, NEW DELHI

Chapter 2
Aquatic Weeds as Potential Future Foods

R.N. Mandal and P. Jayasankar*
Central Institute of Freshwater Aquaculture (ICAR)
Kausalyaganga, Bhubaneswar – 751 002, Orissa
**E-mail: pjayasankar@yahoo.com*

Food crisis is one of the major challenges, human civilization encounters every moment and it will be an inevitable hurdle to feed more than six billion people worldwide in near future. In this regard, the statement of UN Secretary-General Ban Ki-moon to UN's 192 member states, "in a single year, staple foods that feed half of the world more than doubled in price" (Wroughton, 2008), highlighting the severity of the Global food crisis, is relevant. Besides, weather either in single catastrophe or subtle shifts related to climate change has already impacted food production causing structural changes to the agricultural system (Wei *et at.*, 2008). Mitigating this challenge needs to search alternate plant bio-resources which are the potential repository of a variety of food attributes and grow or may be grown with usual sprouting even under inhospitable climatic conditions in fallow wetlands. Such bio-resources include a diverse group of aquatic plant communities which have incredible treasures of nutritional uniqueness to be utilized for the purpose of human benefit, but are still little known. Potential of such aquatic plants may be considered as future foods and may contribute substantially to the food basket of our country, ensuring to food security to a noticeable extent. These additional food items may support people to have physical and economic access to adequate amounts of nutritious, safe and culturally appropriate foods (Anonymous, 2013). In this context, Dr M. S. Swaminathan's view is significant as he mentioned that food security should have three basic criteria: (i) availability of sufficient food, (ii) access to food for all, and (iii) absorption of food to body pertaining to health care (Yadugiri, 2011).

India is endowed with enormous number of wetlands which are habitats of a wide range of biotic resources. These valuable plant bio-resources, treated as aquatic weeds, play multiple roles in aquaculture or in aquaculture related activities, catering to societies in diverse basic needs namely, food crops, cash crops, horticulture crops, biofertilizers, feeds, fodders, green manure, leafy vegetables, ornamental plants, integrated crops, wastewater treatment, and therapeutic substances (Mandal and Saha, 2007). Ironically, these plant bio-resources are usually overlooked despite having unique potentiality which could be harnessed in better way for farmers' welfare. Most of these bio-resources still grow wildly and are utilized traditionally rather than are grown with implementation of modern agricultural methods blended with scientific know-how (Mandal *et al.*, 2012). Importantly, however, a large section of society in rural India manages their livelihood supports by utilizing these bio-resources. This article highlights a glimpse of these aquatic bio-resources with their potential substances.

Classification of Aquatic Plants in View of Future Foods

Aquatic plants considered for possible future foods may be classified into four groups viz; food, feed, fodder and vegetables; all the four items may rationally be put under the term, 'foodstuffs' coined primarily in relation to the sense of both satisfaction of hunger and maintenance of good health through intake of appropriate nutritional requirements essential for respective communities. The term 'food' may be used for human beings and related communities, 'feed' – prepared ones or related foodstuffs used for livestock including fish, 'fodder' – green plant items again used for livestock including fish and vegetables for human beings. Aquatic plants/weeds used as food comprise Makhana, Water chestnut, Lotus, Water Lily and Colocasia. Feed constitutes a variety of food items given to livestock including fish or prepared ones which may have different ingredients of which important ones are wild macrophytes, Azolla, Duckweed, Water spinach and Water chestnut. Fodder again includes a wide range of wild macrophytes, Azolla, Duckweed, Water hyacinth and Water spinach. Quite a number of wild aquatic plants represent leafy vegetables utilized for their unique nutrient constituents, apart from important therapeutic substances.

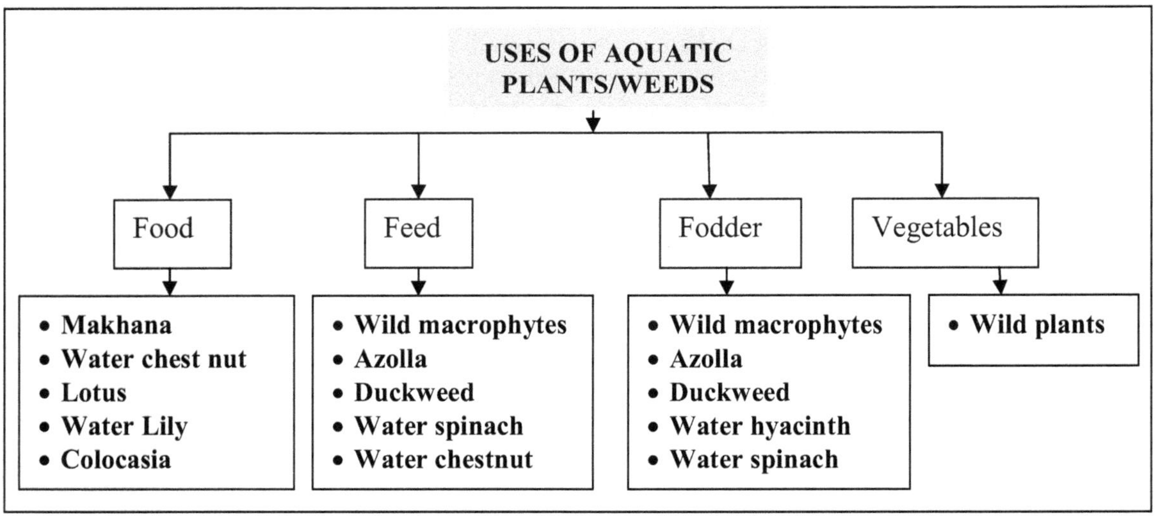

Chart 2.1: Classification of aquatic plants in view of future foods

Food

The term 'food' is considered in the purpose of human consumption as directly or indirectly through various ways of preparation. However, the term 'food' is strictly based on arbitrary choice rather than any scientific reason taken into account.

Makhana

Euryale ferox Salisbery, family-Nymphaceae, locally known as Makhana (Figure 2.1a), is one of the important aquatic plant bio-resources (Mandal and Saha, 2012). It was found growing wildly in fallow wetlands of long range from Kashmir to Manipur alongside Himalayan stretch from Northeast to Far East. Presently, wild genetic stock of this species has been seriously vulnerable due to habitat loss and fragmentation, injudicious exploitation, environmental contamination and climate change. This species is important because it provides high quality nutritive food in addition to important therapeutic substances (Mandal *et al.*, 2010). This species is now widely cultivated in Bihar, India and treated as an important food crop (Jha *et al.*, 1991).

Figure 2.1a: Makhana.

Botany and Ecology

Makhana is an annual aquatic herb with floating leaves. The leaves are prominently nerved and veins appear conspicuous at ventral side. Colour of leaves are green above and purple beneath. The plant is rhizomatous which remains deep rooted with cluster of roots at the bottom. Flowering bud initiates during March-April, blooms afterwards and remains open for 2-4 days. Immature fruit remains above water surface until it gets fully matured. Mature fruit looks globose in shape with persistent sepals at the apex. It sinks under water while it is about to ripen. One mature fruit contains 30-40 seeds each of which looks oval in shape with black hard seed coat. Fruit coat ruptures after ripeness and then bursts, resulting in the spread-over of seeds in bottom. This process continues during June to August (Mandal *et al.*, 2010).

In floodplain areas, closed types of water bodies with standing water depth in the range of 1.0-2.0 m are found as suitable habitats for makhana vegetation. Makhana plant grows well in the condition of neutral pH of water bodies and soil sediment having C/N ratio nearly 10 ideal for plant growth and increase of biomass (Dutta *et al.*, 1984). In addition, soil sediment has high concentration of Fe, Zn and Co all of which act as micronutrients for plant growth (Jha *et al.*, 1991).

Cultivation Practice and Harvest

During pre-monsoon, seeds get germinated and sprouting takes place afterwards at the time of rainy period. Seedlings take about four months to become mature. Farmers apply fertilizers if required. Usually much care is not required for Makhana cultivation. However, harvest is very much tedious as well as cumbersome task which requires skilled labours along with effort of typical drudgery (Mandal *et al.*, 2010).

Importance as Food

Seed of Makhana is edible. Seeds are ground to flour which is mixed with dal (pulse grain) or vegetables for consumption. This preparation is more nutritious and also has higher calorific value

Figure 2.1b: White puffs from Makhana.

than that of puff ones. Traditionally, people of North India, Bihar in particular, take this preparation as one of their items in daily meal. However, common practice for use of makhana is to prepare white puffs from seeds through a series of cumbersome tasks (Mandal *et al.*, 2010). The white puffs (Figure 2.1b) are marketed as either puff grain or puff flour. Not only puff is eaten as delicious food but also treated as medicated one.

Table 2.1: Nutritional values: (Proximate composition of 100 g edible portion of major food items: comparative evaluation)

Food	*Moisture (g)*	*Protein (Nx6.23) (g)*	*Fat (g)*	*Minerals (g)*	*Crude Fibre (g)*	*Carbo-hydrate (g)*	*Energy Kcal*	*Ca (mg)*	*P (mg)*	*Fe (mg)*
Makhana	12.8	9.7	0.1	0.5	–	76.9	347	20	90	1.4
Rice	12.6	8.5	0.6	0.9	–	77.4	349	10	280	2.8
Wheat	12.2	11.8	1.5	1.5	1.2	71.2	346	41	306	5.3
Pulse	9.8	17.1	5.3	3.0	3.9	60.9	360	202	312	4.6
Potato	74.7	1.6	0.1	0.6	0.4	22.6	97	10	40	0.48

Source: Gopalan *et al.*, 2007 after modification.

Proximate composition of edible part of makhana (Table 2.1) seems to be almost similar to that of staple foods like rice, wheat and has more caloric value than that of common pulse grain and vegetable with particular reference to potato. In addition, it contains a good proportion of sugar, ascorbic acid and phenol which are the major nutrient constituents present in other edible fruits (Sikka *et al.*, 1978;

Sikka *et al.*, 1979; Gopalan *et al.*, 2007). Despite its lower protein contents, though nearly equal to protein content of rice and wheat, makhana has been analyzed as nutritionally superior as its essential amino acids indices (EAAI) in the raw and puff parts of makhana were found to be 93 per cent and 89 per cent, respectively (Nath and Chakrabarty, 1985). Comparatively, other foods show following lower values: rice (83 per cent), wheat (65 per cent), Bengal gram (81.55 per cent), Soy bean (85.6 per cent), cow's milk (88.8 per cent), and mutton (87.2 per cent), except fish (89.2 per cent) (Anonymous, 1952).

Therapeutic Value

Makhana has unique therapeutic constituents which are remedy to several human ailments like culinary, digestive, renal and reproductive problems (Dragendroff, 1885; Stuart, 1911). The seeds are used for the treatment of stomachache, articular pain, seminal loss, diabetics, spleen, and gonorrhoea. The puffs are easily digestible and good for human health because of its low fat content. In addition, the different parts of the entire plant are used for medicinal purposes as remedy for rheumatism, polyurea, spermatorrhoea, parturition and bile disorder (Chopra *et al.*, 1986; Jha *et al.*, 1991).

Water Chestnut

In recent past an effort is initiated to bring economically important aquatic plants under commercial cultivation programme so as to exploit their potential through utilization of large scale fallow wetlands, particularly in South Asian countries. Water chestnut (Figure 2.2a),(*Trapa bispinosa*, family Trapaceae), is one such species of important aquatic plant communities, potentiality of which is yet to be explored. Of late, its utility as cultivable species has been recognized because of its edible fruits, kernel of which is tasty, nutritious (Mandal *et al.*, 2010) and medicinal too (Chopra *et al.*, 1986).

Figure 2.2a: Water chestnut.

Botany and Ecology

Trapa is an annual herb with buoyant rosette of leaves around a central stem. Leaves have spongy inflated petioles which enable the rosette to float. A central floating stem is surrounded with numerous fibrous roots. Flowers are white and are born at leaf axis of younger leaves just above the water. Fruits develop above, but move downward as meristem elongates, and thus become matured into water. The single seeded matured fruits are woody and bear two pointed spines.

Water chestnut is grown in nutrient rich water body with pH ranging between 6.5 and 8.5. Climatic conditions are recorded as rainfall ranging between 150 and 300 cm annually, temperature from 18°C to 38°C and humidity between 60 and 98 per cent (Mandal *et al.*, 2010).

Cultivation Practice

At the onset of monsoon during late June, lands are prepared for plantation of trapa seedlings. Depth of water is maintained in the range of 1.0-1.5 m. Seedlings are planted keeping distance of

1.25cm between two individual ones and the distance between two rows of seedlings generally remain 75.0cm. Seedlings take three months to flower, accompanied fruiting afterwards.

During late September first harvest takes place, followed by second one after 3-4 days. First harvest yields around 200 kg of fruits, and the second one 250 kg of fruits. Individual plant may produce 5-6 fruits/harvest. Yield continues up to 3 months from initial harvest and then the entire plant dies out (Mandal *et al.*, 2010).

Figure 2.2b: Red fruits of Water chestnut.

Culinary Importance

There are two colours of fruit-coat: red and green (Figure 2.2b). Red fruits are more acceptable than green fruits as per culinary preference. Fresh kernel is edible directly or eaten after boiling in certain occasions. Flour of dry kernel after grinding is usually treated as substitute of cereal flour. Fresh kernels are used for preparing desserts and other Bengali delicacies (polauw, payesh, etc.).

Table 2.2: Nutritive values: (Proximate composition of 100 g edible portion).

Food Items	*Moisture (g)*	*Protein (N×6.23) (g)*	*Fat (g)*	*Minerals (g)*	*Crude Fibre (g)*	*Carbo-hydrate (g)*	*Energy Kcal*	*Ca (mg)*	*P (mg)*	*Fe (mg)*
Fresh kernel	70	4.7	0.3	1.1	0.6	23.3	115	20	150	1.35
Dry kernel	13.8	13.4	0.8	3.1	–	68.9	330	70	440	2.4

Source: Gopalan *et al.*, 2007 after modification.

Besides these substantial nutrient constituents (Table 2.2), kernel contains comparatively higher amount of Zn (1.4-1.56 mg/100g), which has multiple roles in metabolism. Natural source of Zn is poor and therefore, it is supplemented from other sources or in the form of chemical medicine. In such condition, consumption of kernel provides a certain amount of Zn essential in human health. Other micronutrients are available in kernel with respective concentration/100g: Mg 72µg, Cu 1.31µg, Mn 0.85µg, and Cr 0.011µg (Gopalan *et al.*, 2007); all of which, in low concentration, are beneficial to human health.

Therapeutic Value

Water chestnut is of substantial medicinal value: nut has cooling effect, useful for remedy of diarrhoea, dysentery and bilious affections. White soft kernels ground and mixed with milk are useful for remedy of gastrointestinal trouble and leucorrhoea. Kernel is also used for cooling of bile effect. It is reported useful also for remedy of memory loss and improving thinking skill (Chopra *et al.*, 1986).

Lotus

Lotus (*Nelumbo nucifera*) of the family Nelumbonaceae, is treated as a sacred flower (Figure 2.3a) in India. Its flower is valued for its cosmic beauty and versatile qualities. Lotus flower creates sanctity at shrines across all the religions and is in great demand as horticultural crop. Historically, lotus is grown in the far East for 5000-7000 years (Wong, 1987). It is extensively acquainted with people of Asia and now its appreciation has gradually percolated to the West. In Buddhism, lotus symbolizes purity (Shen-Miller, 2002). Virtually, all parts of lotus plant are used for multipurpose benefits of human need (Shen-Miller, 2002; Mandal *et al.*, 2007).

Figure 2.3a: Lotus flower.

Botany and Ecology

Lotus is a rooted floating leaved aquatic plant. It propagates through runner – a propagating part that can penetrate vertically about 1.0 m down into bottom and spread horizontally about 1.5 m distance to develop new shoot and foliage. Erect and stout leaves supported with long petiole place its canopy over the water surface, sometimes even above the water surface. Its petiole is adapted to increase its length up to 3.0 m proportionately with increasing water level. In India, it grows well in all types of wetlands: beel, jheel, karanjali (shallow wetlands), ditches, ponds, reservoirs and lakes. Tropical climate is suitable for its growth and development.

Cultivation Practice

Growth of lotus plant cycles through two distinct periods: (i) Active period and (ii) Dormant period. Total period of cultivation covers 6-7 months which commence at mid of March and ends at mid of October. The entire period of cultivation including harvest of flower is divided into three phases: (i) 1st phase (ii) 2nd phase and (iii) 3rd phase. Usually, 1st phase starts at mid of March and continues up to May, 2nd phase from June to July and 3rd phase from July to mid of October.

Importance as Food

People use almost the entire parts of lotus: rhizome (Figure 2.3b) and roots are used for food after boiling; starchy rhizomes are eaten raw, roasted (Figure 2.3c), as pickles, and as dried slices, fried as chips (Orozco-Obando, 2009). Thalamus is used as vegetable directly or being blended with other vegetables. Tender leaves are also used as vegetables and sometimes mixed with flour to make delicious lotus bread (Orozco-Obando, 2009). Seeds are an oriental delicacy and are eaten raw, roasted, boiled, pickled, candied, ground as meal and roasted to make puff (Billing and Biles, 2007). Seeds are even eaten as a snack, and the stem and roots are made into soups and other dishes (Evans *et al.*, 2004). Petioles are cut into number of pieces and mixed as one of the ingredients in preparation of pickle. Pickle having lotus petiole is more in demand than that without petiole. Tender lotus leaves are used as fish fodder, particularly for grass carp (*Ctenopharyngodon idella*).

In addition to the above substantial nutrient constituents (Table 2.3), rhizome contains different micronutrients in respective concentrations (mg/100g): Mg, 168; Cu, 1.22; Na, 438; P, 3007; S, 258 and Ca, 444; all these nutrients are important with regard to human health.

Figure 2.3b: Rhizomes of lotus.

Figure 2.3c: Roasted and cooked rhizomes.

Table 2.3: Nutritional value: (Proximate composition of 100 g edible portion).

Parts	*Moisture (g)*	*Protein (N×6.23) (g)*	*Fat (g)*	*Minerals (g)*	*Crude Fibre (g)*	*Carbo-hydrate (g)*	*Energy Kcal*	*Ca (mg)*	*P (mg)*	*Fe (mg)*
Seeds (green)	84.6	3.9	0.7	1.1	0.9	8.8	57	49	151	
Seeds (dry)	10	17.2	2.4	3.8	2.6	64.0	346	36	294	2.3
Petiole (dry)	9.5	4.1	1.3	8.7	25.0	51.4	234	405	128	60.6
Rhizome	85.9	1.7	0.1	0.2	0.8	11.3	53	21	74	0.4

Source: Gopalan *et al.*, 2007 after modification.

Therapeutic Value

Therapeutically, flowers are cooling agents, used as astringent in diarrhoea, cholera, fever and recommended as cardiac tonic. Seeds are used to check vomiting, given to children as diuretic, form a cooling medicine for skin diseases and leprosy, and considered as antidote to poison. Petioles are considered as astringent and cooling, useful in burning sensation of body, and bleeding piles. Root in powder form is prescribed for curing piles, dysentery, and dyspepsia, is used as a paste in skin affection and ringworm (Chopra *et al.*, 1986). Honey, if it has major proportion of nectar of lotus flower, is treated as good as 'Amrita' in Ayurveda. It is believed to be so precious that people who are suffering from many common ailments can recover after its intake (Bhattacharjee, 1995; Mandal *et al.*, 2007).

Note: In mythological view, 'Amrita' means the food items by which intake, human beings become immortal; 'Ayurveda' is the indigenous therapeutic system based on medicinal plants practised in India.

Possible Benefits

- ☆ High market value of flowers and 'lotus honey'
- ☆ Petals used for extraction of scent
- ☆ Thalamus used as vegetables
- ☆ Petioles used for preparation of pickles
- ☆ Leaves used as plates
- ☆ Seeds, rhizomes and roots used as food and medicine
- ☆ Lotus cum fish culture – an integrated approach of aqua-farming
- ☆ Organic aqua farming due to periphyton assemblage
- ☆ Maintenance of aesthetic beauty
- ☆ Maintenance of ecological balance through accumulation of pollutants

Water Lily

Water Lily (Figures 2.4a, b) has a worldwide distribution. It has a variety of colour and some species have fragrance. It grows in temperate, tropical and sub-tropical climates. It is mostly used as water gardening plant in western country because of its attractive and beautiful flower.

Figure 2.4a: Water lily.

Botany and Ecology

Nymphaea under the family Nymphaeaceae has four common species like *Nymphaea alba, N. rubra, N. pubescens* and *N. stellata* growing in India. It is a perennial plant growing up to 245.0cm deep. The leaves develop directly from root stock on long petioles measured 12.5-35.5cm across, orbicular in shape, cleft towards the middle on one side and smooth along the outer margins. Long flowering stalk also develops directly from rootstock and bears a single flower which opens up during morning and closes during afternoon. Each flower is about 10.0-17.5cm across. In the middle of the

Figure 2.4b: Water lily.

petals, there are many yellowish stamens surrounding single pistil. Each flower lasts for only 4-5 days and produces globose shaped fruit having enormous number of seeds.

Importance as Food Item

All the parts of *Nyphaea* are used as food stuff for human beings: Rhizomes (Figure 2.4c), stalks and thalamus are eaten either raw or prepared for culinary items (Figure 2.4d). Seeds are eaten raw or fried to make puffs which are used as good as rice puff. Rhizomes are used after boiling which are tasty to eat. Beside human beings, all the parts of this flora are very much used for fodder, a common practice eastern part of India.

Table 2.4: Nutritional value: (Proximate composition of 100 g edible portion).

Parts	*Moisture (g)*	*Protein (N×6.23) (g)*	*Fat (g)*	*Minerals (g)*	*Crude Fibre (g)*	*Carbo-hydrate (g)*	*Energy Kcal*	*Ca (mg)*	*P (mg)*
Rhizome (red)	49.1	4.1	0.3	1.6	1.5	43.4	193	65	217
Rhizome (white)	62.5	3.1	0.3	1.3	1.1	31.7	142	76	220
Seeds	10.0	8.3	1.0	0.9	4.2	75.8	346	20	110
Flower	90.8	1.6	0.6	0.7	0.9	5.4	33	29	18

Source: Gopalan *et al.*, 2007 after modification.

The nutrient profile (Table 2.4) is important in respect of human health. People collect this wild flora from wetlands (Figure 2.4e) and make use in preparation of food items.

Figure 2.4c: Rhizomes as food.

Figure 2.4d: : Stalks and thalamus as food items.

Figure 2.4e: Water lily in wetlands.

Therapeutic Uses

Rhizome and stalk of *N. alba* are used as astringent, slightly narcotic and administered in dysentery. Infusion of flower and fruit is given in diarrhoea. Root contains alkaloid nymphaeine and several astringents. Flowers and rhizome yield two alkaloids both of which have sedative actions in small doses. Powdered roots of *N. pubescens* are used for remedy of piles, dysentery and dyspepsia. Flowers are used for preparation of cardiac tonic. Powdered roots of both *N. rubra* and *N. stellata* are used for remedy of piles, dysentery and dyspepsia. Flowers are prescribed in palpitation of heart (Chopra *et al.*, 1986).

Colocasia

Colocasia esculanta (Figure 2.5a) under the family Araceae is native to wetlands of Malaysia. The specific epithet, esculenta, means edible in Latin. In India, it was reported being cultivated in wetlands before 5000 BC. It has important food value with rich nutrient status.

Figure 2.5a: Water Colocasia.

Botany and Ecology

C. esculanta is a rhizomatous herbaceous plant with variable shape and sizes of rhizomes. Petioles grow up to the height ranging 0.5-1.5m; Spathe is 30cm long, leaves are measured to be 45 × 25 cm, dark green in dorsal surface while light green in ventral, triangular-ovate, mucronate, basal lobes with sub-rounded. Tropical climate and wet habitats are ideal for growth of this species, but sometimes cultivars of this species are grown in rather semi wet soil.

Cultivation Practices

Rhizomes are planted in mud soil. Usually, fertilizer is not required for its cultivation because organically loaded soil is the facilitator of its growth. However, for mass cultivation practice, fertilizer is now applied. It takes 3-4 months to harvest edible rhizomes.

Importance as Food Item

All parts of this plant are edible and contain substantial nutrient properties (Tables 2.5–2.7). In raw form, it is not advisable to eat due to presence of calcium oxalate which is toxic and develops itching inside mouth. Toxin can be minimized by way of cooking and thus tubers become palatable. Corms of small round variety are peeled and boiled before their consumption. In wild ones, runner – propagating unit, is also edible and used for cooking.

In addition, its leaves (green variety) contains the following/100g: β-carotene 15,700μg, Mg 32mg, while tuber is analyzed to have Mg 28mg, Na 9.0mg, K 550 mg and Mn 0.28mg; all these are essential for human health for protecting body from different ailments.

Table 2.5: Nutritional value: (Proximate composition of 100 g edible portion).

Parts	*Moisture (g)*	*Protein (N×6.23) (g)*	*Fat (g)*	*Minerals (g)*	*Crude Fibre (g)*	*Carbo-hydrate (g)*	*Energy Kcal*	*Ca (mg)*	*P (mg)*	*Fe (mg)*
Tuber	73.1	3.0	0.1	1.7	1.0	21.1	97	40	140	0.42
Petiole	94.0	0.3	0.3	1.2	0.6	3.6	18	60	59	2.5
Leaves (black)	78.8	6.8	2.0	2.5	1.8	8.1	77	460	125	0.98
Leaves (green)	82.7	3.9	1.5	2.2	2.9	6.8	56	227	82	10.0

Source: Gopalan *et al.*, 2007 after modification.

Table 2.6: Vitamins content: values as per 100gm of edible portion.

Parts	*Carotene (μg)*	*Thiamine (mg)*	*Riboflavin (mg)*	*Niacin (mg)*	*B_6 (mg)*	*Folic Acid (μg)*	*Vitamin. C (mg)*	*Choline (mg)*
Tuber	24	0.09	0.03	0.4		16	54	12
Petiole	104	0.07	0.07	0.1			3.0	
Leaves (black)	12000	0.06	0.45	1.9			63.0	
Leaves (green)	10278	0.22	0.26	1.1			12	

Source: Gopalan *et al.*, 2007 after modification.

Table 2.7: Amino acids present in *Colocasia* (mg) per gm N.

Parts	*N g/100gms*	*A*	*H*	*L*	*T*	*Ph*	*Tr*	*M*	*C*	*Th*	*Le*	*Is*	*Va*
Tuber	0.48	470	110	300	110	320	230	080	160	280	510	270	380
Leaves	1.09	250	130	260	040	180		060		150	360	310	280

A: Arginine; H: Histidine; L: Lysine; T: Trysophan; Ph: Phenylamine; M: Methionine; C: Crystine; Th: Threonine; Le: Leucine; Is: Isoleusine; Va: Valine.

Source: Gopalan *et al.*, 2007 after modification.

In terms of nutrient status, Colocasia (Figure 2.5b) is one of the important foodstuffs; however, its proper use as potential food is yet to be explored. It grows abundantly in wetland areas; populace use to make it as their common food item and take varieties of its culinary preparations for its easy availability. The huge amount of its use as food items is not probably recorded in account since it is treated as a non-conventional crop.

Therapeutic Value

Colocasia is used for remedy of bowel complaints. Petiole juice is styptic, stimulant, rubifaceant, and corm juice used in alopecia and in scorpion sting (Chopra *et al.*, 1986).

Figure 2.5b: Rhizomes of colocasia.

Feed

A wide range of aquatic plants/weeds are used as ingredients for preparation of foodstuff known as 'feed'.

Wild Macrophytes

Wild aquatic macrophytes contain substantial amount of nutrients (Table 2.8). The proximate composition is: moisture ranges 84.1-95.9 per cent, dry matter 4.1-15.9 per cent, crude protein 8.7-26.8 per cent, crude fat 2.2-5.1 per cent, carbohydrate 9.3-35.6 per cent, ash 8.0-25.3 per cent, and crude fiber 15.0- 28.1 per cent along with caloric content (2.47- 4.2 kcal/g). They are blended with other feedstuffs and fed to livestock including fish. It is experimented that about 50 per cent of plant protein is converted to fish protein and even 30 per cent inclusion of macrophytes in the preparation of feeds has shown better balance of amino acid with particular reference to fish species. More and above, herbivores have shown satisfactory growth in respect of food intake, feed conversion ratio and relative growth rate while feeding upon such feeds (Mandal *et al.*, 2010).

Table 2.8: Nutritive status of a few aquatic macrophytes (dry weight basis).

Species	*Ash (per cent)*	*Protein (per cent)*	*Lipid (per cent)*	*Carbohydrate (per cent)*	*Calorific Content (Kcal/g)*	*Reference*
Carex hudsonii	3.0	–	–	–	4.05	Fischer, 1973
Ceratophyllum demersum	20.6	21.7	5.97	27.9	3.71	Boyd,1968
Chara sp.	35.8	17.5	1.63	23.8	2.58	Boyd, 1968
Eichhornia crassipes	1.7	16.0	3.4	28.1	3.82	Boyd, 1969
Elodea densa	21.1	20.5	3.27	29.2	3.35	Boyd, 1968
Glyceria fluitans	10.0	–	–	–	4.11	Fischer, 1973
Hottonia palustris	10.0	–	–	–	4.06	Fischer,1973
Hydrilla verticillata	2.8	17.5	3.47	32.6	3.46	Boyd,1969
Juncus effuses	13.0	–	–	–	3.93	Fischer,1973
Lactuca sativa	15.0	–	–	–	3.86	Fischer, 1973
Lemna minor	17.2	20.4	3.8	–	–	Banerjee and Matai,1990
Myriophyllum spicatum	40.6	9.8	1.81	18.8	2.47	Boyd, 1968
Najas guadelupensis	18.7	22.8	3.75	30.56	3.55	Boyd, 1968
Nelumbo nucifera	9.9	21.9	2.8	–	–	Banerjee and Matai,1990
Nymphae spp.	14.1	16.8	2.6	–	–	Banerjee and Matai,1990
Ottelia alismoides	16.5	14.3	2.9	–	–	Banerjee and Matai,1990
Potamogeton crispus	16.0	10.9	2.85	37.2	3.61	Fischer, 1973
Trapa natans	13.6	10.8	5.1	–	–	Banerjee and Matai,1990
Typha latifolia	6.9	10.3	3.91	32.2	3.61	Boyd, 1968
Wolffia arrhiza	14.5	21.5	5.50	10.6	–	Culley and Epps, 1973

Azolla

Azolla (Figure 2.6a) is used an ingredient for preparation of livestock feed, particularly fish feed. It improves the water quality by acting as Controlled Ecological Life Support System (CELSS) because of its unique photosynthesis oxygen releasing capacity (Liu *et al.*, 2008), while being cultivated in water bodies.

Botany and Ecology

The genus Azolla belonging to family Azollaceae comprises 6 species like *Azolla pinnata, A. nilotica, A. caroliniana, A. filiculoides, A. microphylla,* and *A. rubra;* among which *Azolla pinnata* occupies a prominent place for its unique role in agriculture and also in aqua-farming in India (Mandal *et al.*, 2012). *A. pinnata* has its main axis with alternate leaves and adventitious roots at regular intervals. The entire vegetation appears to be a mat and floats freely on water surface.

Its growth usually appears to be lush green because of its association with cyanobacteria, even in unproductive habitats. It can double its biomass in 3-10 days. In tropical rice fields, azola has a standing crop of 8-10 t/ha on wet weight basis, with fixing of nitrogen @ 2-3 kg/ha/day. Azolla grows in the varied range of ecological factors: pH, 4.0-9.0; temperature, 20-33°C; humidity, 30-95 per cent.

Cultivation Practices

A simple method of *Azolla* nursery for its large scale production/ha in field condition is described as (Mandal *et al.*, 2012):

- Nursery field is to be thoroughly prepared with having inlet and outlet for water supply and drainage.
- Water is to be manipulated at a depth of 10 cm usually and may vary depending up on the situation of the field.
- 10.0 kg of fresh cattle dung mixed in 20litre of water is sprinkled in each plot, followed by inoculation of 8.0 kg of *Azolla* in each plot.
- Single Super Phosphate (100 g) is applied in three split doses at intervals of four days as a top dressing fertilizer.
- Fifteen days after inoculation, Azolla is harvested; about 40-55 kg of fresh Azolla is obtained from each plot per harvest.

Figure 2.6a: *Azolla*.

Figure 2.6b: *Azolla* based diets.

Composition and Utilization of Azolla

Proximate composition of *Azolla pinnata* is mentioned on dry matter basis (per cent): dry matter, 6.6; crude protein, 21.6; ash, 15.4; crude fat, 3.8; crude fiber, 16.6; calorific value, 4.2 kcal/g, in-vitro digestibility as 77.7. *Azolla* based diets (Figure 2.6b) exhibited quite encouraging result while fed to juvenile tilapia (*Oreochromis nilotica*). *Azolla* after being mixed with rice bran or yeasts or purified enzymes might improve its ingestion and digestibility. Prepared diets with even 15 per cent incorporation of dry Azolla meal with different percentage showed excellent growth for tilapia while fed. It yielded positive growth due to presence of its some W-6 fatty acids while blended in the diet of rohu (*Labeo rohita*) (Santiago *et al.*, 1988; Mohanty and Dash, 1995).

Duckweed

The term 'Duckweed' manifests an integral relationship occurring in nature between biologically two distinct organisms: one is duck – water fowl and other weed - aquatic tiny flowering plant. Duck prefers eating these soft floating plants while foraging to water bodies where these plants grow naturally. Probably, the relationship in between duck and weed evolved as the euphony of the term 'duckweed' for simple namesake, but later it has become the synergy of ecological integration between two distinct organisms (Mandal *et al.*, 2010).

Botany and Ecology

Duckweed comprises four genera *viz., Spirodella, Lemna, Wolffia,* and *Wolfilfia* belonging to family Lemnaceae. Three of them serve as fodder for many species of fish and aquatic birds. It can grow in any type of organic waste-laden wetland distributed in tropical and subtropical countries. With reference to worldwide distribution, *Spirodella* has its greatest diversity in South Africa, *Lemna* widely distributed in North America and South-East Asia, *Wolffia* found mainly in America, South-East Asia and Australia.

1. *Spirodella polyrrhiza*: It is the largest one (each frond 6.0mm x 1.5mm; the individual canopy covers 11.0mm area in diameter (Figure 2.7a) among duckweeds species. Individual leaf is obovate in shape along with round tip and soft in nature; lower side of each leaf has pink colour due to presence of anthocyanin pigment. Each *Spirodella* plant has several roots that become tangled with having conspicuous root cap.
2. *Lemna minor:* Each new leaf emerges from a cleft or scale in the margin of the last leaf. A fast growing plant, very often, has 3 or 4 attached fronds (each frond 2.5mm x 1.5mm; the individual canopy covers 4.0mm area in diameter (Figure 2.7b). Each leaf is oblong –spherical in shape and soft in nature and pendulous roots are seen beneath at the center axis.
3. *Wolffia arrhiza:* It has, very often, single frond oval in shape. It is called water meal because without microscope it looks like dot, <0.3mm (Figure 2.7c). Floating on the surface, it is the smallest flowering plants without root.

Cultivation Practice: A Few Tips

Cultivation of duckweed requires necessary care along with a few techniques. All the species of duckweed consist of parenchymatous tissues; so they are soft, delicate and may get damaged easily. Therefore, handling duckweeds needs special attention.

Chart 2.2 shows the steps of duckweeds cultivation at the level of 500kg duckweeds production/ day/hectare water body (Naskar *et al.*, 1984; Landolt and Kandeler, 1987).

Half-one meter water depth

↓

Application of 10kg cattle dung

↓

5 days required for precipitation of cattle dung

↓

Application of duckweed inoculum

↓

Application of Urea, TSP and MP @ ratio of 10: 2: 2kg

↓

Application of crude sea salt 4.5kg

Chart 2.2: Steps of duckweed cultivation.

(TSP: Triple Super Phosphate; MP: Muriate of Potash)

Figure 2.7a: Duck weed (*Spirodella polyrrhiza*).

Figure 2.7b: *Lemna minor.*

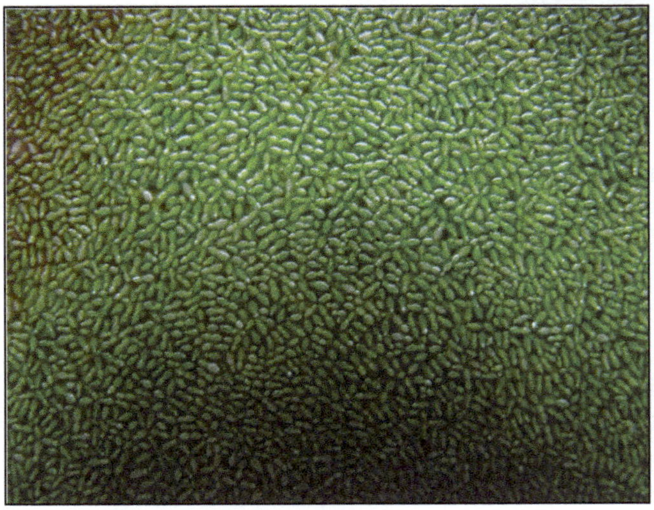

Figure 2.7c: *Wolffia arrhiza.*

Precautions: Though duckweed species tolerate a wide range of temperature, yet moderate range of temperature like 22-28°C and 10–15 lux light intensity are ideal for duckweeds growth. Hot temperature may warm the upper surface; in such condition water exchange is essential after one-week interval, else their growth may get restrained. pH should be in the range of 6-8.

Duckweeds as Feed: An Evaluation

Their rich nutrient properties (Table 2.9) and softness have been graded as easily digestible food. Duckweeds contain 91-94 per cent moisture. The proximate analysis of three duckweeds shows that protein content on dry weight basis varies 20.4-25.3 per cent, where *L. minor* contains maximum crude protein - 25 per cent. However, crude protein content of duckweed may be increased by adding additional nutrients in culture system. All the three species of duck weeds have negligible amount of fiber ranging 7.6 and11.7 per cent; considered to be suitable as fish feed, when blended with other ingredients for preparation of pelleted feeds of fish.

Nutrient Status of Duckweeds

Table 2.9: Proximate composition (per cent) of duckweeds (dry weight basis).

Item	*W. arrhiza*	*L. minor*	*S. polyrrhiza*
Moisture	94.7±0.07	94.0±0	91.0±0
Dry weight	5.3±0.07	6.0±0.3	9.0±0.4
Crude protein	20.4±0.36	25.3±0.9	23.8±0.8
Crude fat	4.63±0.13	4.5±0.5	3.8±0.5
Crude fiber	11.6±0.17	7.6±0.5	11.7±0.6
Ash	17.6±1.52	17.6±0.6	18.3±1.0
Calcium (Ca) ppm	0.16±0.03	2.0±0.02	1.47±0.47
Phosphorus (P) ppm	0.41±0.05	1.0±0.04	0.18±0.04

Source: Chareontesprasit and Jiwyam, 2001; Hassan and Edwards, 1992; Kalita, *et al.*, 2007.

Lemna minor as a Food to Fish

L. minor shows gross energy 357.8±1.6 (kcal/100 g) which is valued as 78.4±1.5 mg protein/kcal, considered to be substantial amount in terms of quality of food consumed by fish. Besides, it contains vitamin E (26.6±0.3 mg/100g), vitamin C (3.8±0.2mg/100g) and carotenoid (0.1±0.01 mg/100g); all these are very essential for fish in respect of suitable growth and good health (Kalita *et al.*, 2007).

Anti-nutrient Properties

Duckweeds contain negligible amount of antinutrients. With reference to *L. minor*, (other two species are yet to be analyzed), it has negligible amount of antinutrients (g per cent) like; Trysin inhibitor 1.47±0.5, Calcium oxalate 3.5±0.7, Tannin 0.9±0.3 and Phytate 0.004±0.003 (Kalita *et al.*, 2007).

Water Spinach

Water spinach (Figure 2.8), *Ipomoea aquatica* under the family Convolvulaceae, is known as Morning glory – a name derived because of its beautiful flowers. It is sold in tightly packed bunches in Asian markets and considered as one of the cheapest nutrient sources preferred by all groups of consumers – rich and poor. Therefore, it is cultivated as an important crop of high nutritive value essential from human health (Mandal, *et al.*, 2008).

Botany and Ecology

Figure 2.8: Water spinach.

Ipomoea aquatica is treated as a common perennial/annual aquatic plant which exhibits its variable habits in accordance with nature of habitats: it floats over the water surface in case of full water course area, and may creep on the ground in marshy wetland. It has funnel shaped; white, pink or pale lilac colour; 18cm long x 9 cm across; erect with solid peduncle. A single stem can grow and extend its apical shoot up to 20m. long, along with profuse branching as its apical shoot may advance 12 cm per day under suitable environment. This flora is basically a vine, which may form dense masses of tangled vegetation, thus developing impenetrable canopies over the water surface. Tropical and subtropical climate are ideal for its proliferation and organically enriched wetland acts as facilitator for its high proliferation (Mandal *et al.*, 2008).

Cultivation Practice

I. aquatica is propagated by seeds, or by stem cuttings. Packed viable seeds are now available in market. These seeds are sown over the marshy places during peak winter period. Farmers collect stem fragments by chopping long stem into several pieces, each of which must have at least single node from which roots initiate. In case of full course water bodies, long stem can be put at the edge of the ponds, and then fixed it with bamboo pole so that wind or water wave cannot take the vine away. It then propagates rapidly by sprouting its branches and shoots.

First time harvest may take place one month after seed sowing and then it is done at weekly interval. As many times as cuttings are to be done, biomass of vegetation is found increasing as proliferated with numerous branching. It is estimated that about 60 kg of yield (wet basis) is harvested per 100 m^2 during 1st time cutting, which then induces crop to sprout a number of branches which effect the biomass production more. Consequently, 2nd time harvest doubles the production as compared to 1st yield. Gradually, production of yields increases up to 4th successive harvest, and then it declines (Mandal *et al.*, 2008).

Importance as Food

The nutrients analysis of Water spinach shows high amount of crude protein and organic matter as 32.2 per cent and 70.0 per cent, along with low amount of crude fibre as 10.0 per cent, which is found beneficial as an ingredient for fish feed preparation. Its efficacy as an ideal food component is reflected with its rich nutrients status: crude lipid, 6.0 per cent; ash, 30.0 per cent; total carbohydrate (NFE +crude fibre), 31.8 per cent; crude fibre, 10.0 per cent; vitamin E, 28.5 per cent; vitamin B1, 87µg; Nicotinic acid, 0.6mg; Riboflavin, 120µg; vitamin C, 137 mg/100g; along with rich carotenoids: β-carotene, xanthophylls and taraxanthin. There are several minerals like K, 41.4mg per cent; Mg, 31.0

mg per cent; Zn, 1.7 mg per cent; Cu, 0.1 mg per cent; Ca, 2.0 mg per cent; Na, 5.0 mg per cent; P, 1.0 mg per cent; along with gross energy content, 337.9 kcal/100g; and protein/energy value, 95.3 mg protein/kcal (Gopalan *et al.*, 2007).

Fish Feed

When formulated feed using Water spinach was given for 60 days to the fingerlings of three carp species *viz., Labeo rohita, Catla catla,* and *Cyprinus mrigala* to evaluate its efficacy as a fish feed, it revealed a significant result in terms of growth performance as feed conversion efficiency (FCR) ranging 1.8 – 2.5, gain in body weight (BWG) ranging 18.2 – 33.3g, protein efficiency ratio (PER) ranging 0.7 – 1.0, and protein retention and specific growth rate (SGR) ranging 0.7– 1.0 per cent per day (Mandal *et al.*, 2008).

Water Chestnut

Trapa leaves containing 1.4g/100g of Zn is an important ingredient to perform multiple roles in metabolism and thereby used for preparation of fish feed. With these potential biochemical properties, *Trapa bispinosa* plants has been experimented as fish feed to IMC (Indian major carps) and proven as valuable supplement if blended with conventional foodstuff (Kalita *et al.*, 2007).

Proximate composition of Water chestnut (*Trapa bispinosa*) leaves shows the following components (dry matter basis): moisture 7.6 per cent, organic matter 87.0 per cent, crude protein 11.4 per cent, crude lipid 8.0 per cent, ash 13.3 per cent, crude fibre 4.2 per cent, total carbohydrate 67.3 per cent, gross energy 3.47kcal/g, vitamin E 61.3mg/100g, vitamin C 3.6mg/100g, carotenoid 0.2mg/100g, thiamine 0.05mg/100g, riboflavin 0.07mg/100g, Niacin 0.6mg/100g, Zn 1.4mg(per cent), Mg 25.1mg(per cent), Cu 0.1mg(per cent), Ca 2.0ppm, Na 5.0mg(per cent), K 27.7mg(per cent), P 0.9g/kg. Besides, it contains few potential fatty acids (g per cent): Myristic acid 0.007, Palmitic acid 0.02, Palmitoleic acid 0.1, Stearic acid 0.003, Linoleic acid 0.2, Arachidic acid 0.02, Arachidonic acid 0.007, Behenic acid 0.003, Nervonic acid 1.2. This unique nutrient status of Water chestnut is found to be worthy in preparation of any kind of feed items (Gopalan *et al.*, 2007; Kalita *et al.*, 2007).

Fodder

The term 'fodder' is conceived when green plant materials are used directly as food for livestock including fish. There are diverse communities of macrophytes growing in natural ecosystem; importantly, most of which are contributors as fodder to livestock including herbivorous fish.

Wild Macrophytes

A wide array of diversified aquatic macrophytes (Table 2.10) (recorded about 47 species) serves as potential fodder to about 40 fish species belonging to two major families *viz.*, Cyprinidae and

Table 2.10: Diversification of macrophytes used as fodder (Mandal *et al.*, 2010).

Plant Groups	*Species*		*Genera*		*Families*	
	Number	*(Per cent)*	*Number*	*(Per cent)*	*Number*	*(Per cent)*
Algae	1	2.12	1	2.63	1	3.7
Bryophyta	2	4.25	2	5.26	2	7.4
Pteridophyta	2	4.25	2	5.26	2	7.4
Monocotyledon	30	63.82	22	57.89	12	44.44
Dicotyledon	12	25.53	11	28.94	10	37.03

Chichlidae found in tropical and sub-tropical countries. These macrophytes represent different life forms such as emergent, free floating, rooted floating and submerged and belong to several families; of which major ones are Amaranthaceae, Araceae, Azollaceae, Cabombaceae, Ceratophyllaceae, Characeae, Convolvulaceae, Haloragaceae, Hydrocharitaceae, Lemnaceae, Lentibuliaraceae, Menyanthaceae, Najadaceae, Nymphaeaceae, Polygonaceae, Pontederiaceae, Potamogetonaceae, Salvinaceae, Trapaceae, and Typhaceae.

Azolla

Composition of Azolla appears to be rich because of its 21.6 per cent crude protein with EAA (Essential Amino Acid) having lysine rich, along with arginine and methionine. Azolla having lysine rich protein usually becomes tasty as fresh food to animals including fish (Cagauan and Pullin, 1994). Azolla is also a major source of β-carotene content (Table 2.11) while used as fodder. Importantly, the β-carotene content of leaf protein is a major source of pro-vitamin-A used for treatment of cancer and respiratory diseases. It is reported from farmers that eggs which are laid by poultry birds and duck while feeding upon Azolla fodder (Figure 2.9) have much yellow colour in yolk. It is also reported that farmers use to feed Azolla to cows for obtaining more quantity of milk (personal conversation with farmers).

Table 2.11: A comparative account of β-carotene content among common leaf vegetable.

Name of Common Leafy Vegetables	*Carotene Amount (μg/100g)*
Amaranthus gangeticus (Tender)	5,520
Amaranthus paniculatus	14,190
Bethua leaves	1,740
Cabbage	120
Carrot leaves	5,700
Colocasia leaves (black variety)	12,000
Colocasia leaves (green variety)	10,278
Coriander leaves	6,918
Cow pea leaves	6,702
Curry leaves	7,560
Drumstick leaves	6,780
Ipomoea leaves	1,980
Lettuce	990
Mustard leaves	2,622
Radish leaves	5,295
Spinach	5,580
Azolla pinnata	**63,200***

Source: Dewanji, 1993; Gopalan *et al.*, 2007.

Figure 2.9: Duck feeds Azolla.

Duckweed

Three genera of duckweed like *Lemna minor*, *L. gibba*, *Spirodella polyrrhiza* and *Wolffia arrhiza* occupy an important place as fodder given to poultry along with herbivorous fish with particular reference to grass carp (*Ctenopharyngodon idella*) (Azim and Wahab, 2003). Besides, duckweeds are found as preferred fodder to a wide range of herbivorous fish like silver berb (*Puntius gonionotus* and *P. jerdoni*), tilapias (*Oreochromis niloticus*, *Tilapia rendelli* and *T. zillii*), rohu (*Labeo rohita*) (Singh *et al.*, 1967; Gaigher *et al.*, 1984; Kalita *et al.*, 2007) because of their high nutrient contents and low amount of fiber and fat.

Water Hyacinth

Water hyacinth is used as fodder (Figure 2.10) for both non-ruminant and ruminant animals like pig, cattle, buffalo and horses. Mainly tender leaves are used for this very purpose before development of its fibre content. The plant leaves are chopped and mixed with other materials like banana stalks and vegetable wastes. The total ingredient is then boiled together in which the proportion of water hyacinth is about 10 per cent (Mandal *et al.*, 2011).

Figure 2.10: Water hyacinth as fodder.

Nutrient Value

The fresh water hyacinth contains following nutrients (per cent): Moisture 75.5, organic matter 3.5, nitrogen 0.05, ash 1.0, phosphorus 0.06 and potash 0.02. In addition, water hyacinth exhibits rich profile of amino acids along with a few volatile fatty acids (Trivedi and Thomas, 2005).

Amino acid contents (per cent): Arginine 2.98, Histidine 1.9, Glycine 5.14, Alanine 5.59, Serine 3.85, Aspertic acid 17.37, Glutamic acid 9.29, Proline 4.73, Methionine 0.73, Cystine 11.6, Phenylanine 4.72, Tyrosine 2.98, Threonine 4.32, Lysine 5.34, Isoleusine 4.32, Valine, 0.27 and Leucine 7.2 (Trivedi and Thomas, 2005). All these important amino acids, if extracted through easy method which need to be developed, will certainly be beneficial to enrich different common food items by way of their value addition.

Fish Fodder: An Evaluation

Tilapia (*Oreochromis niloticus)* feeding upon diet containing 25-75 per cent of semi-composted form of Water hyacinth (*Eichhornia crassipes*) showed better enhancement of growth as compared to control diet having higher protein and energy level; even no significant change occurred in fish physiology afterwards (Edwards *et al.*, 1985). Inclusion of 10 per cent water hyacinth in semi-composted form also favoured the considerable growth rate of other herbivorous fish like *Puntius javonicus*, *Cyprinus carpio* and *Oreochromis mossambicus* (Hutabarat *et al.*, 1986) *Heteropneustes fissilis* and *Clarias batrachus* while feeding upon diet containing water hyacinth leaf meal (Nimat and Jafri, 1984; Rath and Datta, 1991).

Water Spinach

This species can play a significant role as fodder used for livestock including fish, especially for grass carp, *Ctenopharyngodon idella.* This species is known being a voracious grass eater, which can consume every day the amount of grass equal to its body weight. In such condition, *Ipomoea* vegetation may be put and maintained in the periphery of water bodies where grass carps are cultured. In view of nutrient conversion, maintenance of *Ipomoea* vegetation in any water bodies may also fetch two benefits at a time: (i) water spinach eating grass carps are to be treated as an organic product that will have more consumer preference, and thus may generate profitable revenue. (ii) Water spinach eating grass carp indirectly benefits the health of consumers, as *Ipomoea* is a source of rich nutrients which ultimately reach human beings via fish. It is also used as a fodder given to cattle, pig and goats, which prefer to relish it. Chopped and boiled, it is given to cows to get a maximum amount of milk (Mandal *et al.*, 2008).

Vegetables

There is a major concern about dwindling of wetlands by human interference. Experts opine that loss of wetlands brings a number of inevitable treats towards environment, one of which is loss of wetlands biodiversity. Wetlands are known repository of many worth biotic resources, which serve human civilization since immemorial time. Ironically, there is not so much strong hue and cry about disappearance of a few little known wetlands flora; which importantly support livelihood to considerable population in rural areas. These are mostly seasonal and a few are annual, having dense population, thus sharing a major part in floral community of aquatic vegetation. Not that they are merely leaf vegetables eaten but that they are potentially nutritious and medicinal too. Probably, therefore anybody could find them as one of the daily items seen in full meal in rural areas. Without manuring and less care, they grow wildly in fallow wetlands. Treated as non-conventional leaf vegetables they are more organic than other leaf vegetables cultivated. Rural masses are acquainted with their taste along with traditional knowledge based medicinal importance to some extent, but are not much aware of their nutritional excellence (Mandal and Mukhopadhyay, 2010).

Botanical Description

1. *Ipomoea aquatica* Forsskal: Perrenial/annual herb with long stem trailing on mud or floating over water surface. Leaves are elliptic in shape; flowers are large and showy: corolla pink, purple and white.
2. *Centella asiatica* (L.) Urban: Creeper with numerous roots developed from lower nodes. Leaves are orbicular, reniform with short peduncle. Inflorescence is umbel shaped, comprising 3-6 flowers axillary (Figure 2.11a).
3. *Hygrophilla auriculata* Heine: Erect herb with long spines in each node. Leaves are sessile and oblong-lanceolate. Flowers are white-purple (Figure 2.11b).
4. *Bacopa monnieri* (L.) Wettstein: Prostrate but branches are erect. Leaves are fleshy, dark green, oblong and sessile. Flowers are white and solitary.
5. *Mullugo cerviana* L.: Erect annual herb. Leaves are opposite at whorl. Flowers are dull white at terminal cyme.
6. *Polygonum plabejum* R. Br. Prodr.: Prostrate and diffusely branched herb. Leaves are small, but are larger than internode. Flowers are whitish purple.
7. *Enhydra fluctuans* Lour.: Profusely branched perennial herb. Leaves are opposite with margin distinctly dented. Flowers are white capitulate (Figure 2.11c).
8. *Marsilea minuta* L.: Perennial fern. Leaves are quadrifoliate; sori are black and oval (Figure 2.11d).

Figure 2.11a: *Centella asiatica*

Figure 2.11b: *Hygrophilla auriculata* .

Figure 2.11c: *Enhydra fluctuans*.

9. *Alternanthera sessiles* (L.) R. Br. Prodr.: Decumbent herb with purple stem. Leaves are lanceolate. Flowers are white cluster at axile.
10. *Alternanthera philoxeroides* (Mart.) Griseb.: Perennial creeper with erect branching. Leaves are opposite, lancolate, light green. Flowers are while capitulate at axile.

Figure 2.11d: *Marsilea minuta.*

Practical Uses

Both medicinal values and nutritive attributes of these wild leafy vegetables appear to constitute important properties which can benefit human health (Tables 2.12 and 2.13). Probably, therefore rural populations are dependent on these leafy vegetables for their daily consumption. Traditional knowledge that have been inherently passing through generation, surprisingly resemble also to modern therapeutic values. Nevertheless, these species are neglected and any conservation policy is yet to be formulated to protect them. These wild vegetables are full of valuable nutrients, which can protect human health, particularly rural population which usually suffer from malnutrition. It is a fact that a large section of rural populations are not having a considerable amount of nutritious food on daily basis. In this circumstance, they have at least a minimum amount of nutrients by eating these wild leafy vegetables and thus get nourished. In such case these wild vegetables, which are having substantial amount of valuable nutrients, may be substitute to fill up the nutritional vacuum at rural areas to some extent.

Conservation Issues

Destruction

Indiscriminate exploitation, particularly *Centella asiatica, Hygrophilla auriculata, Bacopa monnieri, Mullugo cerviana, Enhydra fluctuans,* and *Marsilea minuta* by local people who ignorantly harvest and supply them to pharmaceutical agencies, apart from their harvest for regular consumption is a great concern. In return they are offered a negligible amount of money. The rate at which wetlands are being dwindled is a seriously alarming matter for survival of these little known plant communities. Because once their habitats are lost, they will be under inevitable threats. Eventually, the effect will obviously bring to have their disappearance.

Prioritization of Conservation

Probably, there is a gap of consciousness between traditional eaters and those who think about their conservation. Traditional eaters are not in a position of thinking towards disappearance of these wild leafy vegetables. They think if these are available, they will collect them. They are not much aware of the impact of their disappearance. It is not expected so! Environmental conservationists know about their utility, but probably they suffer from lack of information about the data of proper record pertaining to the account of their use by rural population: how many number of peoples per annum

Table 2.12: Summary of usage of following wild leafy vegetables.

Scientific Name	*Family*	*Parts Used*	*Purpose of Uses*	*Status*	*Traditional Knowledge*	*Medicinal Importance*
Ipomoea aquatica	Convolvulaceae	Entire plant	Leafy vegetable	Wild and partially cultivated	Use for remedy of constipation	Considered wholesome for females suffering from nervous and general debility.
Centella asiatica	Apiaceae	Leaves	Leafy vegetable and medicine	Wild	Remedy for chronic amoebiosis and blood purification	Useful tonic in diseases of skin, leprosy, nerves and blood tonic for improving memory, syphilic skin disease for internally and externally.
Hygrophilla auriculata	Acanthaceae	Leafy twigs	Leafy vegetable and medicine	Wild	Use for remedy of anemia, especially female	Tonic for jaundice, dropsy, rheumatism, anasarca and disease of the urinogenital tract.
Bacopa monneiri	Scrophulariaceae	Leafy twig	Leafy vegetable and medicine	Wild and non conventional management	Increase memory, especially children	Tonic useful for asthma, epilepsy, insanity, hoarseness.
Mollugo cerviana	Ficoidacoae	Leafy twig	Leafy vegetable and medicine	Wild	Remedy of skin disease and purification of blood	Febrifuge, used for promoting the flow of lochial discharges and as cure for gonorrhea.
Polygonum plabejum	Polygonaceae	Leafy twig	Leafy vegetable	Wild	Use for remedy of bowel complaints	Given in bowel complaints, dried and powdered taken in internally in pneumonia.
Enhydra fluctuans	Asteraceae	Leafy twig	Leafy vegetable and medicine	Wild	Remedy of skin disease and purification of blood	Laxative, useful in skin and nervous affection, antibilious, and dedulcent.
Marsilea minuta	Marsileaceae	Foliages	Leafy vegetable and medicine	Wild	Remedy for insomnia	Tonic useful for remedy of insomnia.
Alternanthera sessiles	Asteraceae	Entire plant	Leafy vegetable	Wild	Use for remedy of constipation	Not known
Alternanthera philoxeroides	Asteraceae	Entire plant	Leafy vegetable	Wild	Use for remedy of constipation	Not known

Source: Chopra *et al*., 1986.

Table 2.13: Nutritional status of following wild leafy vegetables (contents/100g).

Scientific Name	*Moisture (g)*	*Protein (g)*	*Fat (g)*	*Minerals (g)*	*Fiber (g)*	*Carbo-hydrate (g)*	*Energy (Kcal)*	*Ca (g)*	*P (g)*	*Fe (g)*	*Carotene (µg)*	*Ribo-flavin (mg)*	*Vitamin C (mg)*
Ipomoea aquatica	90.3	2.9	0.4	2.1	1.2	3.1	28	110	46	3.9	1980	0.13	37
Centella asiatica	92.7	0.9	0.3	1.3	2.1								
Hygrophilla auriculata	87.2	3.0	0.4	2.8	1.4	5.2	36	330	21				
Mollugo cerviana	91.7	2.4	0.4	1.0	2.2	2.3	22	370	67	12.3			
Polygonum plabejum	83.2	3.2	0.7	3.9	2.1	6.9	46	194	48				
Enhydra fluctuans	94.1	1.2	0.25	1.0	1.4								
Marsilea minuta	86.9	3.7	1.4	2.1	1.3	4.6	46	53	91				
Alternanthera sessiles	77.4	5.0	0.7	2.5	2.8	11.6	73	510	60	16.7	1926	0.14	17
Alternanthera philoxeroides	85.7	3.3	0.8	2.7	–	7.5	50	322	29	16.8			

Source: Gopalan *et al.*, 2007.

feed upon these wild leafy vegetables, and how much amount of them are utilized annually. People use them as a non-conventional vegetables, and to some extent know about their medicinal importance through traditional knowledge.

Steps to Conserve them

How the wild leafy vegetables to be conserved is a critical task, though not impossible. There are suggestions which may be implemented under conservation programme. A through survey is essentially required to generate data about their record of use: (a) total area of their coverage, (b) locality where they grow, (c) climatic condition they grow in, (d) phenological record (e) method of reproduction, (f) varieties of usage, (g) their amount of exploitation/annum and ways of exploitation, (h) status in red data book, (i) details on medicinal and nutritive value, (j) present status in 'Ayurveda therapeutic use.

Aquatic Plants and Wetlands Environments

India, with its annual rainfall of over 130cm, varied topography and climatic regimes support and sustain diverse and unique wetlands (Prasad *et al.*, 2002) which are usual habitats of 20 per cent of the known range of biodiversity (Deepa and Ramachandra 1999). However, if a little effort is put in the perspective of utilization of large scale fallow wetlands, poor masses can earn their support based livelihood from harnessing the potential of these aquatic plants. It is important that these aquatic plants grow in the lean period cycled in between two major crop seasons - 'Kharif' and 'Rabi-'. In this lean period, farmers usually face hardship in earning livelihood due to lack of sufficient agricultural activities. These valuable aquatic plants are usually found growing in the lean periods and thereby may appear to be conducive to support farmers having substantial earning.

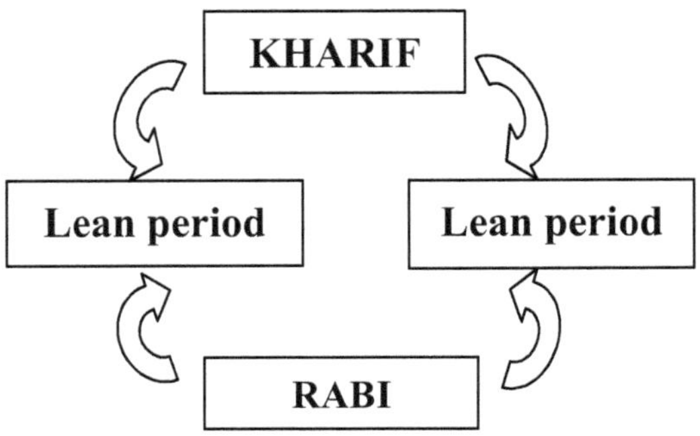

Figure 2.12: A conceptual model on crop cycle.

This wide range of wetlands may have great impact on human society to the great extent as:

Water Conservation

Conservation of freshwater is one of the important issues worldwide. Freshwater in near future can seriously be in crisis because of global warming and climate change. In this respect, maintenance of wetlands along with their biodiversity with particular reference to aquatic plants may protect freshwater from the loss due to excessive evaporation. Canopy coverage of aquatic plants hinders light penetration inside the water and protects it from huge amount of water loss. This kind of conservation programmes through maintenance of aquatic plants may fetch advantages like storage of freshwater and production of high quality nutritive food.

Conservation of Wild Fish Species

During post monsoon period, water bodies covered by partial aquatic vegetation become suitable place for shelter of a wide range of wild fish species which are now threatened due to injudicious catch, fragmentation of habitats, loss of suitable niche and water pollution. In fact, wild fish species pass their most important event of life phases in that shelter before taking part in breeding, spawning and rearing of progenies. In this respect, these water bodies and their aquatic vegetation serve the important repository of gene pool for air breathing cat fishes, non air breathing cat fishes, climbing perch, along with a variety of minnows and small prawns.

Conservation of Aquatic Ecosystem

Submerged parts of aquatic plants act as suitable substrata around which a variety of periphyton algae assemble and then act as food sources for zooplankton, insects and other aquatic organisms. A grazing food chain initiates and propels from this primary food source, upon which zooplankton, aquatic insects and other organisms graze and feed. These secondary food sources like zooplankton, insects and lower organisms are, in turn, fed by wild fish species and thus, a well balanced food web is found prevalent. Besides, a huge amount of periphyton assemblage enhances productivity of water bodies by addition of O_2 through photosynthesis.

Epilogue

Nurturing aquatic plants as possible future crops needs 3P: Protection, Propagation and Proliferation. Protection includes both maintenance of natural habitats of aquatic plants and conservation of its wild genetic stocks. Propagation requires production and development of aquatic plants of cultivable varieties through application of advanced biotechnology. Proliferation certainly indicates cultivation of aquatic plants with aiming to bring as much as fallow wetlands for their sustainable utilization.

It may be perceived that if one-fifth area of the total fallow wetlands are utilized in the purpose of harnessing potential of nutrient substances of selective aquatic plants, it can surely bring about a remarkable change in the scenario of future crops. It may also augment the higher scale of revenue generation compared to the present context; else a huge amount of nutrients which aquatic plants contain will get lost without proper utilization. The country has tremendous human resources of invigorating energy who need to be engaged, in such case a concerted effort is required to harness potential of both fallow wetlands and their aquatic plant resources.

Acknowledgement

We acknowledge Dr. H. K. De, CIFA; Dr. C. Biswas, CRIJAF and Dr K. Ghosh, University of Burdwan for their valuable suggestions. Thanks are due to the Director, CIFA for facilities and to Dr. Warner Orozco Obando, USA, for kindly providing some of the images included in this book.

References

Anonymous, 2013, Food security Network, Healthy food for all, http://www. Foodsecuritynews. com/What-is-food-security. htm.

Anonymous, 1952. The Wealth of India, (Publications and Information Directorate, CSIR, New Delhi).

Azim, M. E and Wahab, M. A. 2003. Development of a duck weed-fed carp polyculture system in Bangladesh. *Aquaculture,* **218**: 425-438.

Banerjee, A and Matai, S. 1990. Composition of Indian aquatic plants in relation to utilization as animal forage. *J. Aquat. Plant Manage.*, **28**: 69-73.

Bhattacharjee, S. K. 1995 (1387 Bangabda), *Chiranjeev Banoishadhi* (Glossary of Medicinal plants, Bengali book), Ananda Publishers, Kolkata.

Billing, K and Biles, P. 2007. The Lotus: Know It and Grow It. *International Water Lily and Water Gardening Society*.

Boyd, C. E. 1968. Fresh water plants: a potential source of protein. *Econ. Bot.* **22**: 359-368.

Boyd, C. E. 1969. The nutritive value of three species of water weeds. *Econ. Bot.* **23**: 123-127.

Cagauan, A. G and Pullin, R. S. V. 1994. Azolla in aquaculture: past, present and future. In: Muir, J and Roberts, R. J (eds), Recent advances in aquaculture, Vol.5. pp. 104-130.

Chareontesprasit, N and Jiwyam, W. 2001. An Evaluation of Wolffia meal (*Wolffia arrhiza)* in replacing soybean meal in some formulated rations of Nile tilapia (*Oreochromis niloticus* L.). *Pakistan J. of Biological Sci.* **4**: 618-620.

Chopra, R. N., Nayar, S. L and Chopra, I. C. 1986. Glossary of Indian Medicinal Plants, Publication and Information Directorate, CSIR, New Delhi-12.

Culley, D. D. and Epps, E. A. 1973. Use of duckweed for waste treatment and animal feeds. *J. Wat. Pollut. Control. Fed.* **45**: 337-347.

Deepa, R. S and Ramachandran, T. V. 1999. Impact of urbanization in the interconnectivity of wetlands. Paper presented at the National symposium on Remote Sensing Applications for natural resources: Retrospective and perspective (XIX-XXI 1999), Indian Society of Remote Sensing, Bangalore.

Dewanji, A. 1993. Amino Acid Composition of Leaf Proteins Extracted from Some Aquatic Weeds. *J. Agric. Food chem.* **41**: 1232-1236.

Dragendroff, G. 1885. *Die Heilpflanzen der verschiedenen Volker und Zeiten*, Stuttgart, p. 885.

Dutta, R. N. 1984. Comparative ecological study of Euryale ferox Salisbery in Darbhanga area, Ph D Thesis, Ranchi University, Ranchi, Bihar.

Edwards, P., Kamal, M and Wee, K. L. 1985. Incorporation of composted and dried water hyacinth in pelleted feed for the tilapia *Oreochromis niloticus* (Peters). *Aquacult. Fish. Manage.* **16**: 233-248.

Evans, P., Marschke, M., Paudyal. K and Sap, T. 2004. Flood Forests, Fish, and Fishing Villages. Community Forest Management Trends in Southeast Asia. FAO. 50 pages

Fischer, Z. 1973. Physiology and bioenergetics of grass carp. *Pol. Arch. Hydrobiol.* **20**: 529-557.

Gaigher, I. G., Porath, D and Granath, G. 1984. Evaluation of duckweed (*Lemna perpusilla* and *Spirodella polyrhiza*) as feed for Nile tilapia (*Oreochromis niloticus* x *Oreochromis aureus*) in recirculating unit. *Aquaculture.* **41**: 235-244.

Gopalan G., Rama Shatri, B. V and Balasubramanian, S. C., 2007. Nutritive Value of Indian Foods. Revised and updated by Rao, B. S. N., Deosthale, Y. G and Pant K. C., National Institute of Nutrition, ICMR, Hyderabad, India.

Hassan, M. R and Edwards, P. 1992. Evaluation of duckweed (*Lemna perpusilla* and *Spirodela polyrrhiza*) as feed for Nile tilapia (*Oreochromis niloticus*). *Aquaculture.* **104**: 315-326.

Hutabarat, J., Syarani, L and Smith, M. A. K. 1986. The use of freshwater hyacinth *Eichhornia crassipes* in cage culture in Lake Rawa Pening. Central Java. In: Maclean, J. L., Dizon, L. B and Hossillos, L. V. (Eds.), *First Asian Fisheries Forum*. Asian Fishery Society, Manila, Philippines, pp. 577-580.

Jha, V., Kargupta, A. N., Dutta, R. N., Jha, U. N., Mishra, R. K and Saraswati, K. C. 1991. Utilization and conservation of *Euryale ferox* Salisbery in Mithila (North Bihar), India, *Aquatic Bot.* **39**: 295-314.

Kalita, P., Mukhopadhyay, P. K and Mukherjee, A. K. 2007. Evaluation of nutritional quality of four unexplored aquatic weeds of northeast India for the formulation of cost effective fish feed. *Food chemistry.* **103**: 204-209.

Landolt, E and Kandeler, R. 1987. The family of Lemnaceae - Phytochemistry, physiology, application and bibliography Geobotanischen Instutites der ETH, Vol.4, Stiftung Rubel, Zurich.

Liu, X., Min, C., Xia-shi, L and Chungchu, L. 2008. Research on some functions of Azolla as CELSS system, *Acta Astronautica.* **63**: 1061-1068.

Mandal, R. N and Mukhopadhyay, P. K. 2010. Little-known Treasures of Wetlands Need Conservation. *Science Reporter*. May: 27-29.

Mandal, R. N and Saha, G. S. 2007. Aquatic vegetation – A potential support for rural economy. *Agricultural Situation in India.* **LXIV:** 251-255.

Mandal, R. N and Saha, G. S. 2012. Producing Makhana (*Euryale ferox* Salisbery) for its diverse benefits to depressed economic regions of India. *Water Garden Journal.* **27**: 13-16.

Mandal, R. N., Bar, R., Chattopadhyay, D. N and Datta, A. K. 2011. Assessment of Impact of Water Hyacinth (*Eichhornia crassipes*) on Ecology and Socio-Economy. In: Roy, A. K. (ed.), Evaluation and Impact Assessment of Technologies and Developmental Activities in Agriculture, Fisheries and Allied Fields, New India Publishing Agency, New Delhi, pp. 291-312.

Mandal, R. N., Chattopadhyay, D. N and Mukhopadhyay, P. K. 2012. Potentials of aquatic plant bio-resources. *Water Garden Journal.*, **27**: 12-19.

Mandal, R. N., Chattopadhyay, D. N., Saha, G. S and Mukhopadhyay, P. K. 2010. Water Chestnut cum Fish Culture: A Viable Integration for Rural Aquaculture. *Agrovet Buzz.* **III** (5): 53-56.

Mandal, R. N., Chattopadhyay, D. N., Saha, G. S. and Mukhopadhyay, P. K., 2010. Duck Weeds-Nature's Potential Treasures of Nourishment Source for Production of Herbivorous Fish. *Agrovet Buzz*. **III** (2): 26-31.

Mandal, R. N., Datta, A. K., Sarangi, N and Mukhopadhyay, P. K., 2010. Diversity of Aquatic Macrophytes as Food and Feed Components to Herbivorous Fishes – a Review. *Indian Journal of Fish*. **57**: 65-73.

Mandal, R. N., Pandey, B. K., Chattopadhyay, D. N and Mukhopadhyay, P. K. 2012. Azolla – an aquatic fern of significance to small-scale aquaculture. *Aquaculture Asia.* **XVII** (1): 11-15.

Mandal, R. N., Saha, G. S and Mukhopadhaya, P. K. 2007. Lotus – an aquatic plant of versatile qualities. *Aquaculture Asia.* **X11** (1): 11-13.

Mandal, R. N., Saha, G. S and Sarangi, N. 2010. Harvest and processing of Makhana (Euryale ferox Salisb.) – An unique assemblage of Traditional knowledge. *Indian Journal of Traditional knowledge*. **9** (4): 684-688.

Mandal, R. N., Saha, G. S., Kalita, P and Mukhopadhyay, P. K. 2008. *Ipomoea aquatica* – an aquaculture friendly macrophyte, *Aquaculture Asia.* **XIII** (2): 12-13.

Mohanty, S. N and Dash, S. P. 1995. Evaluation of *Azolla caroliniana* for inclusion in carp diet. *J. Aquacult. Trop.*, **10**: 343-353.

Naskar, K. R., Banerjee, A., Chakraborty, N. M and Ghosh, A. 1986. Yield of *Wolffia arrhiza* from cement cistern with different sewage concentration and its efficacy as carp feed. *Aquaculture.* **15**: 211-216.

Nath, B. K and Chakraborty, A. K. 1985. Studies on the amino acid composition of the seeds of *Euryale ferox* Salisbery. *J Food Sci. Technol.* **22**: 293.

Nimat, R and Jafri, A. K. 1984. Preliminary observations on the use of water hyacinth (*Eichhonia crassipes*) leaf meal as protein source in fish feeds. *Curr. Sci.* **53**: 338-340.

Orozco-Obando, W., Tilt, K and Fischman, B. 2009, Is lotus an ornamental plant or a vegetable? *Water Garden Journal.* **4.**

Prasad, S. N., Ramachandra, T. V., Ahalya, N., Sengupta, T., Kumar, L., Tiwari, A. K., Vijayan, V. S and Vijayan, L. 2002, Conservation of wetlands of India – review. *Tropical ecology.* **43**(1): 173-186.

Rath, S. S and Datta, H. 1991. Use of the water hyacinth, *Eichhornia crassipes* as an ingredient in the feed of *Clarias batrachus*. Proceedings of the National Symposium on Freshwater Aquaculture. Bhubaneswar, India, p. 98-99.

Santiago, C. B., Aldaba, M. B., Laron, M. A and Reyes, O. S. 1988. Reproductive performance and growth of Nile tilapia (*Oreochromis niloticus*) broodstock fed diets containing *Leucaena leucocephala* leaf meal. *Aquaculture.* **70**: 53-61.

Shen-Miller, J. 2002. Sacred lotus, the long-living fruits of China Antique. *Seed Science.* **12**: 131-143.

Sikka, K. C., Gupta, A. K., Singh, R and Gupta, D. P. 1978. Comparative nutritive value, amino acid content, chemical composition and digestibility *in vitro* of vegetable and grain type soybeans, *J Agric Food Chem.* **26**: 312-326.

Sikka, K. C., Singh, R., Gupta, D. P and Duggal, S. K. 1979. Comparative nutritive value of fish protein concentrate from different species of fishes, *J Agric Food Chem.* **27**: 946-949.

Singh, S. B., Pillai, K. K and Chakraborty, P. C. 1967. Observation on the efficacy of grass carp in controlling and utilizing aquatic weeds in ponds in India. *Proc. Indo-Pacific Fish Counc.* **12** (2): 220-235.

Stuart, G. A, 1911. *Chinese Materia Medica*, (Vegetable Kingdom, Shanghai), 558.

Trivedy, R. K and Thomas, S. 2005. Water Hyacinth for pollution control and resource recovery. In: Mathur, S. M., Mathur, A. N., Trivedy, R. K., Bhatt, Y. C., Mohnot, P. (Eds.); Aquatic weeds, problems, control and management, Mimanshu Publication, Udaipur, pp. 155-171.

Wei Xi *et al.*, 2008, Indifference pricing of weather derivatives. 90. *AM. J. Agric. Econ.* 979.

Wong, C. F. 1987. Origin and historical record of *Nelumbo nucifera*, Eds. China Institute of Botany, China lotus, Beijing, *Academia Sinica*, Science Publishers, pp. 1-8.

Wroughton, L. 2008. Developing World leaders urge action on Food crisis, Reuters Africa, Sept. 23, http: / / africa reuters. com / wire / news / usnN23388234. html.

Yadugiri, V. T. 2011. M. S. Swaminathan, *Current Science*, **101**(8): 996-1002.

2014, Future Crops Vol. 2 Pages 51–83
Editor: Professor K.V. Peter
Published by: DAYA PUBLISHING HOUSE, NEW DELHI

Chapter 3
Underutilized Fruits of Indian Arid Zone

Dheeraj Singh, M.K. Choudhary, M.L. Meena and Hari Dayal*
Krishi Vigyan Kendra,
Central Arid Zone Research Institute (CAZRI)
Pali-Marwar – 306 401, Rajasthan
**E-mail: dheerajthakurala@yahoo.com*

India is a natural reservoir of a number of lesser known fruit species. Importance of these fruits lie not only in its nutritive value, but also in the fact that they can be consumed at any time of the day. Minor fruits enter in a limited way in the interstate trade as they are well known locally for their therapeutic, medicinal and nutritive values. All of them are not used as table fruits. For many of them, there are no organized orcharding, no improved or recognized varieties and little is known about their use.

Generally speaking, plants are a fundamental component of the natural resources available to man and the role played by underutilized species is indeed central to reducing malnutrition and empowering the people to allow communities to pursue resources-based rather than commodity-based development (Burgess, 1994; Blench, 1997). The people who benefit from underutilized species in a globalized world are not just the poor but all. The benefits in terms of better balanced diets, diversified income to farmers and related sectors of the society, better maintenance of agroecosystems and greater use of marginal lands along with enhanced preservation of cultural identity (Padulosi, 1999) can be shared by all the humankind. Attention to minor fruits plays a significant role in increasing the income, providing employment opportunities, uplifting of poor, small and marginal farmers and development of value added products.

Plant biodiversity provides the primary source for food, feed, shelter, medicines and many other products and means which make life on Earth possible and enjoyable (WCMC, 1992; UNEP 1995). The number of plant species used by humans around the world is only one third of the number of species which generations of diverse cultures around the world have drawn upon to develop crops which would meet specific needs. Increased reliance on major food crops was accompanied by shrinking of the food basket which humankind was relying upon for generations (Prescott-Allen and Prescott-Allen, 1990). This *nutritional paradox* (Ogle and Grivetti, 1995) has its roots in the agricultural "simplification", a process that favoured a few crops instead of others on the basis of their comparative advantages for growing in a wider range of habitats, their simple cultivation requirements, easier processing and storability, nutritional properties, taste etc. A change in attitude is noticed over the past 5-10 years among policy makers and the public with regard to quality of life as related to the quality of food as well as diverse sources of food. Vitamins and other micronutrients are for instance being searched in plant species with greater emphasis than in the past in recognition of their role in combating diet imbalances. Although 'hidden hunger' affects mainly developing countries, particularly children and older people (FAO, 1997), it is increasingly being recorded among the more vulnerable social groups in developed nations. If the 20th Century witnessed the undertaking of systematic collection to rescue the genetic resources of staple crops (Pistorius, 1997), the 21st Century has started with the awareness on the need to rescue and improve use of those crops left aside by research, technology, marketing systems and conservation efforts. These underutilized crops (referred as minor, orphan, neglected, underutilized, underexploited, underdeveloped, lost, new, novel, promising, alternative, local, traditional, niche crops) are included in world-wide plans of action after having successfully raised the interest of decision makers. Leading international research organisations like the Consultative Group on International Agricultural Research (CGIAR) are also among those taking a keen interest in strengthening the work on these species (Swaminathan, 1999). This global "opening" towards underutilized species is the result of a gradual change of attitude towards biodiversity and plant genetic resources by many countries. Instrumental in this awareness raising process is the 1992 Convention on Biological Diversity and the FAO IV International Technical Conference on Plant Genetic Resources for Food and Agriculture held in Germany in 1996 (UNEP 1992; FAO 1996). The Global Forum on Agricultural Research (GFAR) in 1999 also emphasized the role of underutilized species in raising income of the rural poor (Frison *et al.*, 2000).

This chapter addresses composition, nutritional value, processing and utilization of a few of the neglected and underutilized fruits (Table 3.1) of Indian origin.

Table 3.1: List of common underutilized fruits.

Sl.No.	*Common Name*	*Botanical Name*	*Sl.No.*	*Common Name*	*Botanical Name*
1.	Bael	*Aegle marmalos*	2.	Seetaphal	*Annona squamosa*
3.	Jackfruit	*Artocarpus heterophyllus*	4.	Bamboo fruit	*Bambusa arundinacea*
5.	Chironji	*Buchanania lanzan*	6.	Karonda	*Carrisa carandus*
7.	Lasoda	*Cordia myxa*	8.	Timru	*Diospyros melanoxylon*
9.	Gab/tumbi	*Diospyros embryoteris*	10.	Aonla	*Emblica officinalis*
11.	Loquat	*Eriobotrya japonica*	12.	Baincha	*Facourtia indica*
13.	Banyan tree figs	*Ficus bengalensis*	14.	Pipal tree	*Ficus religiosa*
15.	Gullar	*Ficus glaumerata*	16.	Anjeer	*Ficus carica*
17.	Kaith	*Feronia limonia*	18.	Phalsa	*Grewia subinaqualis*
19.	Mahua	*Madhuca indica*	20.	Rayan/khirni	*Manilkara hexandra*
21.	Molsiri	*Melangii*	22.	Mulberry	*Morus alba*
23.	Senjana/Drumstick	*Moringa oleifera*	24.	Neemphal	*Melia azadirachta*
25.	Cape gooseberry	*Physallis peruviana*	26.	Dates/khajur	*Phoenix dactylifera*
27.	Pomegranate	*Punica granatum*	28.	Jamun	*Syzygium cuminii*
29.	Tamarind	*Tamarindus indica*	30.	Ber	*Zizyphus jujube*

Fruits are important in the human diet, as they act as:

A Source of Food

A number of species are already being used as staple foods, food supplements and source of condiments, spices, thickening agents, flavours, edible oil, etc. An increased interest is recorded among national and international research organisations towards medicinal and aromatic species in view of their role in improving health of poor and their contribution to combat poverty through income generation (Leaman *et al.,* 1999). Attention to underutilized species may also originate from considerations which are not directly related to food security or poverty alleviation, but to the need to safeguard artistic, landscape and cultural values of these species. This is for instance the objective of the international workshop organised by the Italian National Research Council (CNR) in 1997 in Naples (Monti, 1997). In peninsular India, flowers of Mahua and Jack fruit are important food sources during summer before harvest of *kharif* crop. The highest energy available is from tamarind pulp (142 calories) with low moisture and high carbohydrate (34g) content, while *Anona squamosa* supplies half the energy, but has 70. 5 per cent moisture and 11. 8g carbohydrates (Table 3.2). On dry basis, the energy content will be 127, 178 and 179 calories and carbohydrate content 45, 40 and 43g for dates, seethaphal and tamarind pulp, respectively. In fact, on dry basis, there is not much difference between various fruits. Fibre provides good motility to intestine, which helps digestive process. Most of these fruits are good sources of fibre with *Ficus* spp. having the highest amount (3. 2 to 5. 0 g). Fruits with low moisture content will have higher calorific value, mostly from carbohydrates present.

Table 3.2: Comparison of nutritive value of conventional and non-conventional fruits.

	Conventional Fruits					*Non Conventional Fruits*				
	Apple	*Oranges*	*Guava*	*Mango*	*Banana*	*Phalsa*	*Jamun*	*Bael*	*Timru*	*Khirni*
Energy (Cal.)	59	48	51	74	116	72	62	137	112	134
Carbohydrate (g)	13. 4	10. 9	11. 2	16. 9	27. 2	14. 7	14. 0	31. 8	26. 8	27. 7
Fibre (g	1. 0	0. 3	5. 2	0. 7	0. 4	1. 2	0. 9	2. 9	0. 8	–
Minerals(g)	0. 3	0. 3	0. 7	0. 4	0. 8	1. 1	0. 4	1. 7	0. 8	0. 8
Calcium (mg)	10	26	10	14	17	129	15	85	60	83
Phosphorus (mg)	14	20	28	16	36	39	15	50	20	17
Iron (mg)	1	0. 3	1. 4	1. 3	0. 9	3. 1	1. 2	0. 6	0. 5	0. 9
Carotene (øg)	0	1104	0	2743	78	419	48	55	361	495
Vitamin C (mg)	1	30	212	16	7	22	18	8	1	16
Thiamine (mg)	0. 04	0. 08	0. 03	0. 08	0. 05		0. 03	0. 13	0. 01	0. 07
Riboflavin (mg)	0. 03	0. 03	0. 03	0. 09	0. 08		0. 01	0. 03	0. 04	0. 08
Niacin (mg)	0. 2	0. 2	0. 4	0. 9	0. 5	0. 3	0. 2	1. 1	2. 3	0. 7

As a Source of Nutrients

A good balanced diet should supply sufficient amount of major nutrients -protein, fat and carbohydrate- and minor nutrients- vitamins and minerals-. Minerals and vitamins are largely supplied by fruits and vegetables (Table 3.3).

Table 3.3: Major nutrients in edible portion of certain fruits (50 g of edible portion of fruit).

Botanical Name	Moisture (per cent)	Energy (Cal.)	Carbohydrate (g)	Fibre (g)	Fat (g)	Protein (g)
Tamarindus indica	20. 9	142	34	2. 8	0. 05	1. 6
Manilkara elangii	54. 7	80	18	2. 2	0. 50	0. 9
Bambusa arundinacea	56. 3	76	17. 1	2. 0	0. 05	2. 0
Phoenix dactylifera	59. 2	72	16. 9	1. 9	0. 20	0. 6
Aegle marmalos	61. 5	68	15. 9	1. 5	0. 15	0. 9
Manilkara hexandra	68. 6	67	13. 9	–	0. 20	0. 3
Feronia limonia	64. 2	67	9. 1	2. 5	1. 85	3. 6
Feronia indica	67. 8	57	11. 4	2. 4	0. 90	0. 9
D. embryopteris	69. 6	56	13. 3	0. 8	0. 05	0. 7
Mangifera indica	73. 6	56	11. 4	–	0. 80	0. 7
Diospyros melanoxylon	70. 6	56	13. 4	0. 4	0. 10	0. 4
Ficus religiosa	62. 4	55	10. 6	5. 0	0. 85	1. 3
Annona squamosa	70. 5	52	11. 8	1. 6	0. 20	0. 8
Ficus bengalensis	74. 1	36	5. 9	4. 3	1. 00	0. 9
Emblica officinalis	81. 8	29	6. 8	1. 7	0. 05	0. 3
Physallis peruviana	82. 9	27	5. 6	1. 6	0. 10	0. 9
Ficus cunia	;79. 4	26	5. 4	3. 2	0. 30	0. 6
Moringa oleifera	86. 9	13	1. 8	2. 4	0. 05	1. 3

Diet of an adult should include at least 30 g and for children at least 50g of fruits in addition to leaf vegetables and milk. Ascorbic acid (Vitamin C) is needed to develop resistance against diseases and other ailments in human body. Protein is essential for growth and maintenance of human body. Although most of the common fruits are low in protein, some of the underutilized fruits and nuts are rich in vitamin C, protein and Vitamin A. Vitamin A is essentia1 for normal vision, development of bones and proper growth.

Vitamins are essential constituents of the diet. Deficiency of one or more vitamins leads to a variety of diseases, which cannot be cured by medicines. Fruits and vegetables supply Vitamins A; B complex and C. Leaf vegetables provide most of the vitamin B complex. The main sources of vitamin C are citrus fruits and aonla. Among fruits, aonla (300 mg) is the best source of vitamin C. Ber (38 mg), Cape gooseberry (25 mg) and mahua (20 mg) come next with values comparable to citrus fruits. Fruits like Cape gooseberry (714 mg), Loquat (280 mg), Khirni (248 mg) and Phalsa (210 mg) are rich in carotene, but are not as high as in apricot (1080 mg), mango (1372 mg) and oranges (502 mg). Niacin level is high in Timru (1150 mg). Comparison of the nutritive values of well known fruits and underutilized fruits (Table 3.4) show that there is not much difference between the two groups. Species linked differences, of course, do exist. Except for the high content of carotene in apricot, oranges and mangoes and of vitamin C in aonla, there are not much differences in the vitamin content of these groups.

Table 3.4: Vitamin content of some fruits (50 g of edible portion).

Fruits	*Vitamin C*	*Carotene*	*Thiamine*	*Riboflavin*	*Niacin*
Aonla	300. 0	4. 5	15	5	100
Timru	0. 5	180. 0	5	20	1150
Gular	2. 5	81. 0	30	25	300
Ber	38. 0	10. 5	10	25	350
Cape gooseberry	25. 0	714. 0	25	10	150
Lokat	0. 0	280. 0	–	–	0
Khirni	8. 0	248. 0	35	40	350
Phalsa	11. 0	210. 0	–	–	150
Mahua	20. 0	154. 0	–	–	–
Senjana	60. 0	55. 0	25	35	100
Jackfruit	3. 5	88. 0	15	65	200
Tamarind	1. 5	30. 0	–	35	350

Calcium, phosphorous and iron are the major nutrients required by the human body in addition to sodium and potassium. Most of the requirements can be fulfilled by vegetables, but inclusion of fruits makes further addition of calcium and phosphorus for growth of bones. The amount of minerals in 50 g of edible portion of fruit is presented in Table 3.5.

Table 3.5: Mineral content of certain fruits (50 g of edible portion).

Fruits	*Mineral (g)*	*Calcium (g)*	*Phosphorus (mg)*	*Iron (mg)*
Tamarindus indica	1. 45	85	55	5. 5
Ficus religiosa	1. 15	145	45	–
Cordia dichotoma	1. 10	20	30	–
Manilkara elangii	1. 15	106	15	–
Moringa oleifera	1. 00	15	55	2. 7
Feronia limonia	0. 95	65	55	0. 3
Ficus bengalensis	0. 95		22	–
Aegle marmelos	0. 85	43	25	0. 3
Buchanania lanzan	*0. 85*	39	14	–
Ficus cunia	0. 80	94	20	–
Bambusa arundinacea	0. 80	5	55	0. 8
Flacoutlia indica	0. 65	50	50	–
Grewia asiatica	0. 55	65	20	1. 6

The highest mineral content is in *T. indica* (1.45g) followed by *C. dichotoma, M. elangii* and *F. religiosa.* All fruits except *C. dicholoma, M. oleifera, B. arundinocea* have higher iron content than other fruits listed here.

A Source of Medicine

They contain pectin and celluloses which stimulate the intestinal activity and protect human body against diseases. Value of underutilized fruits in traditional medicines is well known. These are major sources of raw materials for drugs since antiquity and have provided bulk of products in the traditional system of medicine (Singh and Singh, 2005). In India, the fruits of *Emblica officinalis* (aonla), *Terminalia chebulla* (harar) and *T. bellíricaria* (bahera) are the most common, entering into 219 patented drugs. *Aegle marmelos* (bael) is used in 60 drugs. "Chyavanpras" based on *Emblica officinalis* (aonla) is an excellent example. It consumes 20,000 to 30,000 tonnes of dry fruits of amla annually. Bael (*Aegle marmelos*) pulp contains marmelosin which acts as a laxative and diuretic and in strong doses as cardiac depressent. Jujube (*Zizyphus mauritiana*) fruits are used as an ingredient in the preparation of "joshanda" (an ayurvedic medicine used in chest trouble).

Some of the major underutilized fruits of Indian arid zone are described as under:

Bael (*Aegle marmelos*)

It is an important indigenous fruit of India found naturally in Uttar Pradesh, West Bengal, Tamil Nadu and Bihar. It is quite popular for its curative properties as laxative and therapeutic value. It is a very hardy tree and thrives even under adverse agroclimatic conditions. All the parts of the tree, including stem, bark, root, leaves, and fruit at any stage of maturity and ripening, have some use or other. Various chemical constituents, namely, alkaloids, coumarins, and steroids, are isolated and identified from different parts of the bael tree. Bael pulp contains a balsam-like substance, and 2 furocoumarins-psoralen and marmelosin ($C_{13}H_{12}O_3$). There is as much as 9 per cent tannin in the pulp of wild fruits whereas the rind contains up to 20 per cent. Tannin is also present in the leaves, as skimmianine. The essential oil of the leaves contains d-limonene, 56 per cent a-d-phellandrene, cineol, citronellal, citral; 17 per cent p-cyrnene, 5 per cent cumin aldehyde (Singh *et al.*, 2010).

Table 3.6: Food value per 100 g of edible portion.

Nutrient	*Food Value/100 g*	*Nutrient*	*Food Value/100 g*
Water	54. 9-61.5 g	Protein	1.8-2. 62 g
Fat	0.2-0.39 g	Carbohydrates	28.11-31.8 g
Ash	1.04-1.7 g	Carotene	55 mg
Thiamine	0.13 mg	Riboflavin	1.19 mg
Niacin	1.1 mg	Ascorbic Acid	8-60 mg
Tartaric Acid	2.11 mg		

According to Gopalan *et al.* (1978), it contains (Table 3.6) 61. 5 g of water, 1. 8 g of protein, 0. 39 g of fat, 1. 7 g of minerals, 31. 8 g of carbohydrates, 55 ug of carotenes, 0. 13 mg of thiamine, 1. 19 mg of riboflavin, 1. 1 mg of niacin, and 8 mg of vitamin C/100 g of edible portion. No other fruit has such a high content of riboflavin. Mukherjee and Ahmed (1954) also commented on the high riboflavin content of ripe bael fruit. Tannic acid is the only polyphenolic substance detected in bael fruit. Chemical analysis of bael fruit seeds revealed that the seed contained 62 per cent protein (2 per cent water soluble and 60 per cent water insoluble), 32 per cent oil, 3 per cent carbohydrates, and 3 per cent ash (Banerjee and Maity, 1980). A total of 39 aroma components were identified in bael fruit. Among these components, terpene alcohols and 3-ionone contributed to the aroma of bael fruit (Tokitomo *et al.*, 1982). Marmelosin is the most probably the therapeutically active principle of bael fruit. It was isolated

Figure 3.1: Ripe fruit of Bael (*Aegle marmalos*).

as a colorless crystalline compound. It is present only in the fruit. The content varies from 0. 03 to 0. 37 per cent according to the location of cultivation.

Food Uses

Bael fruit is used widely from time immemorial for processing in mature green form to prepare preserves. To prepare preserves, the hard rind of fruit is removed by a special strong knife. The fruit is sliced into 1.5cm thick pieces, washed in water, pricked with a stainless steel fork, and soaked overnight in cold water. The pieces are then blanched and put in light sugar syrup. The strength of the syrup is gradually raised to 70° Brix.

Very little information is available regarding processing of ripe bael fruit. Extraction of pulp from ripe bael fruit is the main hindrance to processing. Verma and Ahmed (1958) reported that bael fruit powder could also be manufactured successfully. Roy and Singh (1979) discussed a process for successful extraction of pulp from bael fruit. According to these researchers, the bael fruit pulp could be successfully extracted by adding water equal to the pulp (with seeds and fiber), adjusting the pH to 4. 3 with citric acid (titratable acidity 0. 5 per cent), heating at 30°C for 1 minute, and then passing through a pulping machine. Addition of water helped in recovering the pulp. Application of heat coupled with addition of citric acid inactivated the enzymes and also helped in dissolving the mucilage uniformly to provide a homogenous pulp. Bael fruit pulp obtained by the standard method had almost the same consistency and colour as that of mango pulp. Various products can be prepared from bael fruit pulp. Bael fruit nectar is prepared by blending pulp with sugar, acid and water. A composition of 35 per cent pulp, 25° Brix, and 0. 3 per cent acidity gives a highly acceptable product. Similarly, the best composition for bael fruit squash was 50 per cent pulp, 50° Brix, and 1 per cent acidity. Bael fruit slab is prepared by adding 10 per cent sugar with 1500 ppm SO_2 and drying to a moisture level of 14. 5 per cent. Bael fruit toffee is prepared by mixing 40 parts of sugar, 4. 5 parts of glucose, 10 parts of skimmed milk powder, and 6 parts of hydrogenated fat to every 100 parts of bael fruit pulp. The proper moisture content of bael fruit toffee was 8. 5 per cent. Bael fruit powder can be mixed with an equal quantity of milk powder and reconstituted with water as required to make an acceptable beverage (Jauhari and Singh, 1971). Pure and clear bael fruit juice is prepared by enzyme treatment. The storage performance of all these bael fruit products was satisfactory even after 6 months of storage under ambient conditions. Addition of SO_2 not only improves the initial quality of the bael fruit products but also prevents non-enzymatic browning reaction during storage (Roy and Singh, 1979).

Bael fruits may be cut in half, or the soft type is broken open, and the pulp, dressed with palm sugar, eaten for breakfast, as is a common practice in Indonesia. A popular drink (called "sherbet" in India) is made by beating the seeded pulp together with milk and sugar. A beverage is also made by combining bael fruit pulp with that of tamarind. These drinks are consumed perhaps less as food or refreshment than for their medicinal effects. Mature but still unripe fruits are made into jam, with addition of citric acid. The pulp is also converted into marmalade or syrup, likewise for both food and therapeutic use, the marmalade being eaten at breakfast by those convalescing from diarrhoea and dysentery. A firm jelly is made from the pulp alone, or, better still, combined with guava to modify the astringent flavor. The pulp is also pickled.

Bael pulp is steeped in water, strained, preserved with 350 ppm SO_2, blended with 30 per cent sugar, then dehydrated for 15 hrs at 58 - 89° C and pulverized. The powder is enriched with 66 mg/ 100 g ascorbic acid and can be stored for 3 months for use in making cold drinks ("squashes"). The young leaves and shoots are eaten as a vegetable in Thailand and used to season food in Indonesia. They reduce the appetite. An infusion of flowers is a cooling drink.

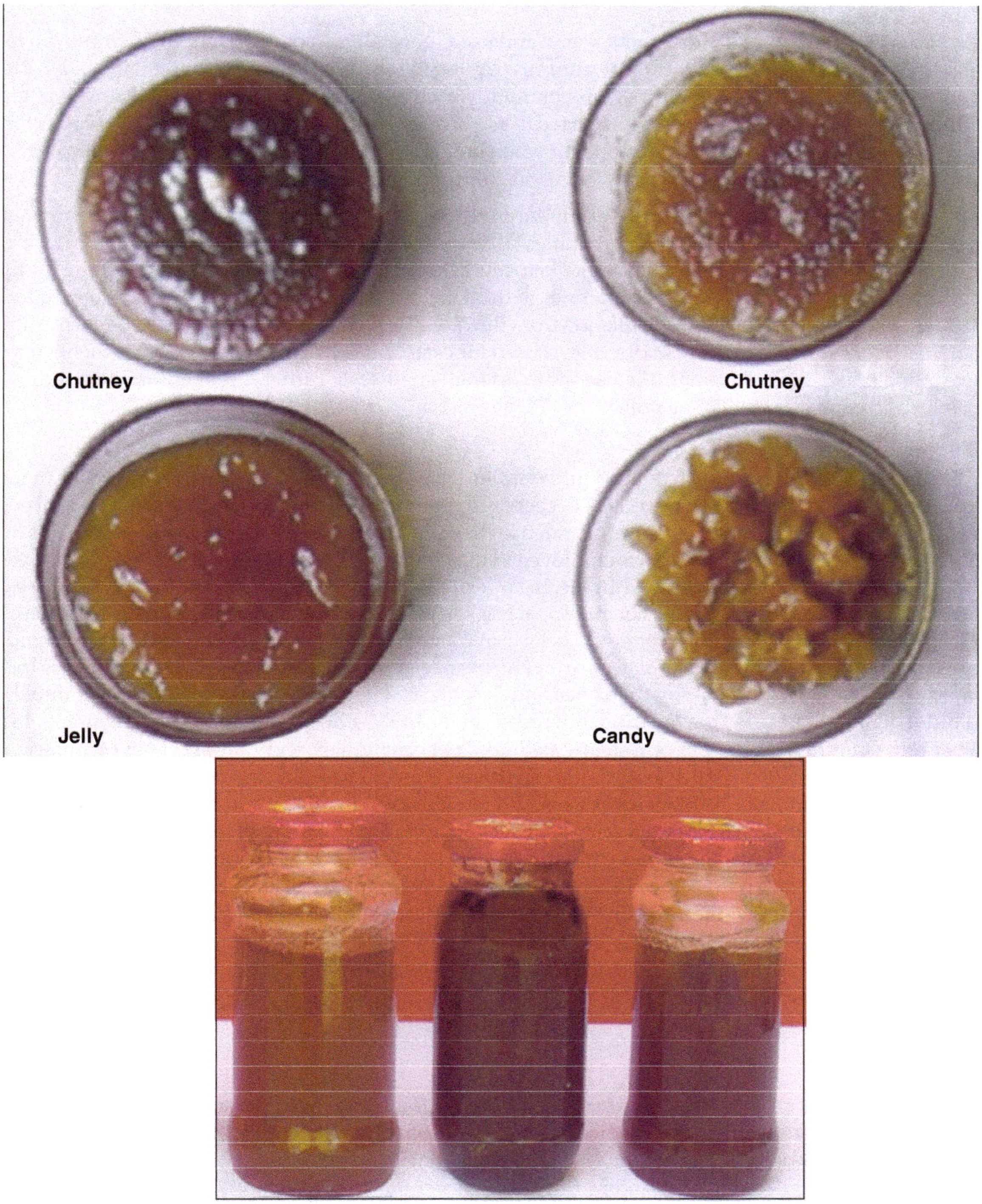

Figure 3.2: Processed products of Bael fruits.

Other Uses

Fruit: The fruit pulp has detergent action and used for washing clothes. Bael fruit is employed to eliminate scum in vinegar-making. The gum enveloping the seeds is the most abundant in wild fruits and especially when they are unripe. It is commonly used as household glue and is employed as an adhesive by jewelers. Sometimes it is resorted to as a soap-substitute. It is mixed with lime plaster for waterproofing wells and is added to cement when building walls. Artists add it to their watercolors, and it may be applied as a protective coating on paintings.

The limonene-rich oil is distilled from the rind for scenting hair oil. The shell of hard fruits is fashioned into pill- and snuff boxes, sometimes decorated with gold and silver. The rind of the unripe fruit is employed in tanning and also yields a yellow dye for calico and silk fabrics. In the Hindu culture, the leaves are indispensable offerings to the 'Lord Shiva'. The leaves and twigs are lopped for fodder. A cologne is obtained by distillation from flowers. The wood is strongly aromatic when freshly cut. It is gray-white, hard, but not durable; is used for carts and construction, though it is inclined to warp and crack during curing. It is used for carving, small-scale turnery, tool and knife handles, pestles and combs, taking a fine polish.

Medicinal Uses

The fresh ripe pulp of high quality cultivars, and the "sherbet" made from it, are taken for their mild laxative, tonic and digestive effects. A decoction of unripe fruit, with fennel and ginger, is prescribed in hemorrhoids. It is surmised that the psoralen in the pulp increases tolerance of sunlight and aids in maintaining of normal skin color. It is employed in the treatment of leucoderma. Marmelosin derived from pulp is given as a laxative and diuretic. In large doses, it lowers the rate of respiration, depresses heart action and causes sleepiness. For medicinal use, the young fruits, while still tender, are commonly sliced horizontally and sun-dried and sold in local markets (Singh and Singh, 2005). They are exported to Malaysia and Europe. Because of the astringency, especially of the wild fruits, the unripe bael is the most prized as a means of halting diarrhoea and dysentery, which are prevalent in India in summer. A bitter, light-yellow oil extracted from the seeds is given in 1. 5 g doses as a purgative. It contains 15. 6 per cent palmitic acid, 8. 3 per cent stearic acid, 28. 7 per cent linoleic and 7. 6 per cent linolenic acid. The seed residue contains 70 per cent protein. The bitter, pungent leaf juice, mixed with honey, is given to allay catarrh and fever. With black pepper added, it is taken to relieve jaundice and constipation accompanied by edema. The leaf decoction alleviates asthma. A hot poultice of the leaves is an effective treatment for ophthalmia and various inflammations, also febrile delirium and acute bronchitis. A decoction of flowers is used as eye lotion and given as an antiemetic. The bark contains tannin and coumarin, aegelinol; also the furocourmarin, marmesin; umbelliferone, a hydroxy coumarin; and alkaloids, fagarine and skimmianine. The bark decoction is administered in malaria. Decoctions of the root relieve palpitations of the heart, indigestion, and bowel inflammations; and overcome vomiting. The fruit, roots and leaves have antibiotic activity. The root, leaves and bark are used in treating snakebite.

Phalsa

Phalsa (*Grewia subinaequatis* L.) belongs to family 'Tiliaceae' and is indigenous to India and recognized for its refreshing and medicinal properties (Samson, 1986). In wild form, it is found all along the foot-hills of the Himalayas in Uttar Pradesh, Punjab, Haryana, Rajasthan, Madhya Pradesh, Maharashtra, West Bengal, Bihar and in 'South India over the Western ghats and along the Malabar coasts. In India, this plant grows under natural conditions up to an elevation of 1,000 m. The phalsa is a warm, drought resistant fruit plant and normally sheds its leaves slowly in those areas with mild

Figure 3.3: Phalsa (*Grewia subinaequatis*) fruits.

winter season. The phalsa plant grows vigorously and produces satisfactorily under variable soil types including fine sand, clay or even limestone, when soil fertility is not very poor. Phalsa is often grown in marginal lands close to city markets to facilitate prompt marketing of fruit. In cultivated form, it is commonly grown around cities and towns in Punjab, Haryana, Uttar Pradesh, Rajasthan and Andhra Pradesh. From these locations it could find ready market for disposal. Though the 'phalsa' fruit is extremely delicate in nature, the plant is one of the hardiest fruit trees grown in arid and semi arid regions.

The small fruits, almost round drupes like blueberry and purple, crimson or cherry red in colour when ripe are produced in great numbers in open, branched clusters. Individual fruits measure from 1.0 to 1.9 cm in diameter, 0.8 to 1.6 cm in vertical height, and 0.5 to 2.2 g in weight. Fruits ripen gradually on bushes during summer. While ripening, the fruit skin turns from light green to cherry red or purplish red finally becoming dark purple or nearly black. The ripe fruit is covered with a very thin, whitish blush, and becomes soft and tender. The delicate, fibrous flesh is light greenish-white becoming colored purplish-red from seed reaching near the skin. Overripe fruit flesh attains purple color followed by shriveling of fruit skin due to moisture loss. Large fruits have two hemispherical, hard, buff-colored seeds up to 5 mm in diameter while small fruits are generally single-seeded. The mature but not completely ripe fruits when harvested can be stored upto 48 hours while fully ripe can hardly be stored upto 24 hours at room temperature. The fruits are small in size and ripen gradually on the plant with the result that only a few fruits are ready to be picked in each cluster at a time. Its fruits ripen in summer (May to June) and have an extremely low keeping quality. The ripe phalsa fruits are consumed fresh, in desserts, or processed into refreshing fruit and soft drinks enjoyed during hot summer in India (Salunkhe and Desai, 1984).

Analyses made long ago in the Philippines (Table 3.7) established the following values as reported by Morton (1987): 725 calories/kg edible fruit; moisture, 81.13 per cent; protein, 1.58 per cent; fat, 1.82 per cent; crude fiber, 1.77 per cent; and sugar, 10.27 per cent.

Table 3.7. Nutrient content of phalsa fruits (Yadav, 1998).

Nutrient	*Nutrient Value/100 g Fruit*	*Nutrient*	*Nutrient Value/100 g Fruit*
Calories (Kcal)	90.5	Calories from fat (Kcal)	0.0
Moisture (per cent)	76.3	Fat (g)	<0.1
Protein (g)	1.57	Carbohydrates (g)	21.1
Dietary Fiber (g)	5.53	Ash) (g)	1.1
Calcium (mg)	136	Phosphorus (mg)	24.2
Iron (mg)	1.08	Potassium (mg)	372
Sodium (mg)	17.3	Vitamin A (µg)	16.11
Vitamin B_1, Thiamin (mg)	0.02	Vitamin B_2, Riboflavin (mg)	0.264
Vitamin B_3, Niacin (mg)	0.825	Vitamin C, Ascorbic acid (mg)	4.385

Processing

Phalsa juice has a deep, crimson-red color and a pleasing flavor, which makes the juice very popular. In addition, the juice is extremely refreshing in summer. The popularity of 'Phalsa' fruit is due to its attractive colour and pleasing taste. The phalsa flavor is pleasantly astringent but delicious due to very appropriate sugar-acid blend. It is rated very high in the indigenous system of medicine. Besides this, the juice has a cooling effect especially in hot months. Hence, it is used extensively in summer both as a fresh fruit and also in the form of a home made syrup. Commercial processes like bottling of 'phalsa' juice or syrup, if developed can ensure their ready sale prompting a considerably larger production of this fruit. The pomace (30.92 per cent) left after the extraction of the juice from phalsa fruit is used to extract pigments and the total soluble solids. It was found that addition of 75 per cent water to pomace gave pigments and total soluble solids (Waskar and Khurdiya, 1988). A ready-to-serve (RTS) beverage from phalsa fruit juice was formulated and standardized. It contained 25 per cent juice and a Brix -acid ratio of 45: 1 (Khurdiya and Anand, 1981). The syrup was prepared from phalsa fruit juice in which the clear juice was mixed with an equal amount of sugar and preserved with sodium benzoate (Amba, 1973). A carbonated beverage can also be prepared from phalsa juice. The colour in 'phalsa" fruit is due to water soluble pigments *viz.*, anthocyanins which lend an attractive colour to the juice. Based on paper chramatography and spectral characteristics two anthocyanin pigments *viz.*, delphinidin-3-glucoside and cyanidin-3-glucoside were identified, in the 'phalsa' fruits.

Medicinal and Other Uses

The fruit is astringent and stomachic. Morton (1987) reported that when unripe, phalsa fruit alleviates inflammation and is administered in respiratory, cardiac and blood disorders, as well as in fever reduction. Furthermore, an infusion of the bark is given as a demulcent, febrifuge and treatment for diarrhoea. The root bark is used in treating rheumatism. The leaves are applied on skin eruptions and they have antibiotic action. The fresh leaves are valued as animal fodder. The bark is used as a soap substitute in Myanmar. A mucilaginous extract of bark is useful in clarifying sugar. Fiber extracted from bark is made into rope. The wood is yellowish-white, fine-grained, strong and flexible. It is used for archers' bows, spear handles, shingles, and poles for carrying loads on shoulders. Stems which are pruned serve as garden poles and for basket-making. The flowers contain grewinol, a long chain keto-alcohol, tetratricontane 22-ol-13-one (Lakshmi and Chauhan, 1976). The phalsa seeds produce approximately 5 per cent yield of a bright yellow oil that contains 8 per cent palmitic acid, 11 per cent stearic acid, 13.5 per cent oleic acid, and 64.5 per cent linoleic acid with 3 per cent unsaponifiable (Morton, 1987).

Custard Apple (*Annona squamosa*)

Custard apple (*Annona squamosa* L.) is is one of the most delicious fruit plants of tropical and sub-tropical area belonging to family Annonaceae. It is a drought hardy, tolerant to salinity and saline irrigation water to certain extent and grows well even on shallow soil without much care. Plants shed their leaves during stress period to avoid moisture loss from plants. Its flowering coincides with maximum moisture availability period thus making its survival possible even in extreme conditions. It also sheds off leaves during stress periods. Plant escapes from animal damage due to its hardy nature. Custard apple is mainly found in wild form in Andhra Pradesh, Tamil Nadu, Orissa, Assam and Rajasthan. In India, the fruits are eaten mainly by the lower and medium class people. The fruit is appreciated by all in Central America, Mexico and the West Indies. When fully ripe, it is soft to touch and the stem and attached core can be easily pulled out. The flesh may be scooped from the skin and eaten as such as served with light cream and a sprinkling of sugar. Often it is pressed through a sieve and added to milk shakes, custards or ice cream. A delicious sauce for cake and puddings can be made by blending the seeded flesh with mashed banana and a little cream.

Figure 3.4: Mature fruits of Custard apple (*Annona squamosa* L.).

Composition

The fruit of custard apple is very sweet and delicious hence this fruit is used for table purposes. In recent years it is used in ice cream and puddings. The pleasant flavour and mild aroma are typical of custard apple fruit. Food value of custard apple fruit pulp is given in Table 3.8. The ripe fruits are rich in sugars. Its pulp contains about 73 per cent moisture, 0.8 to 1.5 per cent proteins and 14.5 per cent sugars, 0.3 per cent fat, 1.0 per cent iron and 0.7 per cent mineral matter (Table 3.5). It is quite rich in calcium (0.02 per cent), phosphorus (0.04 per cent) and iron (1.0 per cent). The calorific value is 105/ 100 g of pulp. Glucose is a predominant sugar.

Table 3.8: Food value/100 g of edible portion of custard apple pulp.

Constituents	*Content*	*Constituents*	*Content*
Calories	80-101	Phosphorus	14.7-32.1 mg
Moisture	68.3-80.1 g	Iron	0.42-1.14mg
Protein	1.17-2.47 g	Carotene	0.007-0.018 mg
Fat	0.5-0.6 g	Thiamine	0.075-0.119 mg
Carbohydrates	20-25.2 g	Riboflavin	0.086-0.175 mg
Crude Fibre	0.9-6.6 g	Niacin	0.528-1.190 m
Ash	0.5-1.11 g	Ascorbic Acid	15.0-44.4 mg
Calcium	17.6-27 mg	Nicotinic Acid	0.5 mg

Medicinal Uses

The seeds are so hard that they may be swallowed whole with no ill effects but the kernels are very toxic. The seeds, leaves and young fruits have insecticidal properties. The leaf juice kills lice. The bark contains 0.12 per cent anonaine. Injection of an extract from the bark caused paralysis in a rear limb of an experimental toad. Sap from cut branches is acrid and irritant and can severely injure the eyes. The root bark yielded 3 alkaloids: anonaine, liriodenine and reticuline (muridnine). The leaf decoction is given as a vermifuge. Crushed leaves or a paste of the flesh may be poulticed on boils, abscesses and ulcers. The unripe fruit rich in tannin, is dried, pulverized and employed against diarrhoea and dysentery. The bark is very astringent and the decoction is taken as a tonic and also as a remedy for diarrhoea and dysentery. In severe cases, the leaves, bark and green fruits are all boiled together for 5 minutes in a liter of water to make an exceedingly potent decoction. Fragments of the root bark are packed around the gums to relieve toothache. The root decoction is taken as a fabrifuge.

Processing

Pulp

Extraction of the pulp is a major constraint in processing of custard apple. A centrifugal process is developed to separate the gritty portion from the pulp, which was otherwise objectionable in processed products. Pulp is scooped out from ripe fruit and preserved with potassium metabisulphite. A recovery of 50 per cent of pulp is obtained from the ripe edible fruit. The custard apple pulp can be preserved with a minimum dose of 250 ppm of SO_2 in refrigerator but it requires SO_2 (500ppm) if it is to be stored at ambient temperature. In this way, it has a storage life of about 180 days. The custard apple pulp can be employed for making fruit drinks (Khurdiya, 2001).

Squash

Sugar syrup is prepared by adding cane sugar to boiling water. The syrup is filtered through a muslin cloth to remove impurities. Fruit pulp and the freshly prepared hot syrup are added together in the proportion of proposed recipe on weight basis. The mixture is boiled by adding required amount of citric acid. Finally, the product is treated with potassium metabisulphite (300 ppm) to prevent spoilage. The organoleptic evaluation of the RTS beverage showed that the beverage prepared by using 20 per cent juice was the most acceptable (Kotecha *et al.*, 1995). The organoleptic properties of the custard apple wine were comparable with those of the grape wine properties.

Nectar

Sugar syrup is prepared by adding cane sugar to boiling water. The prepared syrup is filtered through a muslin cloth to remove impurities. The hot syrup is mixed with fruit pulp on weight basis. The mixture is boiled and citric acid added to get a consistent product. The prepared nectar is poured into the pre-sterilized glass bottles (220ml) and sealed airtight by using crown caps. The bottles are sterilized in boiling water for 25 minutes, cooled to room temperature and stored for further use.

Jam

Custard apple jam (pulp 45-50 per cent, TSS 60-68 per cent) with a fixed level of acidity (0.5 per cent) is prepared. The required quantity of pulp and sugar as per the recipe is taken in a steel vessel and heated, stirring continuously. Separately, citric acid is dissolved in a little water and mixed with the pulp. Heating is continued until the required TSS is reached. Chemical preservative, potassium metabisulphite (0.1 g/kg) is mixed into the product after dissolving in a little water. The product is poured into clean and sterilized jam bottles (500 g) and sealed hermetically.

Other Uses

The leaves are used in tanning and they yield a blue or black dye. A fiber derived from young twigs is superior to the bark fiber from *Annona squamosa.* Custard apple wood is yellow, rather soft, fibrous but durable, moderately close-grained, with a specific gravity of 0. 650. It is used to make yokes for oxen.

The seeds of fruit contain 30 per cent oil, used for making soap and paint. The oil has insecticidal properties. When mixed with neem oil, it demonstrated efficacy against Nephotetttx Virus and transmission of rice tungro Virus. The oil cake containing 4 per cent nitrogen is used as manure, beneficial to the crops (Singh *et al.,* 2007).

Jamun (*Syzygium cuminii*)

Jamun is a tall ever-green tree indigenous to India which is generally grown as avenue or as wind break. It is widely grown from Indo-Gangetic plains in North to Tamil Nadu in South. It is also found in the lower range of the Himalayas and *Kumaon* hills. Since Jamun is a hardy fruit crop, it can be grown under adverse soil and climatic conditions. It requires dry weather at the time of flowering and fruit setting but rains are beneficial for better growth, development and ripening of fruit. The jamun trees can be grown on a wide range of soils–calcareous, saline sodic soils and marshy areas.

Composition

The refreshing and curative properties of jamun make it one of the useful medicinal plants of India. Fruits are a good source of iron, used as an effective medicine against diabetes, heart and liver trouble (Table 3.9). The seed powder of jamun reduces the quantity of sugar in the urine very quickly. Ramanjaneya (1985) reported chemical characteristics of two different types of jamun, one small and the other large. The bigger fruits are suitable for table purposes. The small-sized fruits, which are not suitable for table purposes, are ideal for processing. The sugar content of jamun fruit consists principally of reducing sugars. The high tannin content is mainly responsible for its astringency. Jamun colour is due to anthocyanin pigments (Khurdiya and Roy 1985). Jamun fruit pocesses considerable nutritive value. According to Singh *et al.* (1967), the fruit has 19.7 per cent carbohydrates, 0.7 per cent protein, 0.02 per cent calcium, 0.1 per cent fat, 0.01 per cent phosphorus, 0.4 per cent mineral matter, 0.1 per cent iron, 0.9 per cent fiber, and a calorific value of 83/100 g. Bajpai and Chaturvedi (1990) reported that the vitamins in jamun are thiamine (0.03 mg), riboflavin (0.01 mg), nicotinic acid (0.2 mg), vitamin C (18 mg), and folic acid (3 mg)/100 g of edible pulp.

Figure 3.5: Mature fruits of Jamun (*Syzygium cuminii*).

Table 3.9: Chemical constituents of different types of Jamun fruit.

Parameters	*Small Jamun*	*Big Jamun*
Moisture (per cent)	79.12	81.24
Total soluble solids (per cent)	11.10	12.30
Acidity (per cent)	1.00	1.20
pH	3.35	3.45
Brix/acid ratio	6.94	10.25
Reducing sugars (per cent)	8.41	9.06
Total sugars (per cent)	8.40	9.68
Total anthocyanins (mg/100 g)	242.50	210.00
Tannins (mg/100 g)	428.26	386.25

Source: Ramanjaneya (1985).

Processing

Jamun is exploited commercially for preparation of different products- preserves like jam, jelly, beverages, wine, vinegar and pickles etc. (Shukla, 1979).

Juice

A method of extraction of the *jamun* juice is standardized (Shukla *et al.*, 1991). The maximum yield of *jamun* juice with a high level of anthocyanins and other soluble constituents are obtained by grating the fruit, heating it to 70°C and passing the heated mass through a basket press (Ramanjaneya, 1985). *Jamun* juice, thus, obtained is further heated at 85°C and then cooled to room temperature. Sodium benzoate (500 ppm) is added to the juice before it is stored. Pure *jamun* juice may also be preserved by heat pasteurization. Jamun juice may be concentrated to 60° Brix in a vacuum concentrator. Vacuum concentration with addition of cutback is, therefore, regarded better than the open pan concentration (Ramanjaneya, 1985).

RTS

Jamun juice, being highly acidic in nature, is not consumed as such. A ready-to-serve beverage (nectar) is prepared with 25 per cent juice, 18°Brix and 0. 6 per cent acidity. This nectar is delicately flavoured, which has an appealing colour and when reused is highly acceptable (Khurdiya and Roy, 1985). Storage study of jamun juice and nectar done by Khurdiya and Roy (1985) revealed that at 12 months in cold store, both products were highly acceptable. At room temperature, jamun juice could be stored for up to 8 months and nectar for up to 10 months.

Wine

Jamun fruit is used for preparation of wine in certain parts of India (Bajpai, and Chaturvedi, 1990). Acceptable dry table wine may be prepared from the *jamun* juice (Shukla *et al.*, 1991). In preparation of the *must*, dilution of the whole fruit pulp in a ratio of 1: 1 with water was suitable. Pectic enzyme was beneficial in obtaining a clear product (Shukla *et al.*, 1991).

Pomace Utilization

During extraction of *jamun* juice, a large amount of pomace remains as waste. The pomace contains considerable amount of anthocyanins, tannins and sugars. A method was evolved by addition of 3: 4 of the water to the pomace to obtain pomace extract, which may further be utilized in beverage industry (Ramanjaneya, 1985).

Medicinal Properties

The berry is a powerhouse of health and is officially named as a diabetes-fighter for its hypoglycemic (lowering blood sugar) properties. The seed is used in various alternative healing methods in Unani, Ayurveda and Chinese medicine for digestive ailments. The seeds are very useful and are widely used to treat diabetes alone as well as in combination with other medicines of its class. Seeds are also astringent to bowels, diuretic and stop urinary discharges. Leaves and bark are used for gingivitis and controlling blood pressure. Oral administration of Jamun extracts, from the bark and seeds, boosts insulin levels in diabetics. The berry is also rich in minerals like manganese, zinc, iron, calcium, sodium, potassium and antioxidants (Singh *et al.*, 2010). The jamun fruits are a general tonic, cooling and astringent to bowels. They enrich blood and strengthen liver. They remove bad smell from the mouth and cure biliousness. Very good vinegar is prepared from ripe jamun fruits. This vinegar is an agreeable stomachic and carminative. It is also used as a diuretic. The bark of jamun tree is digestive, astringent to the bowels and anthelmintic. It is good for sore throat, bronchitis, asthma, thirst, biliousness, dysentery, blood impurities and ulcers. The fresh juice of jamun bark alongwith goat milk is given to children in diarrhea. An astringent decoction is prepared from jamun bark and used for gargles and washes. The ash from burnt leaves strengthens gums and teeth. The expressed juice from leaves is used in dysentery alone and in combination with other astringents.

Pomegranate (*Punica granatum*)

Pomegranate is a favourite table fruit in tropical and sub-tropical regions of the world. A native of Iran (Persia), it is found from Kanyakumari to Kashmir. Pomegranate grows well under semi-arid conditions and withstands different soil and climatic stresses. It requires hot and dry climate during fruit development and ripening. It cannot produce sweet fruits unless the temperature is high for a sufficient long period (Samson, 1986).

Figure 3.6: Mature fruits of Pomegranate (*Punica granatum*).

Composition

Pomegranate fruit being rich in minerals, vitamins and nutrients (Table 3.10) finds wide application in traditional Asian medicines both in *Ayurvedic* and *Unani* systems. *Charak*, the great medical physician of ancient India prescribed a large number of formulations using almost every part of this plant in the treatment of dysentery, diarrhoea, stomachache, inflammations, tapeworm, hyrnenoletidosis, dyspepsia, bronchitis and cardiac disorders (CSIR, 1980). The therapeutic properties are reported due to presence of betulic and uroslic acids and different alkaloids, *viz.* pseudopelletierine, pelletierine and other basic compounds (Singh *et al.*, 1990).

Table 3.10: Chemical composition of pomegranate fruits.

Constituents	*Fresh Fruit*	*Dry Fruit*
Moisture (per cent)	78	19
Protein (per cent)	1.6	7.27
Total sugars (per cent)	14.6	66.36
Ascorbic acid (mg/100g)	16.0	72.73
Ash (per cent)	0.7	3.18
Acidity (per cent)	0.58	2.64
Calcium (mg/100g)	10	45
Phosphorus (mg/100g)	70	318
Magnesium (mg/100g)	44	200
Potassium (mg/100g)	133	604
Sodium (mg/100g)	.90	4.09
Iron (mg/100g)	1.79	8.14
Zinc (mg/100g)	0.82	3.73
Manganese (mg/100g)	0.77	3.50
Copper (mg/100g)	0.34	1.55

Processed Products

The fruit, a false berry about the size of a large orange, is filled with shiny red seeds encased in an edible, orange-red pulp. The fruit is mainly used as an ingredient in cooling and refrigerant mixtures and in the preparation of juice, concentrates, condiments (powders and tablets) and pastes (*avaleha*) (Saxena *et al.*, 1984).

The fruit is consumed fresh or can easily be processed into juice, syrup, jams, jelly or even wine. A number of processed products can be manufactured by processing the fruits, which can be preserved for future use (Singh and Maini, 2001).

Fruit Juice

Pomegranate juice makes a delicious drink. On whole fruit basis, the juice yield is about 42 per cent while from grains the yield is about 70 per cent (Phadnis, 1974). The extracted juice can be clarified by heating in a flash pasteurizer at 79-82° C, cooling, settling for 24 hours, racking up and filtering or decanting. The clear juice can be preserved by heat treatment or by using chemicals (600 ppm sodium benzoate). The use of sulphur dioxide is banned for pomegranate due to loss of colour by bleaching action of SO_2. After heating at 80°C, it is filled into bottles while still hot. The bottles are crown corked and pasteurized at 80° centigrade for 30 minutes (Saxena *et al.*, 1984). The juice is highly nutritious and is recommended for patients suffering from gastric troubles. It contains 16.2 per cent Total Soluble Solids (TSS) and 0. 35 per cent acidity and total sugars, 12.93 per cent, reducing sugars, 12.65 per cent, non-reducing sugars, 0.28 per cent and 9.23mg/100g ascorbic acid (Dhumal, 1984). The juice of wild pomegranate is used in the manufacture of citric acid and sodium citrate for medicinal use in Azerbaijan.

Concentrate and Beverage

Fruit juice concentrates, singly or in blended form, are products with great potential on account of their wide acceptability in fruit based beverages. The beverages prepared from pomegranate are highly refreshing and nutritious. A technology was identified using bentolite at 1.0-1.5 g/litre and gelatin at 0.05g/litre as clarifying agents giving better results than milk, charcoal or casein (Gabuniya *et al.*, 1984). The prepared beverage using this technology is excellent in taste and has attractive red colour due to high retention of anthocyanins. The anthocyanins reported in pomegranate have antioxidant properties.

Wine

For preparations of wine, the whole fruits are pressed without crushing or juice may be extracted from pomegranate grains, which gives a yield of 76 to 85 per cent (Adsule and Patil, 1995). Sugar is added to the juice to bring it to 22-23° Brix. Potassium meta-bisulphite is added to the juice to prevent growth of undesirable microorganisms. The juice is fermented with starter wine yeast and the wine is aged and finished in the same manner as the red grape wine. If a sweet table wine is desired, sugar is added to 8-10° Brix. The wine is flash pasteurized at 60°C, bottled hot and sealed. Fortifying the sweetened wine to about 20 per cent alcohol can make wine like port.

Syrup and Jelly

A syrup of 60° brix with an added acidity of 1. 5 per cent as citric acid has a bright purplish – red colour and a delightful taste and flavour (Trapaidze and Abuladze, 1989). It is preserved by pasteurization or by adding sodium benzoate. An attractive jelly can be prepared from pomegranate juice (Phadnis, 1974). Preparation of jelly on a small scale from sweet –sour pomegranates from 'Ganesh' variety has given very good results. (Adsule *et al.*, 1992).

Anar Rub (Pomegranate Jam)

A product known as *anar rub* with fairly good keeping quality can be made by concentrating pomegranate juice, and heating the mixture on a slow fire for a long period. The finished product has a thick consistency and contains 70-75 per cent TSS (Siddapa and Bhatia, 1954). *Anar rub* can be stored for one year and is utilized as a jam, which can be enjoyed with bread, etc.

Anardana

Seeds of the cracked fruits are dried to yield a value added by-product known as *anardana* (Singh *et al.*, 2008) which is used as acidulent and condiment in Indian curries and also used in Ayurvedic and Unani medicines (CSIR, 1980). The use of cross flow drier is recommended to get a uniform, hygienic good quality *anardana* (Pruthi and Saxena, 1984). Out of different varieties screened for preparation of *anardana*, cultivar Bassein Seedless recorded maximum value for acidity, ascorbic acid and anthocyanins and this cultivar recorded maximum organoleptic score showing the suitability for preparation of quality grade *anardana* having bright maroon colour, high acidity and free flowing texture (Singh and Sethi, 2003).

Pomegranate Rind Powder

It is prepared from the waste of fruits. The recovery of pomegranate rind powder was 15. 55 per cent on whole fruit basis and 34 per cent on rind weight basis (Chaven *et al.*, 1995). It is used in medicines, leather and dying industries. It is also used to extract yellow and orange colour used as colouring agents.

Medicinal Properties

Pomegranate is a unique fruit plant producing a wide spectrum of biologically active substances especially important in present-day polluted environment. Pomegranate is loaded with tannins, anthocyanins, polyphenolics and antioxidant vitamins, A, E and C, all of which have a health enhancing effect on the body (Singh *et al.*, 2006). These elements work together to benefit the arteries, plus it keeps the cardiovascular system healthy - the chief health benefit of Pomegranate-. It increases levels of nitric oxide, which improve blood flow to the heart, reduce arterial plaque, reduce systolic blood pressure and help in curing erectile dysfunction. Other benefits include preventing premature aging, stroke, arthritis, Alzheimer's and even cancer. The juice from the fresh fruit is an excellent cooling beverage for alleviating thirst in fevers and sickness. It acts on the liver, heart and kidneys and tones up their functions. It supplies the required minerals and helps the liver to preserve vitamin A from the food. It increases the body's resistance against infections, particularly tuberculosis. It is also of great value in digestive disorders. It is an appetizer, a digestive food item and is useful for patients suffering from colitis and mucous. It binds the stools and tones up the intestines (Singh and Singh, 2005). The chief value of pomegranate is its astringent properties which cause cells to shrink. It is a valuable food medicine for diarrhea and dysentery. The flower buds are also astringent and are useful in chronic diarrhea and dysentery, especially of children. The bark, both of the root and stems of pomegranate tree, is well known for its anthelmintic properties of destroying parasitic worms. The root-bark is, however, preferred as it contains greater quantity of the alkaloid punicine than the stem-bark. This alkaloid is highly toxic to tapeworms. A *sherbet* of the ripe fruit is beneficial in the treatment of typhus, gastric and asthmatic fevers. The root bark is also given as a febrifuge to prevent fevers. The skin of pomegranate fruit is highly beneficial in the treatment of anal itching. The seeds of sour and sweet pomegranate are useful to dissolve gravel in kidneys and bladder. Powder of the dry rind strengthens the gum, stops bleeding, prevents pyorrhea, cleans the teeth and preserves them for a long time

Tamarind (*Tamarindus indica* L.)

Figure 3.7: Ripe fruits of Tamarind (*Tamarindus indica* L.).

Tamarind belongs to Leguminosae family. It is native to tropical East Africa and introduced into India, where it was readily adopted and cultivated (Bueso, 1980). Tamarind is a large, evergreen tree with excellent foliage. It grows to height of 30 m, but mostly it attains a height of about 10 m. The tree grows slowly, taking 10-12 years before it matures and yields fruit.

Composition

The tamarind pod contains about 30 per cent pulp, 40 per cent seeds, and 30 per cent shell (Table 3.11). Baragano de Mosqueda (1980) reported a total sugar content of 31. 01 per cent in tamarind pulp and that reducing sugars contributed about 23. 05 per cent of tamarind fresh weight. Tamarind is a very acidic fruit, with a total acidity range varying from 12. 2 to 23. 8 per cent, mostly as tartaric acid. Almost half of the tartaric acid is in combined form, mainly as potassium bitartarate and to a lesser extent as calcium tartarate. Tamarind pulp is a good source of phosphorus, calcium, and iron. It is an excellent source of riboflavin, thiamine and niacin (Bueso, 1980).

Pulp

Tamarind is valued mostly for its fruit and pulp used for domestic and industrial purposes (Kulkarni *et al.*, 1993). In India, the pulp is eaten raw and sweetened with sugar (Lotschert and Beese,

Table 3.11: Chemical composition of ripe tamarind pulp.

Constituents	*Range*	*Constituents*	*Range*
Water (per cent)	22.6-69.2	Protein (per cent)	1.4-3.3
Lipids (per cent)	0.4-0.81	Carbohydrates (per cent)	59.76-71.8
Total acidity (per cent)	17.1 - 1 8.4	Free tartaric acid (per cent)	8.4-12.4
Minerals (mg/100 g)			
Calcium	54-113.50	Phosphorus	95.40-108
Iron	**0.6-1.0**		
Vitamins (µg/100 g)			
Thiamine	0.15-0.44	Riboflavin	0.16-0.21
Niacin	1.28-2.10		

Source: Pathak *et al.* (1991).

1994). The pulp is usually removed from the pod and used to prepare juice, jam, syrup and candy (Rashda *et al.*, 2000). More commonly, the acidic pulp is used as a favourite ingredient in culinary preparations like curries, chutneys, sauces, ice cream and sherbet in countries where the tree grows naturally (Little and Wadsworth, 1964). The separation of pulp from tamarind seed is rather difficult due to its low water content. In the pulping process, the shell is removed manually, followed by agitation of the seeds in water in order to disperse the pulp and separate it from the seeds. The pulp can also be separated from seeds by covering with water placed in a steam bath for several hours, allowing to stand overnight, filtering through a linen filter, and washing with hot water. The industrial process for the manufacture of tamarind juice concentrate in India is also based on extraction of pulp with boiling water (Nagaraja *et al.*, 1975). Baragano de Mosqueda (1966) described a process for extraction of pulp in which prefermentation of pulp in 1: 2 (fruit water) dilution prior to pulping and heat treatment was used. After removal of the shell, the fruits were soaked for 48 hours to develop aroma and flavour characteristics, to soften the tissues, and to facilitate extraction. Benero *et al.* (1974) reported the mechanical extraction of tamarind pulp from unpeeled fruit.

Tamarind Toffees

In the Philippines, Sri Lanka and Thailand, the fruit pulp after removing the fibres, is mixed with sugar, wrapped in paper and sold as toffees, which are a common sight in front of schools and on urban roadsides. The pulp is also used to make sweet meats mixed with sugar called "tamarind balls" (Purseglove, 1987).

Tamarind Juice

Clarified tamarind juice is one in which the structural and colloidal phases are completely removed by adding gelatine (Baragano de Mosqueda, 1980). The product maintained its distinctive colour and flavor while being absolutely transparent. Benero *et al.* (1974) prepared tamarind beverage containing 9-12 per cent pulp adjusted to 21.5° Brix with sugar. The ranges in composition of beverages containing 9-12 per cent tamarind pulp were 20-21.4° Brix and 0.35-0.44 per cent titrable acidity. The beverage was pasteurized at 85°C, canned in plain tin containers, cooled, dried and stored at 30°C. After 1 year of storage, the beverage was found acceptable by a taste panel.

Tamarind Juice Concentrate

Nagaraja *et al.* (1975) prepared tamarind juice concentrate by clarification of fruit extract and concentration of the clear extract under vacuum to 65-68 per cent total soluble solids. Using this process, an extract is obtained containing 20 per cent soluble solids. The extract is separated from the pulp by sieving and is concentrated under vacuum till the soluble solids reach 68 per cent. The product was filled into sterile glass containers, which sat to jam like consistency on cooling. Yield of the concentrate is about 75 per cent of the pulp used (Kotecha and Kadam, 2003). Tamarind fruit juice is the most valued for its ascorbic acid content, which is about 5.1-6.1 mg/100ml. The recommended concentration of tamarind pulp in syrup is between 20-24 per cent, to produce a beverage with distinctive flavour and acidity (Bueso, 1980).

Tamarind Pulp Powder

Tamarind pulp powder (TPP) is one of the convenience food products prepared by concentrating, drying and milling the pulp into a powder form. Manjunath *et al.* (1991) reported that on an average, the total solid content of tamarind pulp powder varied from 18.6-25 per cent, acidity 8.7-11.1 per cent with an average value of 9.9 per cent (as tartaric acid). The moisture content ranges 3.5-8.8 per cent. Among the minerals, the calcium and potassium contents were high and ranged 74-143 mg and 23.8-27.7 mg. Starch is the major ingredient in tamarind pulp powder (20-41 per cent).

Tamarind Pickle

Pulp is used commercially to prepare tamarind pickle. Pickles are hot, spicy and of salty-sour taste, and can be preserved for several months. Preservation is due to presence of salt, increased acidity and spices. The preparations of pickles are simple and can be done at a household level. Tamarind pickle is available in urban markets of Bangladesh, India and Sri Lanka and is rated as a popular product of tamarind (Singh *et al.*, 2007).

Jam

In making tamarind jam, ripe fruits are shelled and the pulp is boiled for 10 minutes. The pulp is then drained and separated from seeds. For every cup of pulp, two cups of brown sugar are added. The mixture is then cooked and constantly stirred while boiling until it becomes thick in consistency. The resulting jam is cooled, packed in dry, sterilised jars and sealed (Singh *et al.*, 2007).

Syrup

Syrup can be prepared from tamarind pulp using sugar. The recommended concentration of tamarind pulp in syrup is 20-24 per cent, which makes a beverage after dilution with distinct flavor and desirable acidity. Syrup containing 56.7 per cent total solids, 43.8 per cent reducing sugars, 56.54 per cent total sugars, and a total acidity of 1.11 per cent as tartaric acid yielded a refreshing drink after appropriate dilution. Tamarind syrup is made by boiling pulp with required ingredients. To every cup of juice, a half-teaspoon of baking soda is added. The mixture is boiled down to one-half the original quantity, removing the rising scum in the process. The juice is again strained (Kotecha and Kadam 2003).

Candy

Sweetened tamarind fruit is prepared by peeling whole ripe fruits and pouring boiling (62° C) sugar syrup over the fruits. The process is repeated until the fruits are sweet enough. The sweetened fruits are arranged on bamboo racks and dried in sun. Final drying is done in an oven at low temperature. After cooling, the dry fruits are wrapped in cellophane and sold in the market.

Tamarind Seeds

Tamarind seed is a by-product of the commercial utilisation of the fruit. It has several uses. Wasted decorticated kernels contained 46 to 48 per cent of a gel-forming substance. The name jellose is suggested for this polysaccharide as it describes both its jelly forming properties and the carbohydrate character. Seed polysaccharides form a jelly with any kind of sugar without addition of acid (Yoshida *et al.*, 1964).

Medicinal Uses

Tamarind fruits are well known in Europe for their medicinal properties, having been introduced by Arab traders from India. Tamarind products, leaves, fruits and seeds are extensively used in traditional Indian and African medicine. Tamarind cools the system and cleanses the blood. The pulp, rich in vitamins and minerals, is used in Chinese medicine. The bark treats asthma, and a paste from the leaves or the seeds are used to treat boils. The flowers reduce blood pressure. The medicinal value of tamarind is mentioned in traditional Sanskrit literature. The laxative properties of the pulp and the diuretic properties of the leaf sap are confirmed by modern medical science. Several medicinal properties are claimed for preparations containing tamarind pulp, leaves, flowers, bark and root. The American pharmaceutical industry processes 100 tons of tamarind pulp annually and it is a common ingredient in cardiac and blood sugar reducing medicines. Fruit pulp improves appetite and is a gentle laxative. In Sudan dried fruit pulp is taken orally as a purgative, for malaria and for bacterial infections. In Thailand hot water extract of dried fruit pulp is taken orally as an expectorant. In West Indies, fruit pulp is taken orally as a laxative whereas in Peru hot water extract of dried fruit peel is taken orally as a laxative. Tamarind aids in the cure of malarial fever. In Southeast Asia, the pulp is prescribed to counteract the ill effects of overdoses of chaulmoogra (*Hydnocarpus anthelmintica* Pierre), given to treat leprosy, and in Mauritius, the pulp is used as a liniment for rheumatism. The pulp is also effective in ridding domestic animals of vermin in Columbia, through the application of pulp with butter and other ingredients. Powdered seed is used to treat chronic diarrhoea, jaundice and dysentery. It is also used as cardiac and blood sugar reducing medicines. Leaf crushed with salt is used to prevent throat infections, cough, fever, intestinal worms and liver ailments and applied to boils/abscesses to reduce inflammation. Flowers and immature pods are used internally as a remedy to cure jaundice and externally to cure eye diseases and skin ulcers. Bark is used as an astringent for the treatment of diarrhoea, as a cure for asthma and acts as a digestive aid. Roots contain the glycosides: vitexin, isovitexin, orientin and isoorientin which are active components in medicines for leprosy and dysentery (Singh *et al.*, 2007).

Wood Apple (*Feronia limonia*)

Wood apple is a thorny tree of family Rutaceae, commonly found in dry deciduous forests. This fruit is a native of India, and is found throughout the plains of India, particularly in dry situations (Bose and Mitra, 1989). It can be planted under almost all conditions of soil and climate, thrives in adverse climatic conditions, and tolerates salinity and low temperature. It possesses great tolerance to drought. Extensive root system and synchronization of its reproductive phase with high moisture availability make it suitable crop for arid zone. Woodapple starts bearing at an age of 5-6 years. The fruits ripen from November to March. Ripe fruits drop from the trees, and as they have a hard shell outside, the inner pulp keeps well for some more days (Bose and Mitra, 1989). An average tree yields 250-500 fruits/year; bigger and more vigorous trees may yield more. Each fruit weighs about 150-500 g.

Figure 3.8: Ripe fruit of Wood apple (*Feronia limonia*).

Composition

The ripe fruits contain sour-sweet, aromatic pulp, which is about 70 per cent of total weight. Wood apple pulp contains significant amount of proteins. The fruit is a rich source of riboflavin; acidity is about 2. 3 per cent and sugars 7. 2 per cent (Table 3.12).

Table 3.12: Composition of wood apple fruit.

Constituents	*Range*	*Constituents*	*Range*
	(values/100 g edible portion)	*Mineral content*	*(mg/100 g fresh weight)*
Moisture	64.2 g	Calcium	4
Protein	7.1 g	Phosphorus	9
Carbohydrates	17.0 g	Iron	0.5
Fat	0.3 g	Magnesium	41
Total minerals	0.3 g	Copper	0.21
Energy value (kg ca/μg)	74	Manganese	0.18
		Zinc	0.46
Vitamin content	*(Values/100 g fresh edible portion)*		
β-carotene (μg)	61		
Thiamine (mg)	0.04		
Riboflavin (mg)	0.17		
Niacin (mg)	0.8		
Vitamin C (mg)	3		

The acid content of pulp varies from 7.6 per cent in unripe fruits to 2.3 per cent in fully ripe ones. It contains 3.5 per cent pectin and forms an excellent material for making jelly (Bose and Mitra, 1989). The fruit is a tonic, refreshing, antiscorbutic, and alexiformic. It is used as a substitute for bael in the treatment of diarrhoea and dysentery (CSIR, 1980). The pulp is used to treat infections of the gum and throat. The pulp of the ripe fruit is eaten as such or with sugar. It can be used for making sherbet. The fruit is very rich in acid and pectin (CSIR, 1980).

Processing

Extraction of Pulp

The fruit has a thick, hard shell, inside which is a darkish brown and acid-sweet pulp, in which a large number of small seeds are embedded. The wood apple fruits are opened by breaking them and the pulp, along with the seeds and fibre, are separated from the hard shell with the help of a stainless steel spoon. To extract the fine pulp without seeds and fibre, an equal amount of water by weight is added and boiled. The mass is passed through a muslin cloth to separate the seeds and fibre. The pulp thus obtained is homogenous. This pulp is used for preparation of squash, nectar and jelly (Singh *et al.*, 2008).

Squash

Wood apple syrup can also be prepared and used as a drink. Sugar syrup is prepared by adding cane sugar to the boiling water. The strength of sugar syrup is ascertained by testing (30-60°C Brix range). The syrup, thus prepared, is filtered through a muslin cloth to remove impurities. Fruit pulp and the freshly prepared hot syrup are added together in the proportion of the proposed recipe at a

Figure 3.9: Wood apple squash.

temperature of 90°C and poured into pre-sterilized dry glass jars. The filled jars are sealed and stored at room temperature for further use.

Nectar

Sugar syrup is prepared by adding cane sugar to boiling water. The strength of sugar syrup is estimated with help of a hand refractometer. The prepared syrup is filtered through a muslin cloth to remove impurities. The hot syrup was mixed with fruit pulp on weight basis. The mixture is boiled and citric acid added to get a consistent product. The prepared nectar is poured into the pre-sterilized glass bottles of 220 ml capacity and sealed airtight by using crown caps. The bottles are sterilized in boiling water for 25 minutes, cooled to room temperature and utilized for further use (Singh *et al.*, 2007).

Jelly

Ripe fruits are used for making jelly. The shell is broken and the pulp is boiled with water, in a proportion of 1: 3 (pulp: water) for about 30 min. Upon cooling, it is strained, and to the liquid is added about its own volume of clean crystalline sugar. The liquid is then boiled to about 105°C and poured hot into sterilized containers for setting. Woodapple jelly resembles black currant or apple jelly. It is clear and bright purple in color, with firm consistency and agreeable flavor (Bose and Mitra, 1989). The filled jars are sealed and stored at room temperature for further use.

Chutney

The acidity of the pulp varies from 7.6 per cent in raw fruits to 6.3 per cent in semi ripe and 2.3 per cent in fully ripe ones. Ripe fruits contain 7.25 per cent total sugars. The fruit thus forms an excellent raw material for chutney making. Fully developed immature fruits are harvested for preparation of chutney. The chutney of this fruit, if standardized, may also be turned into a commercial product.

Other Uses

The fruit shell is fashioned into snuffboxes and other small containers. The trunk and branches exude a white, transparent gum especially after the rainy season which can be utilized as a substitute for adulterant of gum arabic and also for making watercolors, ink, dyes and varnish. It consists of 35.5 per cent arabinose and xylose, 42. 7 per cent d-galactose, and traces of rhamnose and glucuronic acid (Singh *et al.*, 2008).

The wood is yellow-gray or whitish, hard, heavy, durable, and valued for construction, pattern-making, agricultural implements, rollers for mills, carving, rulers, and other products. It also serves as fuel. The wood is also used in making agricultural implements and handles of different tools. The heartwood contains urasolic acid and a flavanone glycoside, 7-methylporiol-D-xylopyranosy l- D-glucopyranoside.

Medicinal Properties

The wood apple has several medicinal properties. The leaves and bark have medicinal values. It is antiscorbutic. It is an antidote for poisons and also helps in curing sore throat. After the rains, the trunk and branches give off a gum called' *Feronia gum'*, which counteracts diarrhoea, dysentery and diabetes. However, this statement should not be understood that the gum totally cures these diseases (Singh *et al.*, 2008). It serves to mitigate the problem. The fruit is much used in India as a liver and cardiac tonic, and, when unripe, as an astringent means of halting diarrhoea and dysentery and effective treatment for hiccup, sore throat and diseases of the gums. The fruit is a tonic, refreshing, antiscorbutic, and alexiformic. It is used as a substitute for bael in the treatment of diarrhoea and dysentery (CSIR, 1980).

The pulp is poulticed onto bites and stings of venomous insects, as is the powdered rind. Juice of young leaves is mixed with milk and sugar candy and given as a remedy for biliousness and intestinal troubles of children. The pulp is used to treat infections of gum and throat. The powdered gum mixed with honey, is given to overcome dysentery and diarrhoea in children. Crushed leaf extract is applied on itch and the leaf decoction is given to children as an aid to digestion. Leaves, bark, roots and fruit pulp are used against snakebite. The spines are crushed with those of other trees and an infusion taken as a remedy for menorrhagia. The bark is crushed with that of *Barringtonia* and applied on venom wounds. The unripe fruits contain 0.015 per cent stigmasterol (Singh *et al.*, 2010). Leaves contain stigmasterol (0.012 per cent) and bergapten (0.01 per cent). The bark contains 0.016 per cent marmesin. Root bark contains aurapten, bergapten, isopimpinellin and other coumarins. The fruits are highly acidic and contain 7.6 per cent acidity when unripe and 2.3 per cent when ripe. They also contain 3-5 per cent pectin making it suitable for preparation of quality jelly (Singh *et al.*, 2009). Its fruits are also used in various *ayurvedic* preparations for curing diarrhoea, dysentery, gum and throat problems (Singh and Singh, 2005).

Conclusion

Minor fruits play an important role in food and nutrition security. These fruits are known for their typical flavour and taste (Singh *et al.*, 2008). Consumers today are increasingly conscious of health and nutrition, and there is a tendency to avoid chemicals and synthetic foods, thereby choosing natural foods. In this context, underutilised fruits have unlimited potential. Appropriate process technology needs to be developed to popularize these fruits. Neglected and underutilized crops are essential to the livelihoods of millions of poor farmers throughout the world (Rashda *et al.*, 2000). As noted above, they are part of the (threatened) biological assets of the rural poor. In identifying research and development issues, which should be addressed, it is essential to approach the problem from this perspective. The filiere concept can be considered an evolution of networking concept for plant genetic resources based on more efficient partnership and participatory approaches. The filiere would thus bring about greater participation of local actors to ensure addressing of local needs properly. It would also ensure wider representation and participation of stakeholders of food processing and marketing sectors and policy makers who have traditionally been left aside from PGR activities (Padulosi, 1999).

References

Adsule, R. N. and Patil, N. B. 1995. Pomegranate. (In) *Handbook of fruit Science and Technology,* DK Salunke and SS Kadam (eds), Marcel Dekkar, New York, pp. 455-464.

Adsule, R. N., Kotecha, P. M and Kadam, S. S. 1992. Preparation of wine from pomegranate, *Beverage and Food World,* **19** (4), 13.

Amba, D.1973. Some new fruit beverages, *Indian Hort.* 18: 5.

Bajpai, P. N. and Chaturvedi, D. P. 1990. Jamun, *Fruit of India: Tropical and Subtropical* (T. K. Bose and S. S. Mitra, eds.), Naya Prokash, Calcutta.

Banerjee, N., and Maity, M, 1980. Some physico-chemical studies on a protein rich fruit seed (*Aegle marmelos* Correa), *J. Indian Chem. Soc.* 57(9): 945.

Baragano De Mosqueda, M. 1980 Technology of clarified tamarind juice. In *Tropical and Subtropical Fruits.* Nagy, S. and Shaw, P. E. (Eds.), AVI Publishing, Westport, Conn., 375 p.

Baragano de Mosqueda. M. 1966. Technology *of* clarified tamarind juice, Sp. *Soc.* Cieve. *Natur. La Salle. Memo* (Spanish) 26: 62.

Benero, J. R., Collazo de Rivera, A. L. and George, L. M. I. 1974. Studies on the preparation and shelf-life of soursop, tamarind and blended soursop-tamarind soft drinks, *J. Agr.* Univ. *Puerto* Rico 58: 99.

Blench, R. M. 1997. Neglected Species, Livelihood and Biodiversity in Difficult Areas: How should the Public Sector respond? Overseas Development Institute, Natural Resources Perspective No. 23, UK.

Bose, T. K. and Mitra, M. K. (Eds.) 1989. *Fruits: Tropical and Subtropical,* Naya Prokash, Calcutta.

Bueso, CE. 1980. Soursop, Tamarind and Chironka. (In) *Tropical and Subtropical Fruits.* Nagy, S. and Shaw, P. E. (Eds.), AVI Publishing, Westport, Conn., 375.

Burgess, M. A. 1994. Cultural responsibility in the preservation of local economic plant resources. Biodiversity and Conservation 3: 126-136

Chaven, UD. Adsule, RN. and Kadam, S. S. 1995. Physico-chemical properties of pomegranate rind powder, *Beverage and Food World,* **22**(1), 36.

CSIR 1980. *The Wealth of India: Raw Materials,* Council of Scientific and Industrial Research, New Delhi.

Dhumal, J. 1984. Studies to prolong shelf life of Pomegranate fruits, M. Sc Thesis, M. P. A. U. Rahuri, India.

FAO. 1996. Global Plan of Action for the Conservation and Sustainable Utilisation of Plant Genetic Resources for Food and Agriculture and Leipzig declaration, adopted by the International Technical Conference on Plant Genetic Resources, Leipzig, Germany, 17-23 June 1996. Food and Agriculture Organization of the United Nations, Rome, Italy.

FAO. 1997. Human Nutrition in the developing World. FAO Food and Nutrition Series No. 29. Food and Agricultural Organization of the United Nations. Rome. Italy.

Frison E., H. Omont and Padulosi, S. 2000. GFAR and International Cooperation on Commodity Chains. Synthesis paper presented at the GFAR-2000 Conference held in Dresden, Germany on 21-23 May 2000. (Unpublished)

Gabuniya, N. E., Landkinpanidze, I. D. and Magerammov, M. V. 1984. Technology of obtaining a beverage from pomegranate, *Konsernayai Oveschesushilnaya Pomyshelnnost*, **7**, 27.

Gopalan, C., B. V. Ramasastri, and Balasubramanian, S C 1978. *Nutritive Value of Indian Foods*, National Institute of Nutrition, ICMR, Hyderabad, India, 1978.

Jauhari, O. S., and Singh, R D 1971. Bael, valuable fruit, *Indian Hort.* 16(1): 9.

Khurdiya, D. S. and Roy, S. K. 1985. Processing of jamun (*Syzygium cuminii* Linn.) fruit into a ready to serve beverage. *J. Food Sci. Technol.* 22: 27.

Khurdiya, D. S., and Anand, J C 1981. Effect of extraction method, container and storage temperature on phalsa fruit juice, *Indian Food Packer* 35: 6.

Khurdiya, D. S. 2001. Postharvest management of underutilized fruits for fresh marketing. Winter school on exploitation of underutilized fruits. pp. 266-274. 3rd-23rd Oct.2001.

Kotecha, P. M. and Kadam, S. S. 2003. Preparation of ready-to-serve beverage, syrup and concentrate from tamarind. *Journal of Food Science and Technology*, Mysore, 40: 1, 76-79; 15.

Kotecha, P. M, Adsule, R. N. and Kadam, S. S. 1995. Processing of custard apple: Preparation of ready-to-serve beverage and wine. Indian Food Packer. 49(4): 5-11.

Kulkarni, R. S., Gangaprasad, S. and Swamy, G. S. K. 1993. Tamarind: Economically an important minor forest produce. *Minor Forest Products News*, 3(3): 6.

Lakshmi, V. and J. S. Chauhan, J S. 1976. Grewinol, a keto-alcohol from the flowers of *Grewia asiatica*. Lloydia 39: 372–374.

Leaman, D. J, H. Fassil and Thormann, I 1999. Conserving medicinal and aromatic plant species: identifying the contribution of the International Plant Genetic Resources Institute. Study commissioned by the International Development Research Centre (IDRC). International Plant Genetic Resources Institute, Rome, Italy.

Little, E. L. and Wadsworth, F. W. 1964. *Common Trees of Puerto Rico and the Virgin Islands, Agriculture*. Handbook 249, Washington DC. US Department of Agriculture: 548 p.

Lotschert, W. and Beese, G. 1994. *Tropical Plants*. Collins Photo Guide. Harper Collins Publishers: 223.

Manjunath, MN., Sattigeri, V. D., Rama, Rao S. N, Udah Rani, M. and Nagaraja, K. V. 1991. Physico-chemical composition of commercial tamarind powder. *Indian Food Packer*, 45: 39-42.

Monti, L., 1997(Ed). Proceedings of the CNR International Workshop on "Neglected Plant Genetic Resources with a Landscape and Cultural Importance for the Mediterranean Region", 7-9 November 1996, Naples, Italy. 150 pp.

Morton, J. F. 1987. Phalsa. (In) Fruits of warm climates. Julia Morton, Miami, FL. p 276–277.

Mukherjee, B., and Ahmed, K.1954. Riboflavin *Pakistan J. Bioi. Agr. Sci./: 47-51* (1954).

Nagaraja, KV., Manjunath, M. N. and Nalini, M. L. 1975. Chemical composition of commercial tamarind juice concentrate. *Indian Food Packer*, 29: 17.

Ogle, B. M. and Grivetti, L. E. 1995. Legacy of the chamaleon: edible wild plants in the Kingdom of Swaziland, Southern Africa. A cultural, ecological, nutritional study. Part II- Demographics, species availability and dietary use, analyses by ecological zone. Ecology of Food and Nutrition.17: 1-30.

Padulosi, S. 1999. Criteria for priority setting in initiatives dealing with underutilized crops in Europe. *In* Implementation of the Global Plan of Action in Europe. Pp. 236-247.

Pathak, R. K., Ojha, C. M. and Dwivedi, K. 1991. Adopt patch budding *for* quicker multiplication in tamarind, *Indian Hort.* 36: 17.

Phadnis, N. A. 1974. Pomegranate for dessert and juice, *Indian Hort*, **19**(3), 9.

Pistorius, R. 1997. Scientists, Plants and Politics – A history of the Plant Genetic Resources Movement. International Plant Genetic Resources Institute, Rome, Italy. 134 pp.

Prescott-Allen and Prescott Allen 1990. How many plants feed the world? Conserv. Biol. pp. 90-93.

Pruthi, J S. and Saxena, A. K. 1984. Studies on *anardana*, Journal of food Sci and Tech., 41 (4): 31.

Purseglove, J. W. 1987. *Tropical Crops. Dicotyledons*, Longman Science and Technology: 204-206.

Ramanjaneya, K. H. 1985) Studies on some aspects of jamun (*Syzygium cuminii* Skeels) and its processing, Ph. D. thesis, I. A. R. I., New Delhi.

Rashda, Ali., Faiza, Rab., Herregods, M., Nicolai, B., Jager-A-de and Roy, S.K. 2000. Research needs and new product development from underutilized tropical fruits. Proceedings of the XXV International Horticultural Congress. Part 8. Quality of horticultural products: storage and processing, new outlooks on postharvest biology and technology, potentiality of processing of underutilized fruits of the tropics, Brussels, Belgium, 2-7 August, 1998. Acta-Horticulture, No. 518, 241-248;

Roy, S. K. and Singh, R. N. 1979. Studies on the utilization of *bael* fruit (*Aegle mannelos* Correa) for processing. Preparation and preservation of *bael* products. *Indian Food Packer.* 33 : 9-14:

Salunkhe, D. K. and Desai, B. B. 1984. Phalsa. (In) Salunkhe and Desai (eds.), Postharvest biotechnology of fruits. Vol. 2. CRC Press, Boca Raton, FL. p. 129.

Samson, J. A. 1986. The minor tropical fruits. p. 316. (In) Tropical fruits. Longman Inc., New York.

Saxena, A. K., Mann, J. K. and Berry, S. K. 1984. Pomegranates post-harvest technology, chemistry and processing, *Indian Food Packer*, **41**(94), 31.

Shukla, J. P. 1979. Ph. D. thesis. CSA University of Agriculture and Technology, Kanpur.

Shukla, K. G., Joshi, Yadav, M. C. S. and Bisht, N. S. 1991. *Jamun* wine making: Standardization of a methodology and screening cultures. 1. *Food Sci. Tech.*, 28(3): 142.

Siddapa, G. S. and Bhatia, B. S. 1954. The identification of sugars in fruits by chromatography, *Indian J Hort*, **11,** 19.

Singh D, Lobsang Wangchu and Prerak Bhatnagar 2006. Effect of Processing and Packaging material on Anthocyanins Retention in Pomegranate. *Indian Journal of Arid Horticulture*, 1(2): 17-20.

Singh D, Lobsang Wangshu and Kavita A. 2008. Quality analysis of different samples of *Anardana. Journal of Medicinal and Aromatic Plants* (JMAPS), 30, 60-64.

Singh D. and Maini S. B., 2001, Studies on physico-chemical characteristics of important pomegranate cultivars, *Journal of Farming System Research and Development*. 1(3): 16-18.

Singh D. and Singh R K 2003. Processed Products of pomegranate. *Natural Product Radiance A Bimonthly Digest on Natural Products*. National Institute of Science Communication, CSIR, New Delhi, 3 (2). 66-68.

Singh D. and Singh. Ranjay K. 2005. Fruits Diversity to Combat the Human Healthcare: Exploration of Tribal Wisdom Singh, Hand Book of Ethno-medicine, Institute of Ethno-medicines, Munchen, Germany, pp 437-449.

Singh D. Wangchu L and Moond S. K 2007. Processed Products of Tamarind. *Natural Product Radiance A Bimonthly Digest on Natural Products*. National Institute of Science Communication, CSIR, New Delhi. Vol. 6 (4), 2007, pp.315-321.

Singh Dheeraj, Lobsang Wangshu and Prahalad, V. C.2008, Processing and marketing feasibility of underutilized fruit species of Rajasthan, India. Studies on Agricultural and food sector in Central and Eastern Europe. Agri-Food business: Global challenges –Innovative solutions, (Ed.) Thomas Glauben *et al.*, Leibniz Institute of Agricultural Development in Central and Eastern Europe. Salle, Germany. pp 70-81.

Singh Dheeraj, M. Choudhary, V. C. Prahalad and Kavita A. 2009, Value addition to forest produce for nutrition and livelihood. *Indian Forester*, Vol 135 (9): 1271-1284.

Singh Dheeraj, Surendra Kumar Moond and Lobsang Wangchu. 2007. Medicinal value of tamarind (*Tamarindus indica*).(In) Phytopharmacology and therapeutic values I RMPA Vol.19, *Studium Press LLC, USA*, pp 293-304.

Singh, D.2007. Enhancement of nutritional availability through processing and value addition of lesser known fruit species in arid and semi-arid regions *Environment and Ecology*. Kalyani-741235, West Bengal

Singh, D. and Sethi, V. 2003. Screening of pomegranate genotypes for the preparation of quality grade *anardana*. *J Food Sci Technol*, 40 (2) 236-238.

Singh, D., S. Bhushan, L. Wangchu, Kavita, A. and Moond, S. K. 2010, Natural Antioxidant Photochemicals in Fruits, Berries and Vegetables and their Degradation Status during Processing. In: Comprehensive Bioactive Natural Products, vol. 4: Antioxidants and Nutraceuticals. (Eds.): V. K. Gupta and Verma, A. K. *Studium Press LLC, USA*, pp: 99-131.

Singh, R. P., Gupta, A. K. and Bhatia, A. K. 1990. Utilization of wild pomegranate in north-west Himalayas- Status and Problems, *Proc Nat Semi Production and Marketing of Indigenous Fruits*, New Delhi, pp. 100-107.

Singh, S., Krishnamurthi, S. and Katyal, S. L. 1967. *Fruit Culture in India.* ICAR, New Delhi, p. 255.

Swaminathan, M. S. 1999. Enlarging the Basis of Food Security: the Role of underutilized Species. International Workshop held at the M. S. Swaminathan Research Foundation. 17- 19 February 1999. Chennai. India.

Tokitomo, Y., Y. Shimono, A. Kobyahi, and Yamanishi, T, 1982. Aroma components *of* bael fruit (*Aegle marmelos*), *Agr. Bioi. Chem.* 46(7): 11-18.

Trapaidze, T. G. and Abuladze, L. S. H. 1989. Pomegranate cultivars resistant to cracking. Subtropicheskie Kultury, (2): 95.

UNEP. 1992. Convention on Biological Diversity. United Nations Environment Programme.

UNEP. 1995. Global Biodiversity Assessment. United Nations Environment Programme University Press, Cambridge, UK.

Verma, U. P., and Ahmed, S 1958. Drying of bael (*Aegle marmelos*) pulp, *Indian Food Packer* 12(7): 7.

Waskar, D. P., and Khurdiya, D. S. 1988. Studies on the utilization of phalsa pomace, *Abstracts of Int. Food Convemion and Exhibition* (IFCON-88), CFTRI, Mysore, India, p. 24.

WCMC. 1992. Global Biodiversity: Status of the Earth' Living Resources – World Conservation Monitoring Center, Chapman and Hall, London, New York. 5494 pp.

Yadav, A. K. 1998. Phalsa. Commodity sheet FVSU-004 of the Fort Valley State Univ., Ag. Res. Sta.

Yoshida, T., Yokoo, Y. and Koyama, Y. 1964. Jelly strength of tamarind polysaccharide jelly, *Nippon Shokuhin Kogyo Gakkai-Shi J.,* 200.

2014, Future Crops Vol. 2 *Pages 85–102*
Editor: Professor K.V. Peter
Published by: DAYA PUBLISHING HOUSE, NEW DELHI

Chapter 4
Underexploited and Lesser Known Fruits of Mizoram

*T.K. Hazarika**
Department of Horticulture, Aromatic and Medicinal Plants,
School of Earth Sciences and Natural Resources Management,
Mizoram University, Aizawl, Mizoram
**E-mail: tridip28@gmail.com*

Underexploited and lesser known fruits are found in the forests or marginal lands of rural areas without commercial cultivation. These fruits are nutritionally rich and can supplement nutritional requirements especially vitamins and micronutrients to the human beings (Ogle and Grivetti 1985; Maundu *et al.*, 1999). The nutritional qualities of these fruits are comparable and in a few cases even superior to domesticated varieties of high demanding fruits. They provide cheap but quality nutrition for large segments of the population in both urban and rural areas (Dansi *et al.*, 2008). Many of these fruits are still unknown or insufficiently exploited, despite their nutritional values. Most of these fruits have medicinal properties including antioxidant properties. Many of these fruits are used by traditional healers of rural areas as a source of ethno medicine in tribal healthcare system. The vast store of information on indigenous knowledge, practices and technologies is being eroded as a result of rapid urbanization, over-exploitation of resources, unscientific land use, change of lifestyles and behavior (Joshi and Joshi, 2005). Studying indigenous knowledge on plant uses and resource management practices is thus extremely important to support biodiversity conservation programs (Termote *et al.*, 2011). Hence, priority ought to be given to document these useful underutilized and underexploited fruits from the area of high abundance before the inhabitants shift over to modern lifestyles.

These underutilized and underexploited fruits, are known to play an important role in food and nutritional supplement of human beings from time immemorial. These fruits have the ability to grow under stressed and adverse conditions. They have outstanding adaptation capabilities like drought tolerance, salt tolerance and are also resistant to major pests and diseases. A few of these fruits have excellent flavour, juiciness and have very attractive appearance. Many of these species are used as traditional medicinal plants and a few have found important place in the Indian system of medicine and Unani, since time immemorial. These fruits play a major role in the socio-economy of the inhabitants of the areas where they occur to support income and nutrition.

Many of the indigenous fruits remained underexploited due to lack of awareness of their potential, market demand, low and erratic bearing in many cases. Many of them are highly perishable and difficult to store in fresh form. A few are not acceptable as fresh fruit, because of high acidity and/or strong astringent taste. Despite having all the favourable factors, the desired level of development in this indigenous fruit sector is not achieved because of a number of constraints. Although a great diversity of valuable fruit species exist, these fruits are not commercially exploited. All these fruits have unlimited potentialities in the world trade if value addition is done. Organized production and processing for value addition would enhance income of small and marginal farmers and also help in on-farm conservation of valuable germplasm of the region.

Potentialities of Underutilized Fruits in Mizoram

Mizoram, one of the eight states in north eastern hill region, is an expanse of blue green hills and the 23rd state of India in the extreme of Himalayan ranges and covers an area of 21, 087 sq. km and is situated between Myanmar and Bangladesh. It is located in between 21°58´ N and 24 °35´N latitudes and 92°16´E and 93 °29´E longitudes and bounded by Cachar district of Assam in the North, Manipur in the North- East, Myanmar in the East and South and Bangladesh and Tripura in the West (Hazarika *et al.*, 2012). The socio-economic life of rural people depends on their ambient vegetation from where they derive all their material requirements - timber, food, fuel wood, medicinal plants *etc*-. To fulfill their basic requirements, they depend on the wild economic species and to make these resources easily accessible, they cultivate some of these plants in their kitchen gardens (Bhat *et al.*, 2009). Considerable diversity exists among these indigenous fruit species including variation in plant type, morphological and physiological characteristics, reactions to diseases and pests, adaptability and distribution. The vegetation of Mizoram is tropical evergreen and semi-evergreen forest in the lower altitude hills; sub-tropical to montane sub-tropical in the high hills. These forests are covered with a number of underutilized fruits along with other timber and non-timber valuable trees. In this hilly state of North East India, a number of fruits are available which are neither grown commercially on large scale nor traded widely. The tribal people of the state are consuming these minor and lesser known fruits since very ancient time without knowing their nutritive and medicinal value (Lalfakzuala *et al.*, 2007; Lalramnghinglova, 2003). These fruits are consumable but relatively less palatable than other fruits with lesser demand in the market, grown to a limited extent and are not usually cropped in organized plantations with application of inputs and are considered as minor fruits (Mazumdar, 2004). These fruits remain wild or semi–domesticated and traded and consumed locally. Most of them are rich sources of vitamins, minerals and other nutrients like carbohydrates, proteins and fats. Many of these fruits are eaten locally to meet their nutritional requirements, having a broad range of flavour and colour with potential health benefits (Ikram, *et al.*, 2009). They have outstanding capacity like drought tolerance, salt tolerance and also resistance to major pests and diseases. In addition, a few of these fruits have excellent flavour, juiciness and have very attractive appearance. They are characterized by the fact that they are locally abundant but globally rare, and the scientific information and knowledge about them are limited (Gruere *et al.*, 2009). Considerable diversity exists among these fruits in plant type, morphological and physiological characters, adaptability, distribution and reaction to pests and diseases. Apart from nutritional values, they are also used for medicinal purpose, income generation and poverty alleviation programmes in rural areas. Besides, these fruits are also sources of livelihood of people living in rural areas of the region. The tribal people of the state consumed these wild fruits since very ancient time mainly as fruits without knowing their nutritive value.

Table 4.1: Taxonomic description of underutilized and underexploited fruits of Mizoram.

Sl.No.	Scientific Name	Mizo Name	Family	Distribution	Fruiting Time
1.	*Aegle marmelos*	Belthei	Rutaceae	Aizawl	Feb-April
2.	*Anonna squamosa*	Theiarbawm	Annonaceae	Tuikhurhlu	July-Aug
3.	*Artocarpus heterophyllus*	Lamkhuang	Moraceae	Baktawng, Selling, Tlungvel	May-June
4.	*Artocarpus lakoocha*	Theitat	Moraceae	Tanhril,	April-June
5.	*Averrhoa carambola*	Theiherawt	Averrhoaceae	Aizawl, Khumtung, Baktawng	June-Sept
6.	*Baccaurea ramiflora*	Pangkai	Euphorbiaceae	Baktawng	April-June
7.	*Dillenia indica*	Kawrthindeng	Dilleniaceae	Sihphir, Durtlang.	Dec-March
8.	*Dillenia pentagyna*	Kaihzawl	Dilleniaceae	Thingsul-tlangnuam	April-May
9.	*Eleagnus caudata*	Sarzuk	Eleagnaceae	Seling, Tuirial, Thingsul-tlangnuam.	Feb-March
10.	*Elaeocarpus lanceaefolius*	Kharuan	Tiliaceae	Sihphir	Aug-Sept
11.	*Embelia vestita*	Tling	Myrsinaceae	Thingsulthliah	March-April
12.	*Emblica officinalis*	Sunhlu	Euphorbiaceae	Thingsultlangnuam	Dec-Jan
13.	*Euphobia longan*	Theifeimung	Sapindaceae	Darlawng, Khumtung, Baktawng	June-July
14.	*Ficus semichordata*	Theipui	Moraceae	Tanhril.	Jan-Feb
15.	*Garcinia cowa*	Chengkek	Guttiferae	Aizawl	March-April.
16.	*Mimusops elengi*	Ramkelchal	Sapotaceae	Durtlang	Oct-Dec
17.	*Myrica esculenta*	Keifang	Myricaceae	Seling	Feb-April
18.	*Phyllanthus acidus*	Kawlsunhlu	Euphorbiaceae	Thingsulthliah	April-June
19.	*Prunus domestica*	Japan theite	Rosaceae	Thingsulthliah	April-June
20.	*Spondias pinnata*	Tawitaw	Anacardiaceae	Seling, TuikhurhluSihphir.	Oct-Dec
21.	*Syzygium cumini*	Hmuipui	Myrtaceae	Seling	April-June
22.	*Syzygium grandis*	Theichhawl	Myrtaceae	Seling	April-May
23.	*Tamarindus indica*	Tengtere	Caesalpiniaceae	Phulmawi, Darlawng, Thingsulthliah	Feb-March
24.	*Ziziphus mauritiana*	Borai/Kawrsunhlu	Rhamnaceae	Tuirial, Tuikhurhlu	Aug-Oct

Although the genetic resources of these fruits are well adapted to the stressed, arid and semi-arid ecosystems and have high potential for mitigating inevitable climate change scenario, research on genetic improvement of these cultivars is very limited. Since most of these fruit trees have not been cultivated on-farm, there is only scant and dispersed knowledge about their production and nutritional value (Rathore, 2009). Many of these fruits remained underexploited due to lack of awareness of their potential, market demand and low and erratic bearing habit and a few of them get lost due to lack of proper conservation measures. It is of urgent need to conserve available germplasm of these fruits for future generation. There is need to explore and identify minor fruits, areas of their existence and document and evaluate the potential fruits in terms of diversity and availability in the state. Steps should be taken for their genetic resource management, conservation and domestication of elite strains of all the available minor fruits of the state. In-depth study and complete inventorization of these high valued fruits available in rich bio-diversity area is one of the 'Green' prospects of research in Mizoram.

Aegle marmelos (L) Correa.

Aegle marmelos is a medium sized tree attaining height up to 6.5m. with deciduous and hardy nature. This is a thorny tree with dimorphic branches, trifoliate leaves and erenate leaflets. Flowers are whitish-green, scented and bisexual. Fruits are of small to very large size with varying shape from oblong to ovate, round, globose and elongated containing hard outer cover (skull) and soft creamy pink to reddish pulp full of hairy seeds. Seeds have hard and hairy seed coat, flat in shape and mucilaginous.

Tree is of historical importance in Indian culture and various uses are described in the literature. Fruit pulp is consumed raw and also processed to make diverse value added products, most common are sharbat, squash, murabba and jam. Dried pulp is also used in several parts of country. Pulp has several medicinal properties and is commonly used for cure of diarrhoea, dysentery and other stomach ailments. Leaves, root and bark also have medicinal properties and used widely in the Indian System of Medicine. The crystalline substance known as 'marmelosin' from fruits have therapeutic properties. Trifoliate leaves are used in prayer/puja of Lord Shiva.

Annona squamosa

Annona squamosa is a small, semi-deciduous tree, 3-7 m in height, with a broad, open crown or irregularly spreading branches; bark light brown with visible leaf scars and smooth to slightly fissured into plates; inner bark light yellow and slightly bitter; twigs brown with light brown dots (lenticels). Leaves occur singly, 6-17cm x 3-6 cm, lanceolate or oblong lanceolate, pale green on both surfaces and glabrate or nearly so; sides sometimes slightly unequal; edges without teeth, inconspicuously hairy, at least when young, minutely dotted on examination with a lens; thin, dull green to dark green on top surface, and pale blue-green and covered with bloom on underside; apex short or long pointed; base short pointed or rounded; petioles 0.6-1.3 cm long, green, sparsely pubescent. Flowers greenish-yellow, fragrant, on slender hairy stalks, produced singly or in short lateral clusters about 2.5 cm long, 2-4 flowers but not at base of leaves; sepals pointed, hairy, green, about 16 mm long; 3 outer petals oblong, thick and rounded at the tips, fleshy, 1.6-2.5 cm long, 0.6 cm wide, yellow-green, slightly hairy, inside light yellow and keeled with a purplish or reddish spot at the thin, enlarged base; inner petals 3 minute, ovate, pointed scales; stamens very numerous, crowded, white, less than 16 mm long; ovary light green, styles white, crowded on the raised axis. The aggregate fruit formed from the numerous pistils of a flower, which are loosely united, is soft and distinct from other species of the genus. Each pistil forms a separate tubercle, mostly 1.3-1.9 cm long and 0.6-1.3 cm wide. Fruit is round, heart shaped, ovate or conical, 5-10 cm in diameter, with many round protuberances; greenish-yellow when ripe, with a white, powdery bloom; the pulp is white, edible and sweetly aromatic; in each carpel is embedded a seed, oblong, shiny and smooth, blackish or dark brown, 1.3-1.6 cm long, numerous.

Artocarpus heterophyllus Lam.

Trees typically reach a height of 8 to 25 m and a canopy diameter of 3.5 to 6.7 m at 5 years of age. Jackfruit flowers are open-pollinated, resulting in highly variable seedlings. Male and female flowers are borne on same individual trees, but in separate enlarged, fleshy flower clusters which sprout from older branches and from the trunk. The yellowish green fleshy club-shaped male flower spike is borne on a stalk 5 to 10 cm long. Male spikes are on younger branches above female spikes. The numerous tiny flowers, each with a 2-lobed calyx and a single stamen are pale green when young but darken with age. The similarly numerous but slightly larger female flowers are borne in an elliptic or rounded cluster. Each female flower has a tubular hairy calyx, a pistil with a 1-celled 1-ovuled ovary, a slender style, and a broader yellow stigma. Jackfruit flowers are reportedly pollinated by both insects and

wind, with a high rate of cross pollination. The simple, alternately arranged leaves (10 to 15 cm long and 5 to 8 cm wide) are glossy dark green, thick, and leathery. The petioles (leaf stalks) are stout and 1 to 2 cm long. Leaf blades have entire margins and may be oblong to oval or narrow. Leaves are often deeply lobed on young plants and shoots. The fruits of cultivars weigh 10 to 30 kg, although the range for known cultivars extends from 2 to 36 kg. The jackfruit, the largest of all cultivated fruits, is oblong to cylindrical and typically 30 to 40 cm in length. Jackfruits usually weigh 4.5 to 30 kg. The heavy fruits are borne primarily on the trunk and on the interior parts of main branches. Jackfruit is a multiple aggregate fruit. It has a green to yellow-green exterior rind. The hard outer covering is derived from enlarged female flowers. The whitish fibrous pulp within carries many seeds. The acid to sweetish when ripe and banana-flavoured flesh (aril) surrounds each seed. The heavy fruit is held together by a central fibrous core.

Averrhoa carambola L.

A small evergreen tree, generally found all over the state is a slow-growing, short-trunked with a much- branched, bushy canopy, broad and rounded. Mature trees seldom exceed 8-10 m in height and 6-8 m in spread. Flowers are pink white and fruits are star shaped with five prominent ridges. Fruits are ovate to ellipsoid, 6 to 13 cm in length, with 5 (rarely 4 or 6) prominent longitudinal ribs. Slices in cross-section are star shaped. The skin is thin, light to dark yellow and smooth with a waxy cuticle. The flesh is light yellow-to-yellow, translucent, crisp and very juicy without fiber. The fruit has a more or less oxalic acid odour and flavour ranges from very sour to mildly sweet.

Fleshy fruits are quite acidic. They are eaten raw or cooked with sugar. Pickles, jelly and squash can be prepared. It is used as medicine for jaundice and kidney stone. Fruit of carambola is a rich source of reducing sugar, ascorbic acid and minerals like potassium, calcium, magnesium and phosphorous as well as vitamin A (560 IU/100 gm.). Fruits of carambola are widely used for preparation of squash and pickles. The raw fruits are astringent to bowel, stop diarrhoea and vomiting due to biliousness. The ripe fruits, juices and squashes are good for bleeding piles and believed a good remedy for jaundice.

Baccaurea ramiflora Lour.

A slow growing evergreen tree, growing to 25 m with a spreading crown and thin bark common species in the state, middle sized tree, young parts hairy, bark dark grey with vertical lenticels. Leaves simple, 9-25 by 3-9 cm. alternate and spirally-clustered at intervals along the twigs, narrowly elliptic or obovate, apex acuminate, base acute, margin entire or slightly undulate, reddish when young, finely brown-hairy, becoming dark green and shiny above and glabrous when mature. Midrib flat above, prominent below, secondary veins oblique to the midrib, widely parallel, looped and joined at margin, tertiary veins reticulate. Petiole slender swollen at top and base. Flowers small, stipules caducous, come out from old trunk as raceme inflorescens, male and female flowers on different trees.

Males smaller arranged in slender clusters of 10 cm. long, mostly at the end of branches, individual flower with short pedicel. Female slightly bigger, racemes clustered of 30 cm. long on old branches and main trunk. Fruit is berry, ovoid or globose, capsular, yellowish brown, seeds orbicular, embedded in rose coloured pulp (arillus), 2.5-3.5 cm in diameter, with 2-4 large, purple-red seed surrounded by a juicy translucent or pinkish pulp. Average fruit weight is 10-15 g. Fruit pulp is consumed and delicious often sold in the market. Fruit can also be stewed or made into wine; it is also used medicinally to treat skin diseases.

Citrus macroptera

Citrus macroptera is so named because of large wings (-ptera) on the petiole, which is as large as the blade of leaf. The tree is about 5-6 m height. Leaves are large, dual lobbed, dark or deep green, 30-36 cm long and 6 cm wide. Flowers are small, 3.6-5.1 cm diameter. Fruit is 6-7 cm in diameter, has fairly smooth, moderately thick rind and is yellow when ripe. Fruit shape is subglobose, with the size and shape of a grapefruit. Fruit surface is smooth or lumpy aromatic, deep green to yellowish green, rind thick, white, pulp with 9-10 segments, translucent, juicy with numerous greenish wedge shaped seeds. Pulp of fruit is greenish yellow and dry (does not produce much juice). The juice is very sour and somewhat bitter.

The fruit is used mainly for medicine. The juice is used against gastritis and acidity. Fruit is seldom consumed raw or as juice but is instead sliced and cooked, either green or mature. Pickle can also be prepared from it. The oil extracted from the fruit is used in perfumery industry.

Dillenia indica L.

An evergreen tree growing up to 20 m height. Branches generally ascending often fluted at the base and branchlets marked with v shaped leaf scars. Leaves are toothed, 14 - 30 cm by 7 - 12 cm with a conspicuously corrugated surface with impressed veins. Flowers are large, 15–20 cm diameter, with five white petals and numerous yellow stamens. Flowers are fragrant, sepals broadly ovate, concave and pale green. The greenish-yellow fruit is globose with small brown seeds with thick protective covering. The fruit has 5–12 cm diameter with aggregate of 15 carpels, each carpel containing five

seeds embedded in an edible but fibrous pulp. The juicy pulp is aromatic but very acidic. The tree grows in damp places near swamps and river banks.

Fruits are eaten cooked. The fruit pulp is bitter-sour and used in Indian cuisine in curries, jam (ouu khatta) and jellies. It is often mixed with coconut and spices to make chutneys. Excellent jams can be prepared from the fruit. Fruit juice mixed with sugar and water is used as cooling beverage in fevers and as a cough mixture. It is also used as mild laxative. Fruits are rich in protein and vitamin C. Fleshy calyx is eaten raw and cooked as vegetable suited with fish, pulses and root vegetables; made to pickles, jam etc. Mucilage in the fruit is used to wash hair as shampoo and considered good for hair growth.

Elaeagnus latifolia L.

A large woody evergreen scandent shrub with spine; leaves variable, alternate, ovate-oblong, glabrous, clothed beneath with silvery or ferruginous scales. Flowers peduncled, small, scented, perianths covered with ferruginous scales. Fruit about 3 cm long, ovoid, ribbed, red or yellow when ripe, succulant, silky inside. The fruit contains a single large seed that has very low viability. The shrub is found in evergreen tracts of forests.

The ripe fruits are eaten raw with salt and used for making chutney and pickles. Fruit pulp is used for making jam, jelly and refreshing drink. The fruit is a very rich source of vitamins, minerals

and other bioactive compounds. It is a good source of essential fatty acids, fairly unusual for a fruit. It is being investigated as a food capable of reducing incidence of cancer and also means of halting or reversing growth of cancers.

Elaeocarpus floribundus Bl.

It is an evergreen moderate sized tree with spreading crown and clean bole of 12-16 m length. Leaves are ovate - elliptic, often rounded at base, glabrous, with a long pointed tip, and toothed margin, blistered on both surfaces when dry, turning bright red before falling. White flowers are borne in many-flowered racemes, 2.5-12 cm long, in leaf axils. Fruit is light green drupe. Both ends of fruit are pointed, outer surface smooth having a mesocarp fleshy and pleasantly acidic, edible. Stone -3 celled, each having a spindle shaped seed.

The mature fruit is eaten fresh, or cooked and also is favourable as pickle. *Olive oil* is used as edible oil besides it has medicinal values also. It is a good source of Calcium, Phosphorous, Iron α-carotene, Vitamin B and Niacin.

Emblica officinalis

A small to medium sized tree, normally reaching a height of 10-12 m. Small, inconspicuous, greenish- yellow flowers are borne in compact clusters in the axils of lower leaves. Male flowers occur at lower end of a growing branchlet, with female flowers above them. The fruit is round or oblate, indented at base with smooth and shining surface, 6 to 8 faded lines from base to apex giving the impression of ridges and divided segements in the fruit. Fruit is green at maturity and ripe fruit is greenish-yellow. Fruit is juicy, flesh is thick, fibrous or non-fibrous depending on the cultivar. Generally wild fruits are small, hard and bitter in taste. Fruit size, texture, fiber content, taste and acidity vary with the cultivar. Aonla is an important fruit of Indian and consumed fresh, processed and preserved in several ways. Fruits are rich in vitamin C due to presence of leucoanthocanins which retard oxidation of vitamin C.

Aonla is known as a fruit with lot of medicinal values especially increasing immunity and health. Fruit is highly nutritive and is the richest source of vitamin C, next to Barbados cherry. Popular ways of processing to retain nutritive value are Murabba, Chawanparash and Trifala. Several new products are developed - aonla candy, jam, herbal jam, chutney, pickle, squash, juice, sharbat, vinegar etc.

Ficus semicordata Buch.-Ham. *ex* J. E. Sm.

A small to moderate sized deciduous tree, young shoots are hairy, up to 15 m tall, Leaves are carried on 1-1.5 cm long stalks. Leaf blade is variable, mostly elliptic to oblong. Leaves are large and of variable size, thinly coriaceous, glossy green above, pubescent beneath. Leaf margin is entire or coarsely toothed. Fruits are spherical to pear-shaped, 1.2-2 cm in diameter, pink or dull reddish brown with white spots, hairy, on leafless branches. The branches are pendulous and are often prostrate on the ground. Receptacles are in pairs or cluster on old stem even at base of the tree, reddish brown or white when ripe. The edible portion is a multiple fruit botanically known as Siconium consisting of a flesh, hollow receptacle with narrow aperture at the tip and numerous small flowers lining inner surface.

Young leaves and shoots are eaten cooked as vegetable; ripe fruit with aroma is eaten. Fresh figs contain protein (4 g.), Calcium (200 mg.), Iron (4 mg.), Vitamin A (100 IU) and Thiamine (0.10)/100 g of edible portion. Fresh and dry fruit are valued for its laxative property. It possesses various medicinal properties like application for Boils and other skin infections.

Garcinia cowa Roxb. ex D. C.

Plant is small to medium sized (3-5 m), multi branched, evergreen having drooping branches and lanceolated leaves with a girth of 15-30 cm and occasionally 90 cm. Leaves are simple, opposite, 6-15 by 2.5-6 cm, entire, broadly lanceolate to lanceolate elliptic, glossy green. Flowers are yellowish to pinkish, axillary, male and female flowers separated, clustered, sometimes solitary with long pedicels. Fruits developed from superior ovary, are about 5cm in dia., globose, but slightly tapering and somewhat oblique towards the apex. The ripe fruit is dull in colour outside and orange inside. Ripened fruits are pinkish red and fruits are similar to tomato.

The fruit is very acidic. In spite of its sourness, fruits are fondly eaten by local people of Mizoram. The fruits are made into jams and preserves. The thinly sliced fruit pieces are sun dried and used for flavouring various cuisines. It is preserved after slicing and drying in the sun and used as medicine for dysentery. Pickles are also prepared from it. Vitamin C content is 4.05 mg./100 g. Seeds of ripe fruit are sweet and eaten fresh.

Passiflora edulis

It is a perennial, vigorous, climbing, woody vine that produces an edible round or ovoid fruit and has a tough, smooth, waxy dark purple hued rind with faint, fine white specks. Leaves usually 3-5-lobed to halfway, suborbicular -ovate, appressed glandular-pubescent, base cordate, margin subentire to ciliate, apex acute; stipules sub-reniform, deeply cleft into glandular processes. Flowers about 4 cm across, axillary, often solitary; bract and bracteoles 2-3 cm, deeply pinnatesect, glandular -pubescent. Calyx tube short, saucer shaped, lobes ovate-lanceolate. Petals slightly shorter than the calyx lobes. Androgynophore 4-6 mm. Ovary globose. Fruit berry, subglobose, approximately 4 cm across. Inside the fruit is more or less filled with an aromatic mass of double-walled, membranous sacs containing orange colored pulpy juice and as many as 250 small, hard, dark brown to black pitted seeds.

Pulp of ripe fruit is eaten. It can also be taken as "sharbet". The fruit juice derived from 100 g. of pulp (edible portion) is fair to good source of Pro-vitamin A, Ascorbic Acid (30 mg.), Riboflavin (0.1 mg.), Niacin (1.5 mg.) and has a high mineral content (16.3 mg.). Apart from its unique flavour the

comparatively high acid content of passion fruit juice (2.4 per cent -5 per cent) is its most distinctive characteristic.

Phyllanthus acidus (L.) Skeels

The plant is a curious intermediary between shrubs and trees, reaching 2-9 m high. The dense and bushy crown is composed of thickish, tough main branches, at the end of which are clusters of deciduous, greenish, 15-30 cm long branchlets. The branchlets bear alternate leaves which are ovate or lanceolate with short petioles and pointed ends, compound leaf, 2-7.5 cm long leaflets subsessile, obovate, rounded or acute, light green. Flowers in cyathium, axillary, whitish. Flowers are small and pinkish and appear in clusters in 5-12.5 cm long panicles. Fruits are numerous, oblate with 6-8 ribs and 1-2.5 cm wide. Fruit is drupe, globose, pale yellow or light green, waxy, crisp and juicy and very sour. It has only one seed in each fruit.

Mature fruits are acidic, eaten fresh or as pickles. Juice can be used in beverages. When cooked with plenty of sugar, the fruit turns ruby red and produces a kind of jelly. It can also be salted. Fruits

contain 28 Kcal of energy, 91.7 g moisture, 0.7 g protein, 6.4 g carbohydrate, 0.6 g crude fibre, 5 mg Ca, 23 mg P, 0.4 mg Fe, 0.01 mg thiamine, 0.05 mg riboflavin and 8 mg total vitamin.

Spondias pinnata (L f.) Kurz.

A middle sized tree with pleasant aromatic acidic smell. It is deciduous and remains leafless during December to March. Leaves compound, leaflets opposite, 3 paired, flowers uni- or bisexual, greyish-white or pale yellow. Fruit is drupe, about six cm long, acidic and aromatic. It becomes greenish yellow when ripe, 4-10 cm long. It has single seed.

Fruit is eaten raw when ripe; unripe green fruits are made into curry, also form good pickles. Fruits are rich sources of vitamins, very useful for treatment against basillary dysentery and T. B. infection as they are blood purifiers. Fruits are effective against scurvy and rickets. The tender leaves and flower buds are used to prepare a very agreeable curry with fishes.

Syzygium cumini (L.) Skeels

It is a large sized evergreen tree with smooth grey bark and attains a height of 20 m. Leaves are 9-18 cm long, oblong-oval or elliptic, evergreen, opposite, bluntly acuminate. Young leaves pinkish and turn leathery and dark green at maturity. Flowering takes place in February- March and fruiting in May to July. Whitish-yellow flowers emerge in clusters, have a funnel-shaped calyx and 4 to 5 united petals. Drupe emerges in clusters of different sizes. Fruit oblong or ovoid-oblong, at maturity, attain dark purple color, very juicy with 1 or 2 to 5 compressed bold white or green seeds. Seedless fruits are also found. Fruit skin is thin, smooth, shining and pulp is whitish purple. The fruit is astringent and taste varies from light acidic to sweet. Vast genetic variability is present in tree phenology, leaves and fruit characters in India.

It is a very nutritious fruit consumed fresh or processed to various products. Fruit is a good source of iron, minerals, protein and carbohydrate. Fruits are processed for squash, sharbat, syrup, jam, jelly, wine, vineagar and juice. Jamun seeds contain various alkaloids like jambosin and glycoside which inhibit conversion of starch to sugars, therefore, the powdered seeds are useful for sugar patients.

Fruits are important in the Indian System of Medicine and recommended for diabetes, heart and liver problems.

Tamarindus indica Linn.

Tree is medium to large, slow-growing, long-lived, evergreen tree about 25-30 m in height. Leaves are pinnate, dense and feathery and making it an attractive shade tree with an open branch structure. Flowers inconspicuous, five-petalled flowers are borne in small racemes and are yellow with orange or red streaks. Flower buds are pink due to outer color of 4 sepals which are shed when the flower opens. The period from flowering to pod ripening is 8-10 months. Ripe fruits, however, remain on the tree until the next flowering period. Fruits are harvested from April to May. Fruit is a pod, 15 – 36 cm long, brown, irregularly curved pods borne in abundance along the new branches. As pods mature, they fill out somewhat and the juicy acidulous pulp turns brown or reddish-brown. When fully ripe, the shells are brittle and easily broken. The pulp dehydrates to a sticky paste enclosed by a few coarse

strands of fiber. The pods contain from 1 to 12 large, flat, glossy brown, obovate seeds embedded in the brown, edible pulp. The pulp has a pleasing sweet/sour flavor and is high in both acid and sugar. It is also rich in vitamin B and high in calcium. There are wide differences in fruit size and flavor in seedling trees

Tamarind is a multipurpose tropical fruit tree used primarily for its fruits eaten fresh or processed in several ways, used as a seasoning or spice and fruits and seeds are processed for non-food uses. Fruit contains upto 73 per cent edible pulp in which major constituents are tartaric acid and invert sugars. Pulp is a rich source of calcium, phosphorous, riboflavin, niacin and thiamine. Pulp of tamarind is used in several ways as per the locality and food habit of local inhabitants. This is used in preparations of chutney, tamarind powder, puree, juice concentrate, jam, jelly, candies, pickles and fruit leather.

Terminalia chebula Retz.

A medium to tall deciduous tree growing to 30 m tall with a trunk up to 1 m diameter, grown wild in Mizoram but admired by all over India for its high medicinal value. Leaves alternate-opposite, thin coriaceous, ovate or elliptic obovate, entire, apex sub-acute, 7-18 cm long and 4.5-10 cm broad with a 1-3 cm petiole. Flowers in axillary spikes, yellowish white and with offensive smell. Fruit is drupe, obovoid, more or less five ribbed, 2-4.5 cm long and 1.2-2.5 cm broad, blackish. Fruits are tapering towards both ends, green in colour when fresh and black and hard after drying.

It is one of the best known medicinal plants or tanning materials in world. Both unripe and ripe fruits are eaten raw or used as medicine. The fruits are boiled, sliced and sun dried and preserved for medicinal purposes; usually chewed after meal as digestive. It is a major component of many Ayurvedic medicines. It helps to cure gastric trouble, indigestion and beneficial against asthma, piles, worm, colic pain, heart diseases, scabies, jaundice, stone, eye diseases and vomiting.

Ziziphus mauritiana Lam.

Z. mauritiana is a gregarious spiny shrub or a small tree about 8-10 m tall. Branches and branchlets have short stipular spines. The leaves are sessile, alternate, ovate- or oblong-elliptic, 2.5-7.5 cm long and 2-4 cm wide; dense, silky, whitish or brownish hairs on underside and short, downy petioles. On the upper surface, glabrous, dark-green, with 3 conspicuous, depressed, longitudinal veins and there are very fine teeth on the margins. The flowers greenish- yellow, tiny, on short stalk, in leaf axils. The fruit drupe varies in size depending on the wild or cultivated nature of plant, in wild 1.25-2.5 cm long

and in cultivation the fruit reaches upto 5.00 cm in length and 3.5 cm in width. The shape may be oval, obovate, round or oblong; the skin smooth or rough, glossy, thin but tough, turns light-green to yellow, later becomes partial or wholly burnt-orange or red-brown or red. Fully ripe fruits are less crisp and somewhat mealy; overripe fruits are wrinkled, the flesh buff-colored, soft, spongy and musky. There is a single, hard, oval or oblate, rough central stone which contains two elliptic, brown seeds, up to 6 mm long. Flowering is in July to September and fruiting starts from October to January.

Fruit is a rich source of calcium, phosphorous, protein, minerals, vitamin C and vitamin A. Ripe fruit is consumed as a popular dessert and processed for various value added products like murabba, candy, sharbat and squash. Powdered fruits after drying are also consumed. Fruits, seeds and bark are important in the Indian System of Medicine and bark is used as a cure for dysentery and boils and fruit as laxative and aphrodisiac.

Conclusion

All the underutilized fruits are good sources of nutrients, minerals and are available throughout the year. These species also provide ecological security as they have outstanding adaption capabilities to drought, salinity and are also resistant to major pests and diseases as well as ensure sufficient production despite adverse conditions. There is urgent need to take up genetic resources management and improvement of underutilized fruits to ensure food and nutritional security. R and D work is needed to evaluate nutritive and medicinal values including antioxidant properties of the fruits. Exploration of all possible natural populations for genetic diversity, regeneration status in nature, selection of elite strains and documentation and standardization of cultivation practices for popularization of these fruits are urgently required.

Production potential of different species and sustainable harvests of useful parts can boost local economy. It is the right time for selection of elite strains, documentation and standardization of cultivation practices of potential minor fruits of Mizoram to enhance income to farmers by cultivating these fruits. There is need to take up R and D work on genetic resources management and improvement of these fruits through genetic engineering and tissue culture to ensure food and nutritional security of tribal and rural population. Bio-prospecting and value addition, evaluation for medicinal values particularly antioxidant properties of all the available minor and lesser known fruits in Mizoram should also be taken up.

References

Bhat, K.C., Pandey, A., Dhariwal, A.P., Parwar, N.S. and Bhandari, D.C., 2009. " *Tum-thang*" (*Crotalaria tetragona* Roxb. ex Andr.): a little known wild edible species in the north-eastern hill region of India. *Genet Resour Crop Evol.*, 56: 729-733.

Dansi, A., Adjatin, A., Adoukonou-Sagbadja, H., Falade´, V., Ydo- monhan, H., Odou, D and Dossou, B., 2008. Traditional leafy vegetables and their use in the Benin Republic. *Genet Resour Crop Evol.*, 55: 1239–1256.

Gruere, G.L., Nagarajan, O. and King, E.D.I., 2009. The role of collective action in the marketing of underutilized plant species: Lessons from a case study on minor millets in South India. *Food Policy*, 34: 39-45.

http://en. wikipedia. org/wiki

http://link. springer. com/chapter/10.1007/978-94-007-1764-0_4#page-2 Lim, T. K. Edible Medicinal and Non-Medicinal Plants: Vol. 2, Fruits. DOI 10.1007/978-94-007-1764-0-4. 29-30.

http://www. nbpgr. ernet. in/download/UUF-BOOK. pdf

http://www. worldagroforestrycentre. org/sea/products/afdbases/af/asp/SpeciesInfo. asp?SpID=214

Hazarika, T.K. Lalramchuana and Nautiyal, B.P., 2012. Studies on wild edible fruits of Mizoram used as Ethno-medicine. *Genet Resour Crop Evol.*, 59: 1767–1776. DOI 10.1007/s10722-012-9799-5.

Ikram, E.H.K., Eng, K.H., Jalil, A.M.M., Ismail, A., Idris, S., Azlan, A., Nazri, H.S.M., Diton, N.A.M. and Mokhtar, R.A.M., 2009. Ora antioxidant capacity and total phenolic content of Malayasian underutilized fruits. *J. Food Comp. Anal.*, 22: 388-393.

Joshi, A.R. and Joshi, Kunjani, 2005. *Ethnobotany and Conservation of Plant Diversity in Nepal.* Rubrick, Kathmandu, Nepal.

Lalfakzuala, R., Lalramnghinglova, H. and Kayang, H., 2007. Ethno-botanical usages of western Mizoram. *Indian Journal of Traditional Knowledge*, 6: 486-493

Lalramnghinglova, H., 2003. *Ethno-medicinal Plants of Mizoram*. Bishen Singh Mahendra Pal Singh, Dehra Dun, India.

Maundu, P.M., Ngugi, G.W. and Kabuye, C.H.S., 1999. *Traditional Food Plants of Kenya*. National Museums of Kenya, Nairobi.

Mazumdar, B.C., 2004. *Minor Fruit Crops of India: Tropical and Subtropical*. Daya Publishing House, New Delhi.

Ogle, B.M. and Grivetti, L.E., 1985. Legacy of the chameleon: edible wild plants in the kingdom of Zwaziland, Southern Africa. A cultural, ecological and nutritional study. Part IV-nutritional analysis and conclusions. *Ecol. Food Nutr*. 17: 41-64.

Rathore, M. 2009. Nutrient content of important fruit trees from arid zone of Rajasthan. *Journal of Horticulture and Forestry* 1: 103-108.

Termote, C., Van Damme, P. and Dhed'a Djailo, B., 2011. Eating from the wild: Turumbu, Mbole and Bali traditional knowledge on non-cultivated edible plants, District Tshopo, DRCongo. *Genet Resour Crop Evol*. 58: 585-618.

2014, Future Crops Vol. 2 Pages 103–131
Editor: Professor K.V. Peter
Published by: DAYA PUBLISHING HOUSE, NEW DELHI

Chapter 5
Underutilized Spices

S. Datta and A.B. Sharangi*
Department of Spices and Plantation Crops
B.C.K.V.V., Mohanpur– 741 252, Nadia, West Bengal
**E-mail: dr_absharangi@yahoo.co.in*

Only about 30 crop species provide 95 per cent of the worlds' food energy whereas over 7,000 species are known to be used for food and are either partly or fully domesticated. Underutilized crops, usually considered in relation to their end-use, are found in numerous agricultural ecosystems and often survive mainly in marginal areas. Probably most underutilised crops for gums, resins, oils and beverages tend to have less than 5-6 per cent of the species with a single use; those for cereals and medicinals somewhat higher, up to 10-11 per cent with predominately a single use; vegetables and tubers up to 16-17 per cent with a single use; and spices and fruits being special categories with about 25 per cent single use (Williams and Haq, 2000).

The fascinating history of spices is a story of adventure, exploration, conquest and fierce naval rivalry. The addition of spices in the food gives that unique flavor and color to the preparations and one can also get the medicinal values out of it (Sharangi and Datta,2005). Taking the right kind of food with spices enhances health, heals, soothes and refreshes us to a greater extent. The Indian medicine system, the Ayurveda uses these spices to heal certain conditions even now (http://justforhearts.org/all-about-spices/). The widely used spices are onion, garlic, ginger, turmeric, chilli, pepper, cinnamon, cardamom, clove, coriander, cumin, mint, fenugreek, fennel and tamarind etc. Besides these, there are many more spices (Table 5.1) used by the people of different locality. Spices are the plant parts may be the whole plant, bark, stem, leaf, root, rhizome, flower, fruit, and seed. Spices contain alkaloids, flavoproteins, carotenoids, oleoresins, steroids, and oils, which are the sources of flavor, color and stimuli. The commercial products of these spice crops are whole spice, ground spice, spice oil and oleoresin etc., and find applications in flavoring and coloring food, pharmaceuticals, perfumery and cosmetic industries.

Table 5.1: List of some underutilized spices with their main chemical constituents.

Sl.No.	*Name*	*Botanical Name*	*Parts Used as Spice*	*Main Chemical Constituents*
1.	**Ajmund**	*Trachyspermum roxburghianum* (DC.) H. Wolff.	Seed	d-limolene
2.	**Ajowain**	*Trachyspermum ammi* L	Seed	Thymol
3.	**Aniseed**	*Pimpinella anisum* L.	Seed	Anethole
4.	**Asafoetida**	*Ferula foetida* L.	Gum resin	Asaresinotannol
5.	**Black cumin**	*Nigella sativa* L.	Seed	Carvone
6.	**Celery**	*Apium graveolens* L.	Seed, stem, leaf	δ-limolene
7.	**Curry leaf**	*Murraya koenigiii* (L.) Sprengel	Leaf	1-sabinene
8.	**Horse radish**	*Armoracia rusticana* Garten	Root	Allyl isothiocyanate
9.	**Japanese Mint**	*Mentha arvensis* L.	Leaf	Menthol
10.	**Kokam**	*Garcinia indica* Choisy	Rind	Garcinol
11.	**Leek**	*Allium porrum* L.	Swollen leaf base	Quercetin
12.	**Mango ginger**	*Curcuma amada* Roxb	Rhizome	car-3-ene and cis-ocimene
13.	**Opium Poppy**	*Papaver somniferum* L.	Seed	Morphine, codeine, and thebaine
14.	**Paprika**	*Capsicum annuum* L.	Fruit	Capsanthin
15.	**Parsley**	*Petroselinium crispum* Mill	Leaf	Apiin
16.	**Rosemary**	*Rosamarinus officinalis* L.	Leaf	1, 8-cineol
17.	**Sage**	*Salvia officinalis* L.	Leaf	Thujone
18.	**Sweet flag**	*Acorus calamus* L.	Rhizome	Asarone
19.	**Sweet Marjoram**	*Majorana hortensis* Monech	Leaf	Caarvacrol
20.	**Tamarind**	*Tamarindus indica* L.	Fruit pulp	Tartaric acid
21.	**Tarragon**	*Artemisia dracunculus* L.	Leaf	Methyl chavicol
22.	**Tejpat**	*Cinnamomum tamal* (Buch Ham) Nees and Eberum	Bark and leaf	Eugenol
23.	**Thyme**	*Thymus vulgaris* L.	Leaf	Thymol

Ajmund

Ajmund (*Trachyspermum roxburghianum* (DC.) H. Wolff.) belonging to the family apiaceae is an underutilized seed spice. In English it is called wild parsley and '*Radhuni*' in Bengali. It is cultivated throughout South and South West Asia. In India it is mostly cultivated in home stead gardens. It is an annual, erect, aromatic herb that can grow 30-90 cm tall. It is used for preparation of different food items especially for preparation of mixed vegetables. It is used in preparation of soup, sauces and pickles. The fruit yield an essential oil up to 2.5 per cent. The main components of essential oil are δ-limolene, alpha tepinene and d-linallol.

It is commercially cultivated in warm and humid regions. It grows up to an altitude of 750 m above mean sea level. It can be grown any type of soil but the soil should be well drained. Sandy loam to loamy soil with neutral pH are ideal for its cultivation. It can also be cultivated in light soil but sufficient organic matter should be added before its cultivation. Seed is sown in the month of April. It

Figure 5.1: Ajmund plant.

Figure 5.2: Ajmund flower.

Figure 5.3: Ajmund seed.

can also be cultivated as winter crop and in this case crop should be sown in the month of November. Overnight soaking in water and sowing enhance seed germination in the field. It can be sown either by broadcasting or line sowing. In case of line sowing, seed should be sown in 30 cm X 10 cm spacing and the seed rate is 1.5 kg/ha. Farm yard manure or compost at the rate of 15 t/ha should be applied during sowing. Among the inorganic fertilizers, N : P : K should be applied at the rate of 45 : 30 : 30 kg/ ha. Full dose of P and K and one half dose of N should be applied as basal during final land preparation and remaining half nitrogen should be applied at 45 days after sowing. One pre- sowing irrigation is given to enhance seed germination. Apply irrigation during branching, flowering and seed formation and development stages which increased seed yield considerably. Harvesting may be done when the seed changes its colour from greenish to brownish. April sown crop can be harvested in September and winter crop in March – April. Seed yield varies from 0.6 - 0.8 tones/ha depending upon the management practices followed. Powdery mildew and blight are the important diseases affecting the crop and reduce the yield considerably. Among the pests, aphid is important attacking the crop and affecting the crop yield.

Ajowain

Ajowain or Bishop's weed (*Trachyspermum ammi* L.) belongs to family apiaceae and is an annual herbaceous underutilized seed spice. It originated in Mediterranean region and grown in Iran,

Figure 5.4: Ajowan seeds.

Figure 5.5: Ajowan flowers.

Figure 5.6: Ajowain seed.

Afganisthan, Egypt, Pakisthan, North Africa and India. In India, it is mainly cultivated in Gujarat, Rajasthan, Madhya Pradesh, Uttar Pradesh, West Bengal, Punjab, Haryana, Bihar and Tamil Nadu. Essential oil content of seed varies from 3-4 per cent and the main composition of the essential oil is thymol. The other constituents are pinene, cymene, dipentene, terpinene and carvacrol. Ajowan is widely used for seasoning various food preparations. Ajowan is mostly found in pulse dishes (dhal), vegetable dishes and for preparation of pickles. It is also used as preservative and antioxidant. Essential oil is used in perfumery and pharmaceutical industries.

Figure 5.7: Ajowain plant .

Cool weather and cloudiness for about a week after sowing and occasional drizzling during active growth are conducive to successful cultivation of ajowan. Cool and dry climate favours good growth and yield of crop. High atmospheric humidity after flowering invites diseases and insects (Shetty *et al.*, 2012). Except sandy loam soil, it can be grown in all types of soil. It grows the best in loamy or clay loam soil rich in humus and well drained. Sowing is done in October – November in most part of the India, but in some parts, it may be sown in May-June. Seed rate is 3-4 kg/ha. Seeds are sown broadcast or in line. In line sowing, a spacing of 30-45 cm X 20-30 cm may be given depending upon locality and management condition especially irrigation and nutrient management. In most of the areas, farmers grow local cultivar. Pant Ruchika, Gujrat Ajowan-1, Lam Selection-I and Lam Selection-II are a few of the improved varieties released from different research institutes for commercial cultivation. Application of farm yard manure @15t/ha along with N : P : K @ 30 : 45 : 45 kg/ha is beneficial for higher growth and yield of crop. The crop matures 160-180 days after sowing. Fruits become ready for harvesting when, the flower head turns brown. Average yield of ajowain varies from 0.6 -0.8 tones/ha. Collar rot, root rot and powdery mildew are the important diseases which affect yield and quality of crop.

Anise or Aniseed

Anise or Aniseed (*Pimpinella anisum* L.) belongs to family Apiaceae grown for its seed. It is an annual pubescent aromatic herb which attains a height of 60 – 90 cm under cultivation. It is native to Mediterranean region and extensively cultivated in South and South East European countries and also in Egypt, Syria, Iran and South America, Mexico and India. In India, it is grown to small extent as culinary herb or as garden plant. In India, commercial cultivation is confined to Bihar and Uttar Pradesh. To a small extent, it is also cultivated in Gujarat and Rajasthan. Seed contains 2.1 – 2.4 per cent essential oil. The main composition of the essential oil is anethole (84 – 87 per cent) and methyl chavicol (12 – 15 per cent). Fresh leaves are used as a garnishing agent and for flavouring salad. The seeds are used for flavouring food, confectionaries, beverages, and in liquor industries and processed food. Essential oil is used in perfumery, soap, toilet, dental and mouth wash articles preparation. Medicinally, essential oil is used as an aromatic carminative and to relieve flatulence.

The crop is grown in wide range of soil, ranging from sandy loam to clay loam, having a pH of 6.0 – 8.5 and of moderate fertility. A fairly warm weather during sowing is desirable. Low temperature and higher humidity are condusive for vegetative growth, while the crop requires warm sunny weather during seed formation and development stage. EC-22091 is the improved variety of anise developed at

Figure 5.8: Anise plant.

Figure 5.9: Anise seed.

IARI, New Delhi. Seeds are sown at a spacing of 45 cm X 20 cm with a depth of 1 – 2 cm during late October to early November. Like other umbelliferous crops, seed germination is rather problematic and uneven. To obtain uniform germination, the seed should be soaked in water overnight and moist seed sowing is recommended. A light irrigation is to be given to get uniform and higher percentage of seed germination. Irrigation also should be provided at flowering and grain filling stages of the crop. Application of farm yard manure at the rate of 10 t/ha and inorganic fertilizers like N : P : K @ 80 : 60 : 60 kg/ha is beneficial for getting higher yield of the crop. The umbels come to maturity from March to April. The terminal umbels mature first and lateral ones 15 – 20 days later. When 80 – 90 per cent fruits begin to turn grayish green, top of the plants is cut along with branches and tied in bundles. Under favourable climatic conditions and good management practices anise yields 0.7 -0.8 tones/ha. Semilooper and cigarette beetle are the important pests attacking the crop. Among the different diseases, alternaria leaf blight is an important disease that reduces crop yield considerably.

Asafoetida

Asafoetida (*Ferula foetida* L.) belongs to family Apiaceae and is commonly known as hing or hingra. It is the gum oleoresin of the genus *Ferula* that includes 60 different species found mostly in Central Asiatic Countries, Europe and Tropical North Africa. A few major varieties are also grown in Pakisthan, Afganisthan and Iran. Most of the asafetida obtained from the species *Ferula foetida*. The other related species are *Ferula alliaceae* and *Ferula galbani*. *Ferula foetida* is a perennial plant, reaches a height of 1 to 1.5 m and has massive carrot shaped roots. The rhizome produces two types of plants, *viz.* male and female plant. Male plant produces inflorescence whereas female plants do not produce inflorescence. The female plant produces an exudation in the form of very thick and sticky paste sap from the underground rhizome. This exudation is called gum resin which is the commercial part of asafetida. It is extensively used for flavouring curries, sauces, pickles and with other spices. Asafoetida is mainly used to impart taste. The gum resin possesses anti-spasmodic, carminative, expectorants

Figure 5.10: Hing powder and tears.

Figure 5.11: Hing flower.

and stimulant properties. Fried asafoetida powder is used in many ayurvedic preparations to cure convulsions and nervous breakdowns. According to different workers, huge variations in physio-chemical characteristics of different samples of asafoetida are reported. Such variations occur due to certain factors, such as part of the plant from which prepared, the season of collection, method of preparation, the degree of adulteration and nature of adulteration (Pruthi, 1976).

It prefers a rocky dry soil and is usually found above 1350m amsl. The plant possesses a perennial root stock through which they multiply vegetatively. During the onset of spring sprouts come out from the rhizome of the female plants and put forth foliage. It takes nearly one month for green foliage to turn yellow. It is the right stage for tapping (by incision) the rhizomes for asafoetida. Initially, the yellow foliage is removed and then top of the rhizome and brush like mass are covered with loose earth and left for five days. Then brushy mass on the top of the rhizome pulled out and the root and gravel around the rhizome are cleaned, exposing the top of rhizome. After that incision is made at the base. A thick white fluid is discharged from the place of incision and pours slowly downwards. After a few days, end of the root is again cut

Figure 5.12: Hing plant.

off and the accumulated fluid sponged out. This is continued till the plant stops discharging the gum resin. The whole process may continue upto three months. However, the gel dried out and mixed with chemicals to increase the amount and also giving proper shape.

Black Cumin

Figure 5.13: Black cumin flower.

Black Cumin (*Nigella sativa* L.) belongs to family Ranunculace and is an important promising underutilized seed spice. It is native to Mediterranean region and is a highly aromatic spice cum medicinal plant. It is mainly cultivated in Punjab, Himachal Pradesh, West Bengal, Bihar, Odisa and Assam. The commercial product is seed, used as spices and in pharmaceutical industry. Seed emits an aroma resembling strawberries when crushed. The drier tracts of different states are highly suitable for commercial cultivation and seed production. It is an important spice in the Indian kitchen especially for preparation of fast food, oil fry food and for preparation of delicious and tasty dishes. The seed contains about 2.0 to 4.0 per cent essential oil, main component of black cumin oil is carvone. Seed contains about 30 per cent fatty oil used for edible purpose (Pruthi, 2001).

Cool and dry climate is ideal for its cultivation. Sunny weather in day and low temperature in night are highly congenial for vegetative and reproductive growth. Rainfall during flowering is harmful whereas it is beneficial after seed set to get higher yield. Well drained and high to medium land is suitable for its cultivation. Though it can be grown in a wide range of soil, it prefers loamy sand soils. Datta *et al.* (2003) observed that variety, Rajendra Shyama performed the best under West Bengal condition. Banafar *et al.* (2002) found that cultivar NS.- 44 performed the best in Jabalpur, Madhya Pradesh. Sowing time varies from locality to locality. Under South Bengal condition, the ideal time is first fortnight of November while in North Bengal ideal time is mid November. 7.5 -10 kg of seeds are required for sowing depending upon methods of sowing. Higher germination percentage can be

Figure 5.14: Black cumin seed.

Figure 5.15: Black cumin plant.

achieved by soaking the seeds with GA_3 at 100 ppm or 200 ppm under dark condition (Mondal and Maity, 1993). Effect of ethrel on growth and yield of black cumin (*Nigella sativa* L.) under gangetic alluvial soils of West Bengal was studied by Uday Kumar *et al.* (2008). Depending upon the edaphic condition, seed is sown with a spacing of 20-30 cm x 5-10 cm. Datta *et al.* (2001) reported that 45 : 45 : 45 kg NPK/ha was beneficial for higher yield. A fertilizer dose of 30 : 30 : 40 kg NPK/ha produced maximum yield under Assam condition (Luchon and Sarat,2003) whereas a fertilizer dose of 60 kg each of N and P were recommended for Kanpur (Singh *et al.*, 1999). Nataraja *et al.* (2003) reported that application of 50: 40: 30 kg NPK/ha gave maximum yield under Bangalore condition. Irrigation at 35, 60 and 85 days after sowing are beneficial to get higher yield under Raipur condition. The crop should be harvested when plants start yellowing. Delay in harvesting causes shattering loss of seed at field. Generally plants are uprooted or cut just above ground level. The seeds are separated by beating lightly with sticks. Seeds are then cleaned, dried and filled in gunny bags with polylining for storing in cool and dry place. Black cumin produces 9.0-12.0 q/ha under irrigated condition and 5.0-7.0 q/ha under rain fed condition. The important diseases of black cumin are blight and wilt. Generally, pest is not a problem in black cumin cultivation. Sometimes cut worm and fruit borer may cause considerable damage.

Celery

Celery (*Apium graveolens* L.) belongs to family Umbelliferae and is an aromatic herbaceous plant grown for its leaves, seeds and essential oil. Celery seed is known for over 3000 years. It is cultivated in France, Great Britain, Hungary, Japan, India, the Netherlands and the United States. In India, it is mainly cultivated in Punjab, Haryana and Uttar Pradesh for its seed, which has a good aroma and finds a ready export market. The ripe seeds are used as spices to flavour soups, salad, tomato juices and sauces. Celery seed and its extractives are also used in seasoning of meat, flavouring of non alcoholic beverages, baking industry, confectionaries and preparation of chewing gum. Ko *et al.* (1991) reported that celery is widely used in Chinese medicine for its antihypertensive and antiinflammatory activity. The characteristic ally persistent odour of celery is due to essential oil. Celery seed contains 2-3 per cent essential oil of pale yellow colour. The principle component of essential oil is δ-limolene (60 per cent) and d-selinine (10 per cent). In colder climate and hills, celery is a biennial plant and produces seed in second year, but in plains it becomes annual and produces seed in first year.

Celery grows the best on acidic sandy and silt loamy soil. Celery is a moisture loving plant. It thrives the best in relatively cool climate and having moderate and well distributed rainfall during its growing periods. It comes up well in places having low humidity and where plenty of sunshine is available. There are usually self blanched and green cultivars. Michigan, Improved Golden, Cornell-19, Supreme Gold, etc are some of the important self blanched cultivars. Utah, Utah 52-20, Giant Pascal, Emerson Pascal and Fordhook. are the green cultivars. Seeds are sown in March – April and transplanted in May under hilly situation and in plains, seedlings are transplanted in September – October. Seedlings are ready for harvest within 45-60 days when they are 10-15 cm high. Seeds are slow in germination and it takes about 4-6 weeks. Seed rate is about 125-150g/ha. A spacing of 60-75 cm X 15-20 cm is adopted for celery. Vargas Prieto *et al.*(1986) obtained the highest profit from the plot fertilized with N : P_2O5 : K_2O at the rate of 190 : 117 : 52 Kg/ha followed by N : P_2O5 : K_2O at the rate of 130 : 63 : 28 Kg/ha. Calcium and magnesium deficiency are common in celery. Black heart disorder occurs due to calcium deficiency and can be managed by foliar application of calcium nitrate or calcium chloride with appropriate doses. Magnesium deficiency can be controlled by spraying magnesium sulphate (12 kg/ha). Water requirement of celery sharply increased at the start of heart development, 8 weeks after planting, reached maximum at 16 weeks and declined during ripening.

The crop is ready for harvesting in the middle of May and seed yield varies from 9.0-15.0 q./ha. Celery seed usually shatters and to check this, application of polyvinyl acetate is effective as anti shattering agent. Early blight, late blight, damping off and fusarium wilt are the common diseases of celery.

Curry Leaf

Curry leaf (*Murraya koenigii* Sprengel) belongs to family Rutaceae, used for flavouring food for its typical flavour and taste. It is cultivated for its aromatic leaves. It is found throughout the India. Its wild form occurs in Western Ghats, West Bengal, Assam, Orissa, Tamil Nadu, Kerala, Gujarat along with foot hills and plains of the Himalayas from Kumaon to Sikkim. It is cultivated in Tamil Nadu and Andhra Pradesh. Curry leaf is rich in Vitamin A, Calcium (830 mg/100 g dry leaves) and iron (730 mg/100g dry leaves). Curry leaf is traditionally and extensively used in South Indian cooking for preparation of curries, rassam, sambar and chutneys. Leaves, bark and root of curry leaf are used in ayurvedic and yunani systems of medicine.

It does not require exact specific climates and can be grown in dry climate too. Growth of the shoot will be affected if the temperature go below 13°C. It can be cultivated in most of the soil types, but comes up well in light textured red soil. DWD-1 and DWD-2 are the most common varieties of the curry leaf. Good and healthy seeds selected from vigorously growing plants should be used for raising nursery. Young seedling of 3 months age are planted in pit of 30 cm x 30 cm x 30cm size. Normally a spacing of 1.0-1.20 m x 1.0-1.20 m spacing is adopted for cultivation of curry leaf. Better grow and yield can be obtained by applying farm yard manure @ 15-20 kg/plant along with N; P: K @ 150 : 25 : 50 g/plant. During seedling stage, the crop should be irrigated at an interval of 5-7 days and there after 15 days interval. The plant should be maintained as a bush of 1 m height. After each periodical harvest, thorough weeding and irrigation are practiced. At the end of the first year the crop comes to first harvest and the yield is about 400 kg/ha. During 2nd and 3rd year, the crop is harvested at an interval of four months and yield varies from 2000-2200 kg/ha. At the 4th year the crop is harvested at an interval of three month and yield is about 2500 kg/ha. From the 5th year onwards, the crop is harvested at intervals of two and half months to three months and yield varies 3-5 tones/ha. Plants can be maintained upto 20-25 years depending upon management practices. Citrus psylla (*Diaphornia citri*) is a major pest. Aphid also causes considerable damage to the crop. Leaf spot and root rot are the diseases of the crop. Despite pests being major constraints in curry leaf production, they are easily manageable when plant protection measures are taken in time, of course with close monitoring. Preferably, the non residual chemicals like dimetoate, malathion and monocrotophos may be sprayed to bring down pest population well below the threshold level. Even for these chemicals, the farmer must allow a safe period not less than 2-3 weeks for harvesting after spraying pesticides.

Horseradish

Horseradish (*Armoracia rusticana* Garten) belongs to Brassicaceae, a near relative of turnip, cabbage and mustard and originated from middle Europe. It is on of the oldest condiments in the world and its white fleshy and tasty roots are used as commercial part. It is a large leaved hardy perennial herb used as culinary herb in different European countries more than 3000 years. It is a very popular condiment for foods in Germany and surrounding areas. It is grown in in Eastern Europe, U. K., U. S. A., India and Sri Lanka. In India, it is grown in gardens both in north India and hill station of South India (Shastri, 1988). The root contains a pungent, acrid and vesicating volatile oil. The pungency is due to presence of an allyl isothiocyanate and butylthiocyanate similar to mustard oil, occurring in combination with the glucoside 'sinigrin' (Choudhury, 1996). Organoleptocally active substances of horseradish are mainly due to presence of 2-propenyl isothiocyanate and 2-phenyl ethyl isothiocyanate (Taborsky *et*

Figure 5.16: Horseradish plant.

Figure 5.17: Horseradish root.

Figure 5.18: Horseradish flower.

al., 1989). Horseradish is used as an appetizing condiment. It is highly priced as condiment especially with oyster and cold meats. A freshly grated horseradish root, when mixed with vinegar and salt, is much appreciated as an appetizing condiment to enhance the flavour of roasted beef. Inhaling of horseradish essential oil or taking horseradish by mouth promotes blood flow in nasal and sinus tissues and helps possibly relieving respiratory congestion.

Horseradish thrives well in temperate climate and in cool season and at high altitude of tropical countries. It grows the best in deep, rich and moist loamy soil and sunny situation. Horseradish is highly sterile in nature and hence the plant is propagated asexually mainly by root cuttings. Cuttings are made from side roots which are trimmed off in preparing roots for market. Before planting the cutting main field should be well prepared and cuttings are planted at a spacing of 30 cm x 30 cm. Horseradish is generally of two types - Common and Bohemain type-. Among the two types, common type produces better quality roots. Edelkafener, Jugoslavischer, Humberger and Steirischer are a few of the ecotypes of horseradish (Nebel *et al.*, 1988). To grow high quality horseradish, lift and strip the roots two times, first when the biggest leaf is 20 to 25 cm long and again 6 weeks later. Yield of horse radish is about 2 tones/ha.

Japanese Mint

Mint or mentha (*Mentha arvensis* L.) belongs to Lamiaeae and is commercially an aromatic perennial spice commonly known as pudina. It is extensively cultivated in USA, South East Europe and Latin American countries. In India, it is grown all over the country. Its commercial cultivation is restricted to erstwhile UP and parts of Punjab because of high yield and superior quality of oil in terms of its chemical composition. The genus *Mentha* comprises of 40 species. Among the different species of mentha, only 3 species namely *Mentha arvensis* (Japanese Mint), *Mentha spicata* (Spear mint or Pahari pudina) and *Mentha piperita* (Pepper mint) are important and these species are used as spices and condiments. Among the different species of *Mentha,* Japanese mint is the most important one. Mint is used for favouring meat, fish, sauces, vinegar, tobacco and cordials. It is the primary source of menthol, widely used as flavouring agent and as fragrance in confectionaries, toilet articles and in beverages.

Japanese mint is a heavy feeder of major nutrients and flourishes well under humid growing conditions of sub tropical areas. Adequate and regular rainfall during the growing periods and good sunshine during harvesting are ideal for its cultivation (Thomas, 2003). The crop prefers well drained deep, organic soil having capacity to retain moisture. It is grown in wide range of soil varying from sandy loam to clay loam, but medium deep soil, rich in humus is the best suited. The soil should have the ideal pH ranging 6.0-7.5. It is propagated through stolons, suckers and runners. Five hundred kg suckers are required to plant one hectare. The suckers should be cut into pieces of 10-12 cm length and planted in furrows at a spacing of 45-60 cm X 15 cm. MAS-1 and Hybrid -77 are improved varieties of Japanese mint. Farm yard manure (20-25 tones)/ha along with N : P : K should be applied at the rate of 50 : 75 : 37 kg/ha during the time of land preparation. Besides 75 kg of N should be applied in three equal splits. Nair *et al.* (1992) found that application of Fe (5 ppm) increased yield at the second harvest from 1.59 tones/ha to 1.72 tones/ha. Zinc had slight effect on dry matter yield at the second harvest. The crop is harvested when it comes to flowering, which indicates maturity to obtain maximum oil yield and menthol. In the plains of North India, where plant does not flower, crop is harvested at the time when lower leaves start falling. In general, first harvest is taken four months after planting. Subsequently two more harvests can be taken at intervals of 60-75 days. The fresh herbage yield ranges 25-30 tones/ha which in turn gives 120-150 kg/ha essential oil yield. The common practice is to maintain the field for two years in temperate climate. In Northern India the field may be maintained 8-9 months. Powdery mildew and wilt are important diseases infecting the crop. Among the pests, termite attacks the crop during dry months and cut worm and semi looper also cause considerable damage to the crop.

Kokam

Kokam (*Garcinia indica* Choisy) belongs to family Clusiaceae is commonly known as kokam butter tree or mangosteen oil tree or brindonia tallow tree. It has tremendous commercial potential in recent times, because of presence of hydroxyl citric acid (HCA). This chemical controls obeisity in man (Mathew *et al.*, 2007). It is a slender evergreen small tree with drooping branches. It is a dioecious tree growing up to 18 m high. The genus *Garcinia* comprises of 200 old world tropical species of which 20 species are found in India. As a part of traditional material and medica of India, *Garcinia indica*, *Garcinia mangosteena* and *Garcinia morella* are documented in Ayurvedic medicine which are generally known as 'Red Mango'. The fruit is spherical, purple and not grooved having 5-8 seeds compressed in. It is originated from tropical rain forest of the Western Ghats of Kerala and Malaysia. Its commercial cultivation is restricted in the coastal hilly region of Maharashtra and Goa. It is also found in the evergreen forests of Assam, Khasi, Jantia hills, West Bengal and Gujarat. The characteristics aroma and flavour of kokam are due to presence of garcinol. Kokam has commercial importance, since the rind of which is used to impart an acid flavour to curries instead of tamarind. The ripened, rind and juice of Kokam fruit are commonly used in cooking. The dried and salted rind is used as a condiment in curries. It is also used as a garnish to give an acid flavour to curries and for preparing attractive, red and pleasant flavoured cooling syrup. The soft drink and derivative of kokum, popularly known as kokum syrup generally consumed as a natural soft drink to quench thirst during summer. The kernels of *kokam* seed contain about 33 to 44 per cent oil, which is commercially known as "kokam butter". It is considered nutritive, demulcent, astringent and emollient. The rind has antioxidant property. Kokam is well known to counteract the heat. Kokam is especially used with fish curries, three or four skins being enough to season an average dish. It is also used in chutneys and pickles.

It comes up well in hot humid climate under partial shade. It thrives under a mean annual temperature of 20-30°C with 60-80 per cent relative humidity and up to an altitude of 800 m above mean sea level. A well distributed rainfall of 250 -400 cm is ideal for this crop. It can be grown in a variety of soils from sandy loam to clay loam and lateritic soil having good water holding capacity and enough depth. Commercially it is propagated through seed. But kokam is dioecious in nature, as a result upon seed propagation it produces 50 per cent female and 50 per cent male plant. Only female plants produce fruits and male plant can be identified 7-8 years after planting, when producing flower. For this purpose, top working of male plant is done to convert into female plant after retaining about 10 per cent male plant. Inarching and soft wood grafting are also done to ensure female plant and uniformity on yield. Pits of 60 cm x 60 cm x 60 cm are prepared at a spacing of 6 m x 6 m. In the first year, each plant should be applied with 2 kg farm yard manure along with 50 g N, 25 g P and 25 g K and this dose is increased every year and from the tenth year onwards, each plant may be given 20 kg farm yard manure and 500 g N, 250 g P and 250 g K/plant/year. Konkan Amruta is the improved variety of kokam. Irrigation should be given at 7-10 days intervals at early stages of growth and generally drip irrigation is practiced in kokam. In the 10-12 years, in kokam plantation, vegetable and flower crop can be grown as inter crops. Kokam tree starts flowering during November-December and fruits are harvested during April-May. It takes about 7 to 10 years for economical bearing. The fruits are picked when ripe, the rind is then removed and soaked in juice of pulp and then sun-dried. The kokam is difficult to cultivate, usually growing as solitary trees in a tropical forest environment. Yield of kokam varies from 20-30 tones/ha. There are no serious diseases and pests reported to infest kokam plantation. Only leaf minor attacks the plant and feeds the leaf.

Leek

Leek (*Allium porrum* L.) is a non bulb forming member belonging to family Alliaceae. It is eaten as raw or mixed in salad. It is also cooked with other vegetables or used as flavouring agent in soups and stews. It has white narrowly ovoid bulbs and broad leaves. Commercial cultivation is not followed but is mostly grown in kitchen gardens in different parts of India. It is believed originated from the Mediterranean region or Western Asia.

Rapid leaf growth is reported at 21-24°C. High temperature during sowing time inhibits seed germination. The growing soil should be rich in humus. It is either grown directly by seed or by transplanting. Spacing is 20-30 cm x 15-30 cm. Seeds are sown in nursery during October – November. Transplanting is done in 30-40 cm deep trenches. London Flag, Mussel Burg, American Flag Elephant, Goliah, Winterens, Titan and Ficus are some of the varieties of Leek. High rate of potasium stiffens the edible stalk where as high rate of nitrogen produces more tender stalk. Irrigation is beneficial because it requires large amount of water during growth periods. Irrigation is required during dried periods of the year. Along with growth and development, soil is placed around the stem and it is to be done at fortnight intervals. If any flower stalk appears during growth stage, this should be removed to maintain continued vegetative growth.

Mango Ginger

Mango ginger (*Curcuma amada* Roxb.) is a unique spice belonging to the family Zingiberaceae and having mango flavour in its rhizomes with high medicinal importance (Sharangi, 2009). Mango ginger is botanically and morphologically related to neither mango nor ginger, but to turmeric (*Curcuma longa*), but has shorter crop duration of six months. The rhizome has pale yellow central core with less fibre and contains traces of essential oil and curcumin (0.1 per cent). It also possesses a unique flavour of a mixture of mango, turmeric and ginger. It imparts mango flavour mainly due to the presence of car-3-ene and cis-ocimene (Achut and Bandyopadhyaya, 1984;Rao *et al.*1984). Probably due to high usage in the preparations of pickles, sauce and various other culinary formulations, *C. amada* is the second most cultivated Curcuma species after *C. longa* (Syakumar and Sasikumar, 2007). It is mainly cultivated in West Bengal, Kerala, Assam, Tamil Nadu, Odisa. The rhizomes of mango-ginger are used for preparing pickles, chutney, preserves, candy, sauce and salad and in meat and other culinary preparations. Its use in traditional medicine is well reflected in ethno-botanical studies conducted mainly in Indian subcontinent, Myanmar and Thailand. Mango ginger has antibacterial, antifungal, insecticidal, aphrodisiac, antipyretic, anti-inflammatory, anti-mycobacterial, anti-hyper-cholesterolemic and antioxidant properties (Singh *et al.*, 2010). It is useful in biliousness, itching, skin diseases, bronchitis, asthma, hiccough and inflammation due to injuries.

It prefers a warm humid climate and well drained rich loamy soil. The crop comes up well in open conditions, but tolerates low levels of shade and therefore partially shaded situations can also be utilized for its cultivation. It can be well accommodated as an intercrop in coconut gardens and in rotation with other short duration crops like vegetables and also as a crop component in homesteads. Whole or split mother rhizomes or well developed, healthy and disease free ginger rhizomes weighing 15-20 g are suitable for planting and optimum month of planting is May –June. Chatterjee and Chattopadhyay (2009) found that among eight planting dates (17 March to 2 July) planting on 2nd June is suggested for the highest yield and quality of mango ginger under Nadia, West Bengal condition. Amba is a released variety from High Altitude Research Station, Pottangi, Odisa. A spacing of 30 x 30 cm and at a planting depth of 4-5 cm is adopted for planting of mango ginger. Mulch the crop immediately after planting with green leaves or paddy straw and repeat mulching after 50 days.

Remove weeds 45 days after planting and repeat if necessary. Earth up the crop after 60 days of planting. Regarding application of manure and fertilizer, apply 20-30 tonnes of well decomposed FYM/ha and N : P : K at the rate of 30 : 30 : 60 kg/ha. Full dose of FYM and half dose of K should be applied as basal. Half dose of N should be applied at 30 days after planting. Remaining half N and K should be applied at 60 days after planting. A good crop of mango ginger may produce 30-40 tonnes/ha fresh rhizome. Regarding pests, shoot borer (*Conogethes punctiferalis*) damages to the crop. If infestation is severe, spray dimethoate or quinalphos (0.05 per cent). Among the different diseases, appearance of dead heart in the field is the main symptom. To reduce the pest population, pull out the dead hearts with the larvae inside and burn it.

Opium Poppy

Opium Poppy (*Papaver somniferum* L.) belongs to family Papaveraceae and originated from the Western Mediterranean Region. The genus *Papaver* has 50 different species of which 6 species are grown in India. It is mainly grown for its seed and latex. The seed is used as spice for preparation of culinary items and small portion of seed is exported mainly to the Asian and African countries. It is grown in Near East and Asia, most parts of Europe, North and east Africa, Australia, Japan and South America. In India its cultivation is confined mainly to Madhya Pradesh, Rajasthan and Uttar Pradesh. Poppy seed is utilized as spice and as a source of protein and fatty oil. It is nutritive and used in breads, curries, sweet and confectionaries. Poppy seed oil is useful in preparation of linoleic acid, soft soap, ointments and preparation of skin care products. In Europe seed oil is used as a substitute and adulterant of olive oil to which it resembles very much.

It is a crop of temperate climate but can also be grown successfully during winter in subtropical regions. Cool climate favours higher yield, while higher day/night temperature generally affect seed yield. The crop requires deep clay loam soil highly fertile in nature and well drained with an optimum pH of 7.0. The seed is either sown broadcast or lines. Seed should be treated with fungicide, mancozeb. Seed is usually mixed with sand before broadcasting to ensure uniformly spread in the bed. Optimum sowing time is October to early November. Seed rate is 7-8 kg/ha for broadcasting and 4-5 kg/ha for line sowing. A spacing of 30 cm x 30 cm is adopted for line sowing. Germination takes 6-10 days depending upon the soil moisture content. Telia, Dhotia and Ranjihatak are a few of the local races recommended for commercial cultivation. MOP-3, MOP-16, MOP-572, MOP-1186 and UD-177-2 are a few of lines of opium. Farm yard manure @25-30 tonnes/ha is generally applied at the time land preparation. Besides, N: P: K @ 80-90 : 90-100 : 40-50 kg/ha is recommended. Half of N and K and full of P should be applied as basal. Remaining N and K should be applied in two equal splits at 30-40 days and 60 days after sowing. Kharwara *et al.*(1988) found that application of N @150 kg/ha produced significantly higher yield than lower dose of 75 kg/ha. The weeding cum hoeing and thinning should be done 15-20 days after sowing and irrigation. Second weeding should be given 50-60 days after sowing. A light irrigation is given immediately after sowing and another light irrigation given after one week when the seeds start germinating. Chung (1987) reported that irrigation increases total dry matter production, LAI and delayed senescence in dry season. for getting maximum yield, one irrigation each of 5 cm should be applied 50 per cent hooking stage, at 50 per cent flowering stage and at the end of flowering stage. Tomar *et al.* (1990) reported that the greatest reduction in seed yield (294 per cent) was obtained with stress imposed at bud and early capsule stage. When the seeds are fully matured at this stage, crop is harvested. Seed yield varies from 3.5 to 6.0 q/ha. Powdery mildew and downy mildew are important diseases of opium and aphid, thrips, cut worm and capsule borer are the pests.

Paprika

Paprika (*Capsicum annuum* L.) belongs to family Solanaceae. It is a high colour and low pungent chilli. Paprika is one of the most traded in spices and its varieties are native to South America. It is an annual plant grown all over the world for different aroma and taste. It is primarily used as a food colorant in a variety of cuisines. United States, Zimbabwe, South Africa and Hungary are major paprika producing countries. It is also valued for oleoresin, used in the food and cosmetic industries. Paprika is the main ingredient in many spiced sausages, including the Spanish chorizos. Many Spanish, Portuguese and Turkish recipes are incomplete without paprika. It also spices and colours cheeses and cheese spreads and is used in marinades and smoked foods. In India, paprika is used in tandoori chicken, to give it a distinct red colour. Paprika is an emulsifier, temporarily bonding with oil and vinegar to make a smooth mixture for salad dressing. Fresh red peppers have more than seven times as much vitamin C as oranges. It is also an excellent source of betacarotene, that the body converts to vitamin A. There is great demand for natural paprika fruit colour used in processed food in place of synthetic colours. Basically, the colouring matter of paprika is a mixture of carotenoids and red pigments. Capsanthin and capsorubin are the main red pigments in the paprika. Paprika oleoresin has a homogenous dark red colour and is dispersible in water and vegetable oil. One part of oleoresin can replace 50 – 60 parts of ground paprika. It has great demand in the food processing industries. At present, paprika has very little production in India commercially but some special type of Byadagi and Tomato chilli are used for oleoresin extraction. Paprika has great scope to be cultivated in the hill station of North and South India and plains of North and Central India, where mild climate prevails.

In tropical and sub tropical areas of India, it is grown during winter. In hilly areas, the crop is cultivated from March to October. It requires well drained fertile sandy loam to clay loam soil with a pH of 5.5 to 6.8. About one kg of seed raised in 180- 200 m^2 nursery will be adequate for planting of one hectare land. Six – 8 weeks aged seedlings are planted in the main field at spacing of 45-60 cm x 45 cm in hills and 60-75 cm x 45 spacing in plains of India. Apply 20-25 tones/ha farm yard manure at time of land preparation. The recommended dose of N : P : K is 150 : 150 : 150 kg/ha (Anon. 2004). Kt Pl -19 is the improved variety of paprika. Apply irrigation during tenth leaf to flowering stage, flowering, fruiting and after each periodical harvesting if soil moisture is deficient for getting higher yield. Soil moisture deficit at flowering and fruiting stage encourage flower and fruit drop, respectively. On the other hand excess water should always be drained out as the paprika is very sensitive to wilt, water logging even for a shorter periods is detrimental. Fruits are harvested when they reach maturity. Mature red fruits are harvested 70-80 days after planting. Normally, 5-6 pickings are done at an interval of 7-10 days. Common diseases of paprika include seedling damping off, fruit rot, root rot, leaf curl complex and bacterial leaf spot. Blossom end rot and sun scald are the common physiological disorder of paprika. Thrips, mites and aphids are the most common are serious pests of paprika.

Parsley

Parsley (*Petroselinum crispum* Mill) belongs to family Apiaceace and is a hardy, much branched, aromatic and biennial spice. It originated from Mediterranean region and grows wild or is cultivated all over the world. Parsley green leaves are used as salad or for seasoning of different food stuffs. Fresh leaves are also used for garnishing and seasoning of foods. It is also used as an ingredient of soup, sauce and stews. Seed is used mainly as spice. It is used for flavouring and adding taste to different foods. The essential oil in the leaves is superior and used as condiments and seasoning. Essential oil is also used for preparation of gripe water. Parsley tea is used as remedy to rheumatism. The characteristic odour parsley is due to presence of apiin present in all parts of the plant. Fresh leaves are rich in vitamin C, iron, calcium and vitamin A.

Parsley is a cold weather crop and prefers cool climate. In India, it grows better in higher altitude but in subtropical areas it is grown as foliage crop during winter. It grows the best in rich and moist soil suitable for deep cultivation. Sandy loam to medium loam soil rich in organic matter is suitable for its cultivation. Heavy clay soil is not suitable for its cultivation. There are three common varieties of parsley, these are Italian or flat leaf parsley, Curly leaf parsley and Hamburg parsley. Mass curled, Humburg parsley, Pusa Vishwas and Pusa Vikas are mostly grown in India. It is propagated through seed or seedling raised in Nursery (Meena *et al.*, 2007). Seed is sown in March-May in hills and August -September in plains. For transplanted crop, seed (250-300g/ha) is required for raising seedlings in nursery. For broadcast sown crop, 3.5-4.0 kg seed is required for sowing of one hectare, where as line sown crop seed rate is 6.0-8.0 kg/ha. Seed normally takes 15-20 days for germination. Seed soaking for overnight enhances seed germination (per cent) and reduces time of germination. Seeds are sown in a spacing of 30-40 cm x 15-20 cm. In case of transplanted crops, seedlings are planted at a spacing of 40-45 cm x 10-15 cm. Apply 15 tones/ha farm yard manure or compost three weeks before sowing. Apply N : P : K at the rate of 100 : 60-80 : 40 kg/ha. Full dose of P and K and one third dose of N should be applied as basal during the time of final land preparation and nitrogen should be applied in two equal splits as top dressing at 30 and 60 days after sowing or transplanting. One pre- sowing irrigation is given for better seed germination and there after four irrigations at 30-35, 60-70, 80-90 and 110-100 days after sowing or transplanting are beneficial for getting higher yield. The leaves are ready for harvesting in about three months. The outer larger leaves are periodically cut off for several weeks. Depending upon the management practices 3-5 cuttings may be taken. Leaf yield varies from 12-14 tones/ha. If the crop is grown for seed, an yield of 0.5-0.6 tones/ha can be obtained. The major diseases are cercospora blight, septoria blight, parsley mosaic and bacterial leaf spot. Aphid is the major pest of the crop and reduces yield considerably.

Rosemary

Rosemary (*Rosamarinus officinalis* L.) belongs to family Lamiaceace and is an under utilized herbal spice grown for its aromatic leaf and flowering tops. It is a native of the Mediterranean region of Europe, Asia Minor and North Africa. It is cultivated in Yugoslavia, Spain, Portugal and certain parts of Europe and USA. In India, it is cultivated in temperate Himalayan region of North India and Nilgiris in South India. It is an evergreen, perennial dicot shrub. Rosemary is a symbol of faithfulness. In ancient times, people of Greece associated rosemary with memory. During the Medieval time, rosemary was rewarded as an evil protector (Patil and Madhusoodan, 2002). The leaves and flowering tops, on steam-distillation, yield the essential oil. The oil has 1, 8 cineole (20-50 per cent), borneol (20 per cent), camphor, linalool, α-pinene, camphene and β-pinene. The oil is valued for its use in culinary, medicine, perfumery and cosmetic industries. It is also used in flavouring meat, sauces, candy, baked products, perfumery and cosmetics including soaps, cream, lotions, deodorants and hair tonics. The leaves are used in cooking.

It prefers a Mediterranean climate with low humidity, warm winter and mild summer for its successful growth. Climatic conditions of Himalayan region, Nilgiri hills and Bangalore are highly suited for cultivation in India. It can not tolerate frost. It is a very hardy plant and is found growing on rocky terrains in the temperate parts of the World. It prefers well drained calcareous soil. In India, it comes up well in loamy and sandy loam soil with a pH of 6.5 -7.0. There are two types of rosemary under cultivation- 'French Rosemary' and the 'Italian rosemary'-. Rosemary can be propagated by seed or stem cutting. In case of seed propagation, seed should be sown in nursery during September to November. When the seedlings are about 8 to 10 weeks old, they are ready for transplanting to the main field. In stem cuttings, cuttings are ready for planting in the main field 6 -8 weeks after cutting. At time of land preparation apply farm yard manure at the rate of 20 tones/ha and during time of final land preparation apply N: P : K @ 20: 40: 40 kg/ha. After each harvest, N(80 kg/ha) is applied in 4 equal splits as side dressing to promote vegetative growth. Initially the crop should be irrigated once in a week and there after, one irrigation in a week is sufficient for normal growth and development. Depending upon the exposure of plantation, plants start flowering earlier in warmer and low altitude areas and later on high altitude. In some areas, harvesting commences from the second year in August after full flowering, which commences in May-June. Harvesting should begin at the time of 50 per cent blossoming and continue till 75-90 per cent inflorescence emerges and must end when flowers have finished blossoming.

Sage

Sage (*Salvia officinalis* L.) belongs to family Lamiaceae and is an underutilized herbal spice. Dried leaf of sage is the commercial part utilized as spice. It is native to Southern Europe. It is grown and cultivated in Yugoslavia, Portugal, Spain, England,- and Canada. It is also cultivated in India mainly in Jammu and Kashmir. The characteristic aroma and flavour of sage are due presence of essential oil. The main composition of oil is thujone. It is a hardy sub shrub with a height of 15-30 cm and aromatic leaves. It is extensively used in food industry as a standard spice in making stuffing for meat and sausages. Dried and powdered leaves can be utilized as condiment for preparation of cheese dishes, meat and other preparation. Fresh sage leaves are used in salads and sandwiches. Sage oil finds use in perfumes, deodorant and insecticidal preparations. It is also used for adulterating rosemary and lavender oils.

It is a cool season crop and hence its performance in hot and dry climate is not up to mark. Sage thrives on any soil but better quality product is obtained from clayey and loamy soil. Sage plantations are preferably raised from rooted stem cuttings as the crop raised from seed is likely to be a mixture of different types. The cuttings are obtained preferably from broad leaved plants which have the least tendency to flower. In case of seed propagation, seed rate is 1 kg/ha. In case of stem cuttings, basal cutting rooted better than apical cutting. April – May cuttings rooted better than October- November (Raviv *et al.*, 1984). Cuttings are usually planted in the mainfield at spacing of 60-90 cm x 30-45 cm under Indian condition. Bezzi (1987) reported that application of nitrogen (0-150 kg/ha) increased dry matter production, but potassium and phosphorus had no effect on dry matter production in sage. The first year crop is usually light, but after the plants are established, two or more cuttings in a year may be possible. From the second year onwards a crop yield of 1.75 – 2.25 tones/ha may be obtained. Sage plantation should be replanted every 4 – 5 years and by that time they become logy and show sign of die-back.

Sweet Flag

Sweet flag (*Acorus calamus* L.) commonly known as 'Bach or Gorbach in Hindi; as Vacha, Ugragandha or Bhadra in Sanskrit, is an important minor spice cum medicinal and aromatic plant belongs to family Araceae. It is a semi-aquatic perennial herb with long, creeping, much branched, aromatic, rhizomes and fibrous roots which occur widely all over India especially in hilly tracts (Selvi *et al.*, 2003). It is mainly cultivated in the Netherlands, Iran, United Kingdom, India and Sri Lanka. In India, it is common in Kashmir and Kumayun region of the Himalayas. It is cultivated also in Karnataka, Manipur and Nagaland. The fresh rhizomes are used in confectionary and also used as substitute for ginger (Farooqi *et al.*, 2000). In West Bengal, the ground dried rhizome and rhizome powder are used in baits for fishing. Root is used for treatment of Kwashiorkor disease of children. The rhizomes are used as carminative, stimulant and tonic (Jain, 2001). In Ayurveda, it is highly valued as a rejuvenator

Figure 5.19: Sweet flag plant.

for the brain and nervous system and as a remedy for digestive disorders. It is used internally in the treatment of digestive complaints, bronchitis and sinusitis. It ha tonic powers of stimulating and normalizing the appetite. It contains 1.5 to 3.5 per cent essential oil. The essential oil is extracted from the rhizome and utilized in perfumery. Due to presence of acorin in its essential oil, it is commonly a remedy for asthma and chronic diarrhoea.

It is a hardy plant found growing from tropical to sub-tropical climates. Plenty of sunshine should be available to the plant during its growth and after harvesting for drying the rhizomes.

Temperature from 10°C to 38°C and annual rainfall between 70 and 250 cm are the best suited. Sweet flag comes up well in clayey loams, sandy loams and light alluvial soils of river banks. The land should be ploughed twice or thrice prior to onset of rains. The land should be prepared as in paddy fields. It is propagated through rhizomes. Rhizomes obtained from earlier planting are kept preserved in the soil and constantly kept moist. After emergence the rhizomes are cut into small pieces and planted. Sprouted rhizome pieces are planted at a spacing of 30 x 30 cm and depth of 4cm in June -July. FYM @15-20 t/ha along with nitrogen and phosphorus. One third of N along with 50 kg of P and 25 kg of K are the basal requirements. The second dose of N should be given after one month of planting as broadcast and a third dose after two months of planting. Datta *et al.* (2009) reported that the maximum fresh and dry rhizome yield (3013.23 kg/ha, 1389.15 kg/ha respectively) was recorded with 50 t/ha FYM along with100ppm GA_3 followed by application of 50 tones/ha FYM (1342.65 kg/ha in dry form). Regarding the application of N and K, Datta and Dey (2010) reported that combined application of N (90 kg/ha) and K (275kg/ha) recorded the maximum dry rhizome yield (1600 kg/ha) and the highest B: C ratio of 2.06 under *terai* zone of west Bengal. The initial level of water standing in the field should be 5 cm and later increased to 10 cm. Timely weeding and hoeing to control the spread of weeds and to obtain good yield are essential. The uprooted rhizome is cleaned after washing with water and cut into size of 5-7.5 cm length and fibrous roots removed. Mealy bugs and caterpillars are the pests and leaf spot is the common diseases of this crop. The yield is 6 -10 tones/ha of fresh rhizomes.

Sweet Marjoram

Sweet Marjoram (*Majorana hortensis* Monech) is an aromatic perennial herbaceous shrub belonging to the family Lamiaceae and originated from Southern Europe. Marjoram or Sweet Marjoram is characterized by strong spicy and pleasant odour. The typical odour of marjoram is due to presence of essential oil. The oil is colourless or pale yellow to yellow green, with a tenacious odour reminiscent of nutmeg and mace. Sweet marjoram is characterized by a strong spicy, slightly sharp, bitterish and camphoraceous. The Greeks called this plant "joy of the mountain". They believed, it was precious to Aphrodite, Goddess of love, and they used it to crown newlyweds on wedding day. Dried leaves with or without flowering tops in small proportion constitute the spice of commerce. Marjoram grows as an upright compact bush 45-60 cm in height, with a main stem and many softer branches. It grows indifferent parts of Asia, Europe, Africa and America. France and Egypt are the major exporters of marjoram oil in the world. In India, it is being grown in the home gardens of Karnataka, Andhra Pradesh and Tamil Nadu. It is highly esteemed as a food stuff. Fresh leaves are used as garnishing agent in salads and in different food preparations. Marjoram is also used to flavor soups, stews, and sauces. Seeds are used in confectionary industry. Essential oil is used for preparation high grade perfume, soap and liquors industries. Medicinally, it has carminative, expectorant and tonic properties.

Figure 5.20: Marjoram seeds.

It is a warm climate plant and as such sensitive to cold. It thrives well in 22°C day temperature and 15°C night temperature. It requires sunny situation and mid day shade is preferred. Well drained nutrient rich soil is ideal. It has strong flavour when grown in nutrient rich soils. Though ideal pH is 6.9, but it can also successfully grown in saline and alkaline soil. The plant resists drought condition but prolonged drought and drought in early a few months of growth are harmful for the crop It grows

Figure 5.21: Marjoram plant.

Figure 5.22: Marjoram flower.

quickly and pinching should be done to make them bushy. It is propagated by stem cutting. It can also be propagated by seeds and seedlings are transplanted in the main field when they are about 50-60 days old at a spacing of 30 cm x 20-25 cm. Adequate amount of organic manure and NPK@ 240: 40: 80 are to be applied for getting higher yield. The leaves should be harvested just before they begin to flower and if harvested when the flowers have set seed, the taste becomes bitterer. While harvesting, about 10cm of the shoot above the ground should be left for further growth. The subsequent harvest carried out at an interval of 45 days from the previous harvest. After each harvest top dressing with urea is beneficial for getting subsequent herbage yield. In the advance stage of growth pruning is practiced at the end of winter because the herb has a tendency to become hardy. On an average it may yield about 16-18 tones/ha fresh herbage per year. Herbage can be dried properly and stored in an airtight container as whole or crumbled leaves for three months or more.

Tamarind

Tamarind (*Tamarindus indica* L.) belongs to family Leguminosae, one of the important underutilized multipurpose tree spices in the Indian subcontinent. The tamarind fruit pulp is an important culinary ingredient in India since ancient times. It is highly suited for dry land horticulture. It originated in Madagascar. It is now cultivated in semi arid Africa and South Asia. Commercial plantations are reported in Central America and in North Brazil. In India, it is grown almost through out the country, except Himalayan and Western dry region. But its cultivation is concentrated in the States of Tamil Nadu, Andhra Pradesh, Kerala, Karnataka and Orissa. Fruit pulp is used for preparation of beverages, canned turmeric, sauce, jam and jelly. Unripe fruits are used for making chutneys. The typical taste of tamarind is due to presence of tartaric acid.

It can be grown in all types of soil up to an elevation of 1200 m. But it prefers deep alluvial soil. It grows well in tropical climate where summer is hot and dry. The crop is not suitable for temperate or arid climate. It is drought resistant but frost susceptible. Tamarind is propagated mostly through seeds and it takes about 10 years to come to regular bearing and hence vegetative propagation methods like inarching, patch budding or air layering is commercially used. In most of the cases, farmers grow local types available in the locality. Improved varieties like Prathisthan and No.-263 are cultivated in Maharashtra, PKM-1 and Urigam are cultivated in Tamil Nadu. In commercial scale, provided with drip irrigation facilities it can be grown with 5 m x 5 m spacing upto 10-12 years. After that alternate rows are removed to provide 10 m x 10 m spacing. Pits of 1 m x 1 m x 1 m filled with 10 – 15 kg FYM, 100 g single super phosphate and 1.5 kg of neem cake is ideal for better establishment of the crop. Normally no fertilizer management practices are followed in tamarind when it is grown in dry land. It responds well to application of nutrients through inorganic fertilizers. In a first to fifth year grown plant, apply 20 kg of farm yard manure along with N : P : K at the rate of 225 : 160 : 600 g/plant/year. It is gradually increased and nine years and above aged plants, N : P : K at the rate of 900 : 500 : 1800 g/plant/year are applied. Pot watering is done for establishment of young plants during summer. It is normally grown as rain fed crop but irrigation helps in getting early yield. Irrigation at flowering and fruit set stage increases yield. It is essential to keep the plants free from side branches up to 1 m height. The trees come to flower from 3rd year of planting, flowering seen during September – October and fully ripe fruits harvested from March – April. At the early stages, crop yield is low and increases gradually. A fully bearing tree of 12 years produces a yield of 200 kg pod/plant/year. Powdery mildew is the important disease affecting the crop and tree hopper, mealy bug and scale insect are the important pests of tamarind which adversely affects the yield.

Tarragon

Tarragon (*Artemisia dracunculus* L.) belongs to family Compositae and is a well known spice for its unusual flavour. Sometimes, it is also called as French Tarragon. *Artemisia* is a large genus and about 280 species found in the Northern hemisphere. About 34 species are reported in temperate region of North Western Himalayas. Its aroma is warm, aromatic and reminescent of anise. It is cultivated in

France, Spain, temperate zone of USA and colder zone of England. The herb on steam distillation, yields about 0.3 per cent essential oil, responsible for the aromatic aniseed like odour of the herb. Methyl chavicol is the chief constituent of the essential oil (Werker, *et al.,* 1994). Phellandrene and ocimene are additional constituents of essential oil. Both leaves and essential oils are used for flavouring vinegar, pickles, prepared mustard and to a limited extent for flavouring soups, salad, meat dishes, certain cheeses and vegetables. The aromatic leaves are credited with stomachic, stimulant and febrifuge properties.

Tarragon is a xerophytic plant. In central Asia, they are grown in semi desert areas where extremes of temperature, both high and low prevail. It prefers a saline sandy soil. Long days promoted shoot growth where as plants grown under short day produce rosette plant. Under ground buds removed from roots or crowns of tarragon plants in August rooted well in sand in glass house and plantlets are ready for filled planting after 3 weeks. By using bud, crop establishment is much more rapid than the rooted tip cuttings and survival rate was much higher (Lammerink, 1985). Rapid propagation of tarragon can be done by using *in vitro* technique.

Tezpat or Indian Cassia

Tezpat or Indian Cassia (*Cinnamomum tamala* (Buch Ham) Nees and Eberum) belongs to family Lauraeae and is a tree spice. It is a moderate evergreen tree, leaves are glabrous and three nerved from the base. Its dry leaves are used as spice having a warm, powerful taste and odour. It originated in India, mainly found in tropical and subtropical Himalayan region upto an altitude of 2400 m. It is found frequently in North Eastern parts of West Bengal, Khasia hills, Jayanti hills and also in homestead gardens of North Eastern states of India. It is used as flavouring agent in preparation of many vegetarian or non vegetarian dishes. Steam distillation of leaves produces essential oil, the main composition of the essential oil is eugenol. It's leaf essential oil is used in pharmaceutical preparations, soap and cosmetic industries. It is also used in flavouring of all kinds of foods, meat, sausages and sauces.

It can be grown in all types of soil but light soil with plenty of organic matter is the best suited. It prefers humid climate. The seeds are used for propagation. Immediately after harvesting of the seed, they are sown in nursery bed because of recalcitrancy of seeds. Seedlings are raised in beds before onset of monsoon. Usually seeds are sown during March – April and they germinate within one or one and half months. Seedlings are planted in the main field when they are four to five years age. Seedlings are planted at a spacing of 3.0 – 4.5 m x 3.0 - 4.5 m. It takes another 5 – 6 years to grow and starts bearing 10th year onwards. Productivity of Tezpat remains up to 50 -100 years. Natural regeneration helps the farmer to raise new plants at cheap rate. In early stages of growth manuring can be done. No manuring is followed in established garden. Pruning of leaves are done during harvesting stage. The season of harvest is dry days of October to December and in some areas extends up to March. Usually leaves are collected once in a year from young trees and every other year from old and weak trees. Normally yield varies from 8 – 20kg dry leaves/plant/year and the average yield of tezpat is 12 kg dry leaves/plant/year.

Thyme

Thyme (*Thymus vulgarisi* L.) belongs to the Lamiaceae commonly known as thyme or golden thyme. It is an important herbal spice used by man since ancient times. It is an evergreen perennial undershrub reaching a height of 20-30 cm. It is one of the very frequently used spices in fresh or dried form for several European delicacies. It is a common garden plant which survives for many years under good management conditions. The dried leaves and floral tops constituting the thyme of

Figure 5.23: Thyme plant

Figure 5.24: Thyme flower

Figure 5.25: Thyme flower

commerce are known as "Thymi Herba". The dried leaves are curled, are of brownish green colour, usually not longer than 6-7 mm and 2-3 mm broad and marketed in whole or ground form. On steam distillation, thyme produces essential oil and the main composition of essential oil is thymol. Thyme is grown in Europe, Australia and North Africa. It originated in Mediterranean region cultivated in France, Germany, Spain, Italy, Greece, North Africa, Canada and USA. In India, it is found in western Himalayan from Kashmir to Kumayan between altitudes of 1525 to 4000 m. The leaves and flowers are find use as food flavourants and seasoning various food items, especially fish and meat preparations. Thyme oil is used in soap, perfumes and flavouring food products such as sausages, sauces, meat and canned products.

Figure 5.26: Thyme seeds.

The plant prefers a light but fertile and calcareous soil for good and oil content. It can be grown both hill and plains but hilly region is most suitable for its growth. It prefers mild climate for its better growth and yield. Late summer planting is the ideal time for transplanting of seedlings or planting of rooted cuttings. Thyme can be seeds and vegetatively by division of old plants or by cutting or by layering of side shoots in March – April. The seeds are sown directly in rows or the seeds are sown in well prepared nursery beds in good soils. A spacing of 90 cm × 30-45 cm may be maintained in case direct sown crop. While planting the seedlings or rooted cuttings or layer etc. they are planted in a spacing of 60 cm × 30 cm. alight irrigation is provided after planting. Application of farm yard manure and nitrogen is reported to promote the formation of numerous leafy shoots. Frequent irrigation is required during the dry periods. In the hills, in order to avoid frost injury to plants during winter mulching is done. The leaves and flowers which are used for culinary or medicinal purposes are harvested five months after sowing or planting. The leaves and flowers are plucked from plants are dried immediately in shade or in mechanical drier and the dried product should be kept in air tight container to prevent the loss of flavour. Under favourable condition and good management condition, yield (in dry form) of thyme varied from 1.6 – 2.2 t/ha. The yield is comparatively low during the first year. The plants become woody and replanting become necessary after 3 – 4 years. The plant is not damaged by any pest of serious nature but wilt disease is a major problem of this crop.

With shrinking land resources and gradually decreasing land holdings the growers are facing stiff challenges to meet the present day requirements. The major challenges are of productivity, quality, value addition and of equity. So an organized and planned effort to go for high value spice crops may be a meaningful alternative. Latest tools of biotechnology, information technology, analytical chemistry, molecular biology, genetic engineering offers new hopes and reliance on both quantity and quality. It is really a boon that India is rich in genetic diversity in several minor spices. Exploration, conservation, evaluation, cataloguing and utilization of these genetic resources is the need of the hour. Several exporters have predicted that India will have to import over 40 million tonnes of foodgrains by 2030 (Pruthi,2001). Spices, especially the minor spices have also its contribution in this mission. Moreover we must achieve agricultural intensification and diversification in an environmentally friendly manner. Sustainable advances in the production and export of spices can then be achieved (Swaminathan, 1996).

References

Achut, S. G. and Bandyopadhyaya, C., 1984. Characterization of mango-like aroma in *Curcuma amada* Roxb. *Journal of Agriculture and Food Chemistry*, **32** : 57–59

Annonymous (2004. *Crop Production technology of horticultural crops*. TNAU, Coimbatore.

Banafar, R. N. S, Gupta, N. K. and Pathak, A. C., 2002. Suitability of black cumin varieties (*Nigella sativa* L.) for Madhya Pradesh. *Advances in Plant Sciences*, **15**(1) : 165 – 166.

Bezzi, A., 1987. A fertilizer trial on *Salvia officinalis* L. *ItalyMinestero dell Agricultura e delle Foreste*, pp. 315-335

Choudhury, B., 1996. Vegetables. National Book Trust, India (Reprinted). p. 223.

Chung, B., 1987. The effect of irrigation on growth and yield components of popies (*Papaver somniferum* L.). *Journal of Agricultural Sciences*, **108**(2) : 389-394.

Datta, S. and Dey, A. N., 2010. Effect of N and K on growth and yield of Sweet flag (*Acorus calamus* L.). *Green Technology*. **9** : 120-123.

Datta, S., Chatterjee, R. and Ghosh, S. K., 2003. Evaluation of some black cumin accession for yield and quality. *The Orissa J. Hort.*, **31** (1) : 34-36.

Datta, S., Dey, A. N. and Maitra, S., 2009. Effect of FYM and GA_3on growth and yield of Sweet flag (*Acorus calamus* L.) under terai zone of West Bengal. *Journal of Horticultural Sciences.* 4(1) : 59-62.

Datta, S., Poduval, M., Basak, J. and Chattergee, R., 2001. Fertilizer trial on cumin black (*Nigella stiva* Linn.). *Env. and Eco.*, **19** (4) : 920-22.

Farooqui, A. A., Sreerammu, B. S. and Srivasapa, K. N., 2000. Cultivation practices of Sweet.

http://justforhearts. org/all-about-spices/Accessed on 11.08.2012

Jain, S. K 2001. Calamus. p11-12. (In) *Medicinal Plants.* National Book Trust Publishers, India.

Ko, F. N., Huang, T. F. and Teng, C. M., 1991. Vasidalatory action mechanism of apigenin islotaed from *Apium graveolens* in rat thorasic arota. *Biochemicast Biophysica Acta, General Subject*, **1115**(1) : 69-74.

Kraxner, U., Weichmann, J. and Fritz, D., 1986. How should horseradish be manured. *Gemuse*, **22**(9) : 363-364

Lammerink, K. J., 1985. Influence of different storage condition on germination of spice seeds. *Acta Horticulturae*, **253** : 99-105.

Luchon, S. and Sarat, S., 2003. Black cumin (*Nigella sativa* L.) a new aromatic spicy medicinal plants. Abstract of paper presented in *Nat. Sem. On "New perspectives in spices, Medicinal and Aromatic Plants"*, 27-29 Nov. 2003, At ICAR Research Complex, Goa p : 153.

Mandal, A. R. and Maity, R. G., 1993. Effect of some chemicals and micro climate factors on germination of black cumin (*Nigella sativa* L.) seed. *Horticultural Journal*, **6**(2) : 115-120.

Mathew, P. A., Krishnamoorthy, B. and Rema, J., 2007. Potential of kokum and its improvement. *Spice India.* **20**(1) : 18-20.

Meena, R. S., Meena, S. S., Singh, R. K. and Metha, R. S., 2007. Scientific cultivation of parsley. *Spice India.* **20**(4) : 27-30.

Nair, A. K., Subramanyam, A. K., Verma, B. S. and Singh, D. V., 1992. Response of Japanese mint to Fe, Zn at two fertility levels. *Journal of Indian Society of Soil Science.* **40**(4) : 873-875.

Nataraja, A., Farooqui, A. A., Sreeramu, B. S and Srinivasappa, K. V., 2003. Influence of nitrogen, phosphorus, potassium on growth yield of black cumin (*Nigella sativa* L.) *J. Spices and Aromatic crops*, **12** (1) : 51-54.

Nebel, N., Weichmann, J. and Fritz, D., 1986. Yield of different horseradish ecotypes. *Gemuse*, **24** (6) : 272-273

Patil, D. V. and Madhusoodan, K. J., 2002. Rosemary –a herb of mild flavour. *Indian Journal of Arecanut, Spices and Medicinal Plants.* **4**(2) : 80-83

Pruthi, J. S., 1976. Spices and condiments. National Book Trust, India, New Delhi.

Purthi, J. S., 2001. Black cumin. (In) *Minor Spices and Condiments Crop Management and Post Harvest Technology.* ICAR publication, New Delhi, pp : 202-205.

Pruthi, J. S., 2001. Introductory. (In) *Minor Spices and Condiments Crop Management and Post Harvest Technology.* ICAR publication, New Delhi, pp : 35-41.

Rao, S. A., Bandaru, R. and Ramachandran, R., 1989. Volatile aroma components of *Curcuma amada* Roxb. *Journal of Agriculture and Food Chemistry*, **37** : 740–743

Raviv, M., Putievsky, E. and Sanderovich, D., 1984. Rooting stem cutting of sage (*Salvia officinalis*). *Plant Propagator*, **30**(3) : 8-9.

Selvi, B. S., Selvaraj, N. and Raghu, R., 2000. Sweetflag. *Spice India*, **14**(8) : 4

Sharangi, A. B., 2009. Mango-ginger: a two-in-one spice. *Indian Journal of Arecanut, Spices and Medicinal Plants.* **11** (2): 77-79.

Sharangi, A. B. and Datta, S., 2005. Medicinal properties of Spices, *Indian J. Arecanut, Spices and Medicinal plants*,**7**(2);42-49.

Shastri, B. N., 1988. The Wealth of India (Raw Materials). CSIR Publication, Reprinted, Volume-2, p. 260.

Shetty, A. A., Buckseth, T. and Naresh, V. S., 2012. Ajowan a under-utilized seed spice. *Spice India.* **25**(2) : 15-18.

Singh, Kumar, J. K., Saikia, D., Shanker, K., Thakur, J. P. and Negi, A. S., 2010. A bioactive labdane diterpenoid from Curcuma amada and its semisynthetic analogues as antitubercular agents. *European Journal of Medicinal Chemistry*, **45** : 4379–4382

Singh, S. K.; Singh Sardar, Singh, S., 1999. Response of nigella (*Nigella sativa* L.) to nitrogen and phosphorus. *Crop Research*, **18** (3): 478-479

Swaminathan, M.S., 1996. Sustainable food and water security. (In) Interaction as-Food, Agriculture and Environment, Vol I. www. eolss. net/sample-chapters/c09/e4-24-05. pdf.

Syamkumar, S. and Sasikumar, B., 2007. Molecular marker based genetic diversity analysis of *Curcuma* species from India. *Scientia Horticulturae*, **112** : 235–241

Thomas, J., 2003. Mint (Japanese mint). *Spice India.* **16**(3) : 20

Tomar, I. S. and Ramgiry, S.R., 1977. Effect of growth regulators on growth and yield of potato (*Solanum tuberosum* L.). *Advances in Pl. Sci.*, **10**(1) : 51-54

Tomar, S. S., Nigam, K. B., Pachori, R. S. and Kahar, L.S., 1990. Critical irrigation stages in unlanced poppy for seed yield. *Current Research- University of Agricultural Sciences* (Bangalore), **19**(5) : 77-78.

Uday Kumar, J.N., Chatterjee, R. and Sharangi, A. B., 2008. Effect of Ethrel on growth and yield of black cumin (*Nigella sativa* L.) under gangetic alluvial soils of West Bengal. *J. Crop and Weed*, **4**(2): 31-32.

Vargas Prieto, A., Aguillon, G.A., Vasquez Navarro, G. and Chinging, A., 1986. Determination of optimum dose of fertilizer (NPK) for celery in Queretaro region. *Proceedings of the tropical region, American Societies for Horticultural Sciences,* **23** : 190-193.

Werker, E., Putievsky, E., Ravid, U., Dubai, N. and Katzir, I., 1994. Grandular hairs, secretory cavities and essential oil in leaves of tarragon (*Artemisia dracunculus* L.). *Journal of Herbs, Spices and Medicinal Plants,* **2**(30 : 19-32.

Williams, J. T. and Haq, N., 2000. Global research on underutilised crops: An assessment of current activities and proposals for enhanced cooperation. International Centre for Underutilised Crops, Southampton, UK. p-50.

2014, Future Crops Vol. 2 *Pages 133–151*
Editor: Professor K.V. Peter
Published by: DAYA PUBLISHING HOUSE, NEW DELHI

Chapter 6
Cactus Pear

***P.K. Ghosh*[1], *P. Bhattacharjee*[1] *and R.S. Singhal**[*2]**
[1]Department of Food Technology and Biochemical Engineering, Jadavpur University, Kolkata – 700 032, India
[2]Institute of Chemical Technology, Nathalal Parekh Marg, Matunga, Mumbai – 400 019, India
**E-mail: rsinghal7@ediffmail.com*

Cactus pear or prickly pear (*Opuntia ficus-indica* (L.) Miller) is native to Mexico. It is a member of the *Cactaceae* family with 400 species and a large number of varieties. It is an emerging fruit crop which evolved commercially during the second half of the last century. Mexico is the largest producer of cactus pear with 79.4 per cent of the world production, covering 49,165 ha with Italy and South Africa following suit with 12.2 and 3.7 per cent of the production, respectively. Cactus pear is also cultivated in Africa, Australia, the Mediterranean basin, and some parts of Asia (Anorve Morga *et al.*, 2006). In

Figure 6.1: A cactus pear plant with cladode and cactus pear.

view of the changes in global climate, cactus pear is suggested as a promising crop by the Food and Agriculture Organization (FAO) (Barbera *et al.*, 1995).

Cacti are succulent plants which live in dry (xeric) environment and survive long period of drought. Low water requirements and a high water use efficiency ratio make cactus pear suitable for arid regions (Stintzing *et al.*, 2001). The plant consists of 'cladodes' (thick walled succulent stems) which are photosynthetic and have a waxy layer that checks transpiration, and 'glochids' (spines) which are modified leaves which also control transpiration (Figure 6.1). The plant carries out 'Crassulacean Acid Metabolism' (CAM) which allows CO_2 uptake at night, thus minimizing water loss during photosynthesis and consequently the ability to withstand drought. These traits make it highly promising as a crop for soils poor in nutrients and with limited water supply (upto 200 to

Figure 6.2: Popular varieties of cactus pear.
Source: Yahya and Mondragon-Jacobo (2011).

Contd...

Figure 6.2–*Contd...*

300 mm annual rainfall) which are common in less developed countries (Brutsch, 1988). The plant has long roots which run near ground level so that they absorb water when it rains. Cactus pear is consumed as a fruit, while cladodes are consumed as a vegetable.

There are many variants of cactus pear such as 'Reyna' (*O. albi-carpa*), 'Naranjona' (*O. megacantha*) and 'Roja Lisa' (*O. ficus-indica* L.) are the most important. Other varieties include 'Liria' (*Opuntia* sp.), 'Roja Pelota' (*O. ficus-indica* L., Mill.), 'Cardona' (*O. streptacantha*) and 'Camuesa' (*O. robusta* Wendl.) which are cultivated across the world (Figure 6.2) (Yahya and Mondragon-Jacobo, 2011). *Opuntia macorrhiza* is another variant of *Opuntia ficus-indica* having a comparatively lower pH value and lower microbiological spoilage (Moussa-Ayoub *et al.*, 2011).

Botanical Characteristics

The major product of the plant is the cactus pear fruit, a fleshy berry, usually eaten fresh. The fruit varies in shape, size and color, and consists of a thick peel and edible juicy pulp with many seeds

(Figure 6.3) (Osorio-Esquivel *et al.*, 2011). The fruit undergoes several developmental stages depending upon the size and/or color of fruit and pulp. The fruit in 'full coloration' is generally used for consumption as shown in Figure 6.4 (Walker *et al.*, 2011). The fruit weighs between 83 g and 216 g, with an average weight of 100-160 g, depending on origin and cultivar. The pericarp consists of glochids which contains pure crystalline cellulose. The peel (48-52 per cent) and pulp (48-52 per cent) consist of usable part of fruit. The pulp consists of two parts- seeds (4-7 per cent) and edible pulp (44-45 per cent). The pericarp and pulp have soft green, greenish white, canary yellow, lemon yellow, red, cherry red and purple hues (Stintzing *et al.*, 2001).

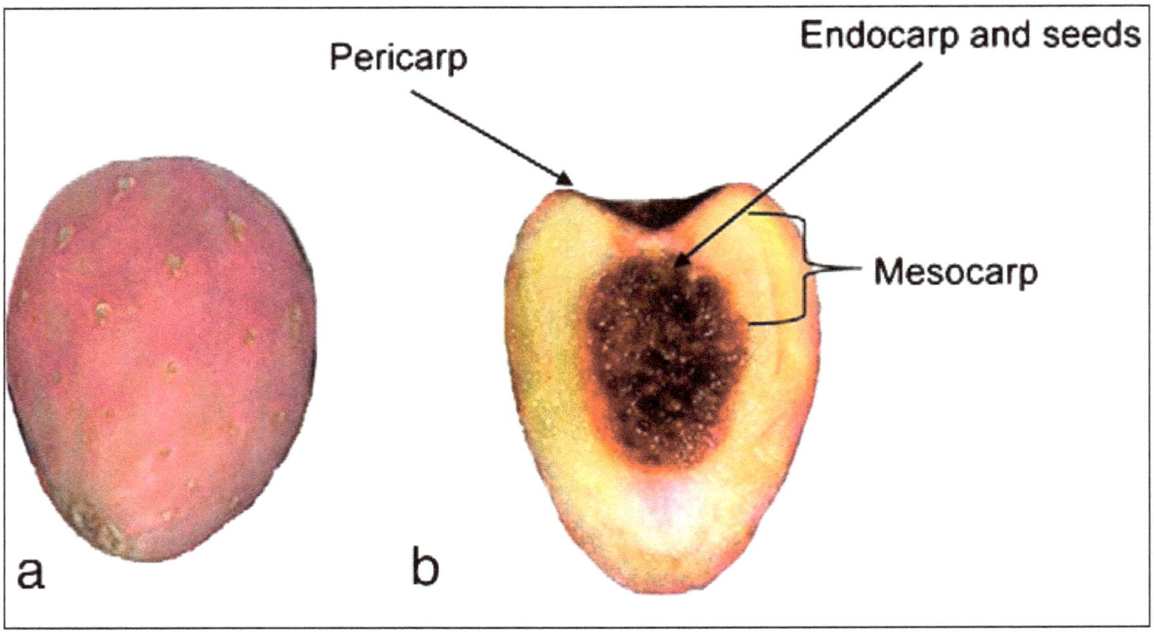

Figure 6.3: A cactus pear fruit.

a. Complete fruit b. Vertical section of the fruit with pericarp, mesocarp, endocarp and seeds.

*Source: Osorio-Esquivel *et al.* (2011).

The stem of the plant 'cladodes' contains about 90 per cent water, dietary fiber and chiefly carbohydrate fraction (40-50 per cent w/w dry basis) with starch and soluble sugars. It also contains proteins (8-13 per cent), vitamins like ascorbic acid, carotenes and minerals, especially potassium and calcium (Ayadi *et al.*, 2009; Rodríguez et *al.*, 2007).

The cactus pear root system is very different from other plants owing to their xeromorphic characteristics which enable the plant to survive prolonged periods of drought. Roots contribute to drought tolerance in three ways. Firstly, by restricting root surface and decreasing their permeability to water; secondly, by rapidly absorbing small quantity of water supplied by light rains via 'rain roots' (which develop within a few hours after a shower and disappear as soon as the soil dries up); via reduction in root surface through which water flows off and by decrease in shoot transpiration consequent to high root negative potential (Sudzuki, 1995).

Figure 6.4: The developmental stages of cactus pear.
From left to right: 1: Small-green; 2: Medium-green; 3: Large-green; 4: Half coloration; 5: Full coloration.
Source: Walker *et al.* (2011).

Genetic Transformation and Tissue Culture of Cactus Pear

Due to general characteristics of cactus pear plant (adaptation to highly contrasting environmental conditions like poor soils and low water availability and high biomass production), genetic transformation is an attractive technology which can be successfully applied to produce recombinant proteins and/or increase their nutritional value in cactus pear. This necessitates need for developing massive propagation conditions that can assure genetic stability of foreign introduced DNA. Since classical cacti breeding methods are slow in developing new cultivars, tissue culture is a suitable alternative for variety development (Figure 6.5) (Angulo-Bejarano and Pardes-Lopéz, 2011). Cruz *et al.* (2009) have reported a 4.1 per cent transformation frequency with biolostic mediated genetic transformation of apical regions of *O. ficus-indica* (L.) followed by micropropagation for shoot regeneration.

Nutritional and Phytochemical Qualities

Cactus pear consists of numerous phytochemicals which are present in the skin, pulp and seeds. The pulp is the main edible portion of fruit which contributes majorly to its nutritional value.

Figure 6.5: (a) Green regenerating compact callus and shoots produced, (b) shoots obtained from regenerating callus (c) shoot proliferation from regenerating calli, (d) plant elongation, (e) extensive rooting in shoots, (f) acclimatization of regenerated shoots, and (g) growth and adaptation of the regenerating plants to greenhouse conditions.

Source: Angulo-Bejarano and Pardes-Lopéz (2011).

Biochemical Composition and Nutritional Quality

The cactus pear fruit consists of about 90 per cent moisture, 40 per cent carbohydrate (5.5 per cent starch), 9 per cent protein (0.024 per cent NPN), 5 per cent lipids, 40 per cent total fibers and 2.5 per cent ash as shown in Table 6.1 (Kossori *et al.*, 1998; Salim *et al.*, 2009). The total fiber composition of the fruit consists of about 50 per cent cellulose, 12 per cent hemicellulose, 40 per cent pectin and 0.05 per cent lignin. The fairly high sugar content (14-18 per cent) and low acidity (0.02-0.15 per cent, pH 5.3-7.1) of the fruit give it a delicious sweet taste, but at the same time makes it very susceptible to microbial invasion limiting its shelf life (Joubert, 1993; Sáenz, 2000).

Table 6.1. Selected nutritional composition of *Opuntia ficus-indica*.

Nutrients	*Skin*	*Pulp*	*Seeds*
Moisture (per cent w/w dry matter)	90.33	94.40	18.05
Carbohydrate (per cent w/w dry matter)	27.6	58.3	1.59
Protein (per cent w/w dry matter)	8.30	5.13	11.8
Fat (per cent w/w dry matter)	2.43	0.97	6.77
Total fibers (per cent w/w dry matter)	40.8	20.5	54.2
Ca (mg/100 g dry matter)	2090	163	258
Mg (mg/100 g dry matter)	322	76.1	208
Na (mg/100 g dry matter)	< 0.85	7.77	< 0.83
K (mg/100 g dry matter)	3430	559	275
P (mg/100 g dry matter)	0.064	0.063	1140
Fe (mg/100 g dry matter)	8.31	16.5	12.1
Cu (mg/100 g dry matter)	< 0.85	< 0.78	< 0.83
Zn (mg/100 g dry matter)	1.70	1.55	4.16
Mn (mg/100 g dry matter)	72.9	6.99	0.83
Ash (per cent w/w dry matter)	3.05	1.00	3.66

Source: Kossori *et al.* (1998); Salim *et al.* (2009).

Table 6.2: Fat soluble vitamins and sterols in pulp and seed oil of cactus pear.

Compound	*Pulp Oil (g/kg fruit)*	*Seed Oil (g/kg fruit)*
α-Tocopherol	0.849	0.056
β-Tocopherol	0.126	0.012
γ-Tocopherol	0.079	0.330
δ-Tocopherol	4.220	0.005
Total vit E	5.274	0.403
β-Carotene	0.420	0.047
Vit. K_1	0.532	0.525
β-Sitosterol	11.2	6.75
δ5 Avenasterol	1.43	0.29
Δ7 Avenasterol	ND	0.05
Campesterol	8.74	1.66
Stigmasterol	0.73	0.30
Lanosterol	0.76	0.28
Ergosterol	ND	ND
Cholesterol	ND	ND

ND: Not detected.

Source: Ramdan and Mörsel (2003).

The pulp and seed oils of cactus pear contain various sterols and fat soluble vitamins which enhance it nutritional potential (Table 6.2) (Ramdan and Mörsel 2003). The seeds constitute about 10-15 per cent of the edible pulp and are discarded as waste after extraction of pulp. However, the seeds contain about 7-15 per cent oil of the total seed weight. The seed oil contains β-sitosterol as the main sterol marker along with tocopherols, avenosterols, stigmasterol, lanosterol, ergosterol, campesterol, cholesterol and vitamins E and K (Stintzing *et al.*, 2000).

The enzymes in the fruit and the cladodes include phosphoenolpyruvate carboxylase, ribulose-1,5-bisphosphate carboxylase/oxygenase, pyruvate and orthophosphate dikinase, phosphoenolpyruvate carboxykinase. These enzymes can be analyzed by SDS-PAGE and visualised in gels using coomassie brilliant blue or transferred to Immobilon-P membrane wherein it can be detected using specific antisera as shown in Figure 6.6 (Walker *et al.*, 2011).

Phytochemical Qualities and Antioxidant Capacity

In cactus pear, phenolics, flavonols, flavonoids and betalains are present (Tables 6.3 and 6.4) which confers it with antioxidant properties (Magloire *et al.*, 2006; Osorio-Esquivel *et al.*, 2011).

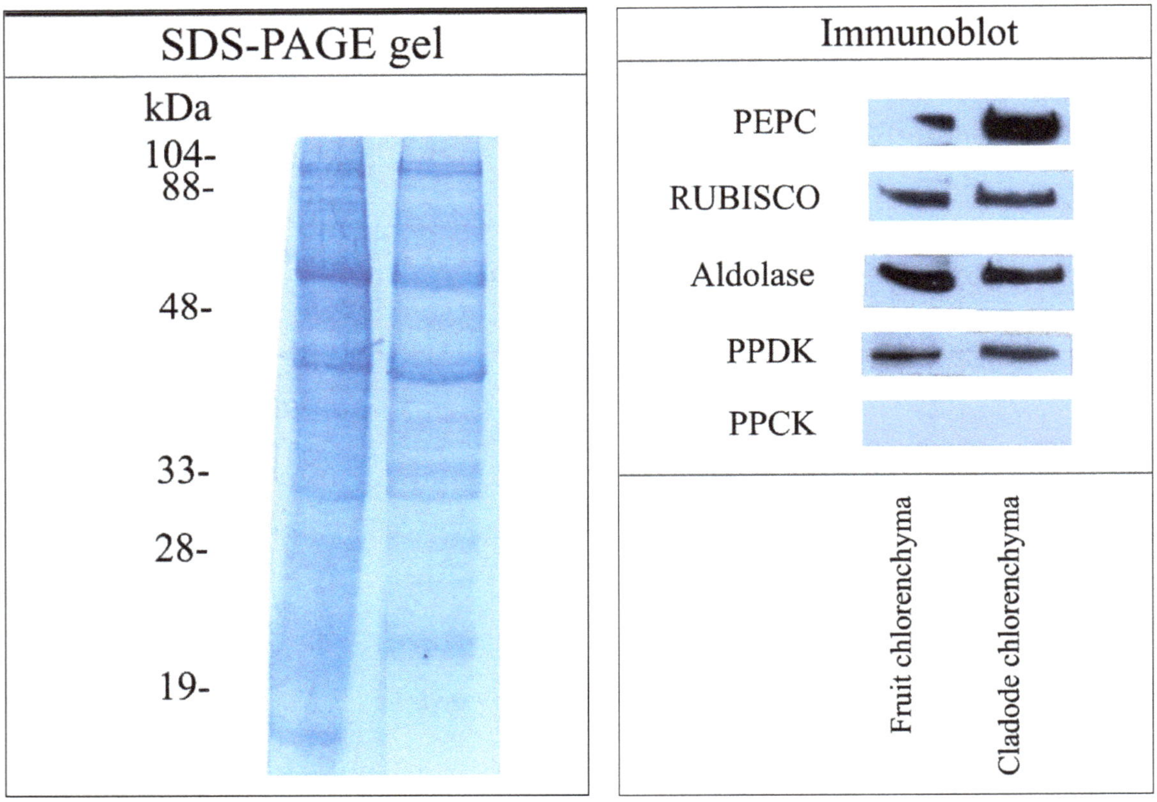

Figure 6.6: Comparison of enzymes abundance and polypeptide pattern in the chlorenchyma of cactus pear fruit and cladode.

PEPC: Phosphoenolpyruvate carboxylase; RUBISCO: Ribulose-1,5- bisphosphate carboxylase/oxygenase; PPDK: Pyruvate, orthophosphate dikinase; PEPCK: Phosphoenolpyruvate carboxykinase.

Source: Walker *et al.* (2011).

Quercetin, kaempferol and isorhamnetin are the major flavonoids in the fruit (Figure 6.7) (Kuti 2004). Cactus pear fruit is a unique source of isorhamnetin glycosides, especially isorhamnetin- 3-O-rutinoside and isorhamnetin triglycosides along with proline and taurine (Galati *et al.*, 2003a). It is also a good source of vitamin C and its total antioxidant activity has been twice than that of pear, apple, tomato and grape and comparable to that of red grape raisin, orange and grapefruit (Livrea and Tesoriere 2006; El-Samahy *et al.*, 2007).

Table 6.3: Phenolics and flavonol content in different parts of cactus pear (µg/g fresh weight).

Phenolics and Flavonols	*Whole Fruit*	*Pericarp*	*Mesocarp*	*Endocarp*
4-Hydroxybenzoic acid	5.67	28.3	12.90	ND
Syringic acid	3.88	4.10	8.70	ND
Protocatechuic	61.30	107.80	49.00	31.10
Rutin	2.67	ND	ND	22.70

ND: Not detected.

Source: Osorio-Esquivel *et al.* (2011).

Table 6.4: Flavonoids distribution in different varieties of cactus pear.

Opuntia Species	*Flavonol Content (µg/g fresh weight)*		
	Quercetin	*Isorhamnetin*	*Kaempferol*
O. lindheimeri	90.5	1.9	1.1
O. strepthacantha	51.0	ND	3.8
O. ficus-indica	43.2	24.1	2.2

ND: Not detected.

Source: Kuti (2004).

Cactus pear contains two types of betalain pigments- betanin (purple-red) and indicaxanthin (yellow) both with radical-scavenging and reducing properties of an antioxidant (Figure 6.8a) (Moßhammer *et al.*, 2005). The other two categories of betalains – the betacyanins (betanin and isobetanin) and betaxanthins (proline-Bx, Isoproline-Bx) are also present in the fruit (Figure 6.8b) (Moßhammer *et al.*, 2006).

The betacyanins in cactus pear can be analyzed by HPLC showing fractions of betanin, isobetanin, betanidin, isobetanidin and phylloactin from pericarp, mesocarp, endocarp and whole fruit (Figure 6.9) (Osorio-Esquivel *et al.*, 2011).

Production and Processing

Cactus pear is usually consumed fresh, but is also put to different traditional and industrial uses. The young parts of the prickly pear cactus, leaves or cladodes are mainly consumed cooked or as a fresh vegetable and also as juice. There are various methods of processing cactus pear depending on end use of product. Cactus pear juice and cactus pear juice concentrate have wide applications in food processing industry.

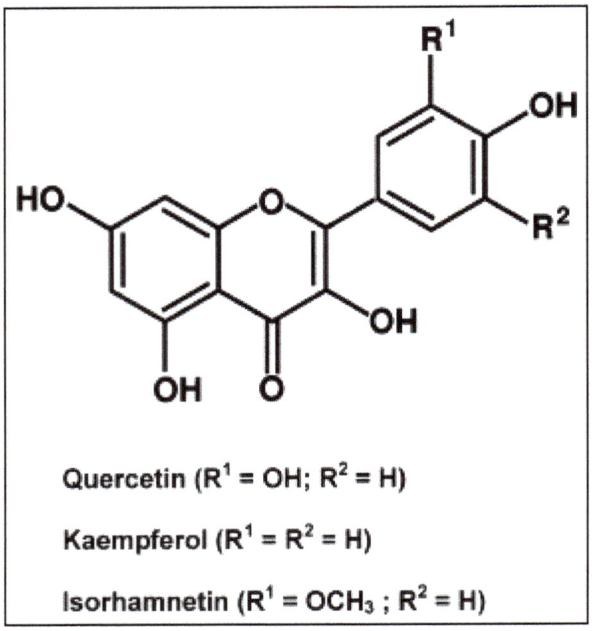

Figure 6.7: Chemical structures of flavonoids (quercetin, kaempferol and isorhamnetin) from cactus pear fruit.
Source: Kuti (2004).

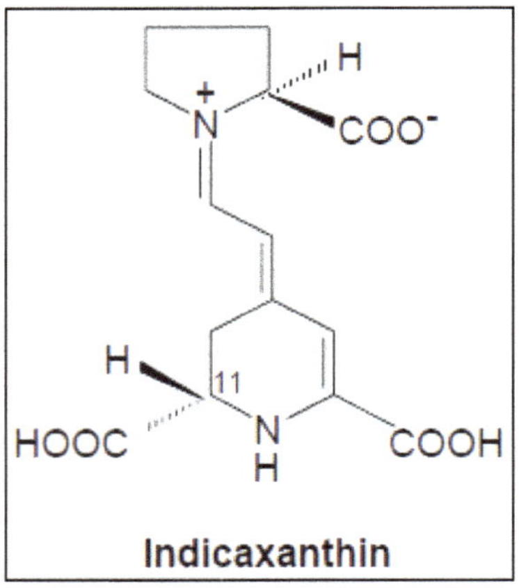

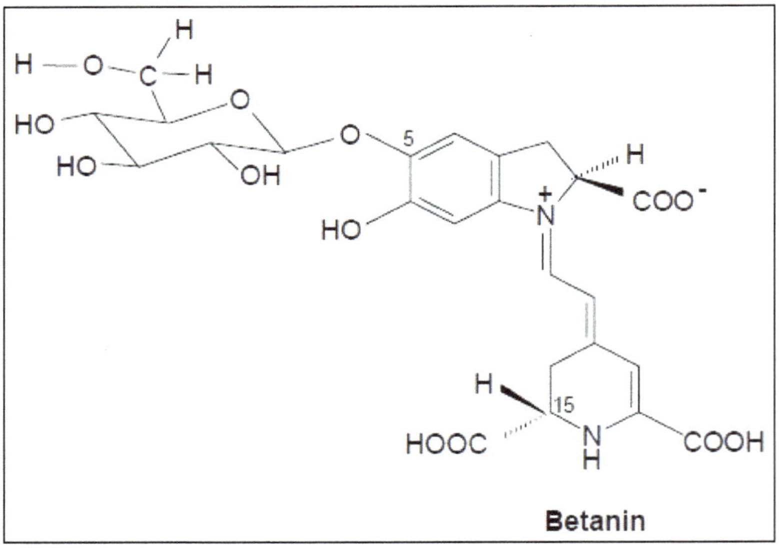

Figure 6.8a: Types of betalains in cactus pear.
Source: Moßhammer *et al.* (2005).

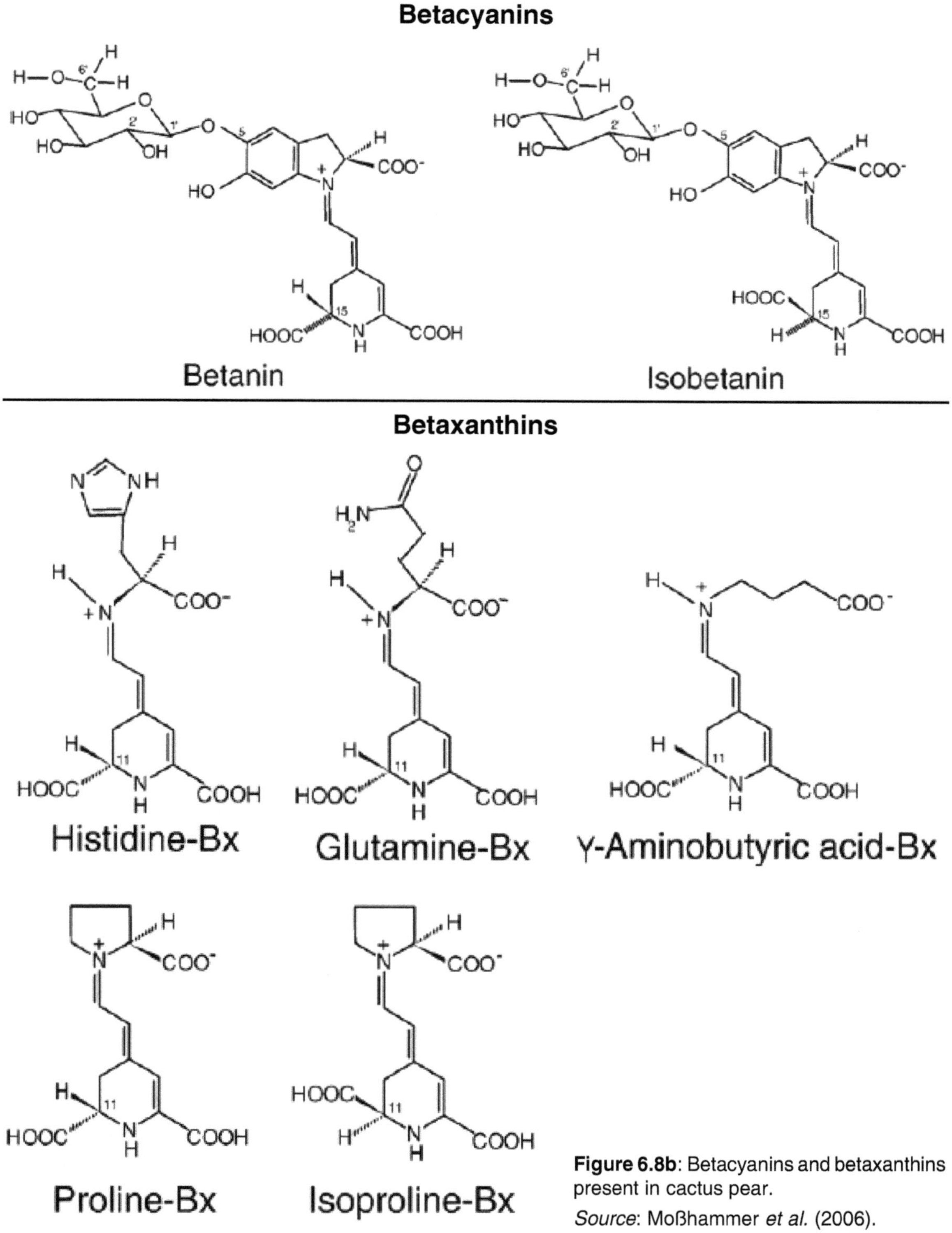

Figure 6.8b: Betacyanins and betaxanthins present in cactus pear.

Source: Moßhammer *et al.* (2006).

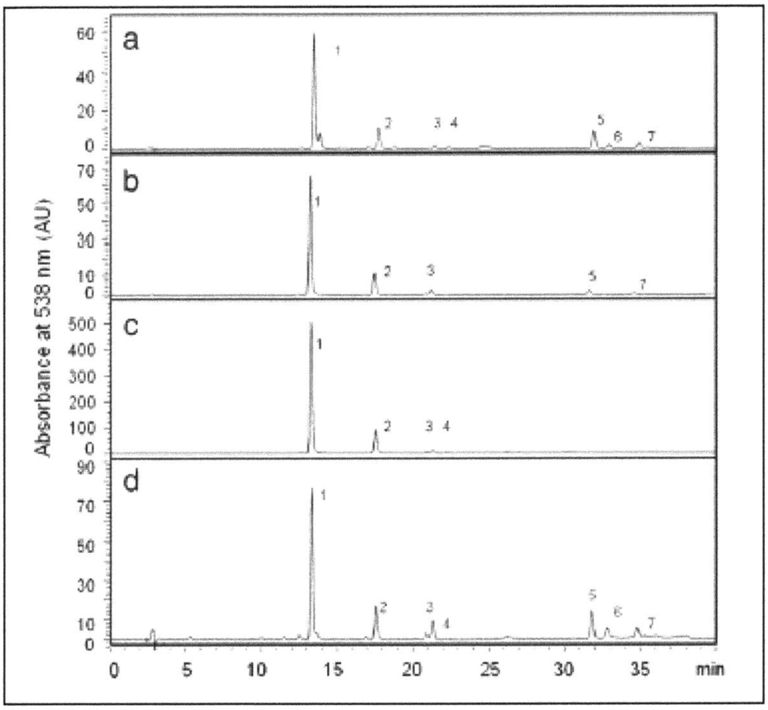

Figure 6.9: HPLC elution profile of betacyanin (λ= 538 nm) from a. Pericarp, b. Mesocarp, c. Endocarp and d. Whole fruit. Peaks 1: Betanin, 2: Isobetanin, 3: Betanidin, 4: Isobetanidin, 5: Unknown betacyanin, 6: Phylloactin, and 7: Unknown betacyanin.

Source: Osorio-Esquivel *et al.* (2011).

Cactus Pear Juice Production

The cactus pears are peeled and sliced with automatic peelers and finished with finishers (removes seeds, mesocarp fibers and residual peel) to obtain pulp (Figure 6.10). Mucilaginous material and residual mesocarp fibers from pulp are subsequently removed by decanter centrifuge. The pre-clarified juice thus obtained is acidified to remove pectic substances by enzymatic treatment with pectinase and citric acid (to avoid thermal sterilization). Thereafter juice is filtered through sheet filters and enzyme is denatured at 92°C in a tubular heat exchanger. Filtration is achieved using sheet filters. Further, pasteurization of the juice is carried out at 92°C at 80 L/h and bottled under steam injection (Moßhammer *et al.*, 2005).

The lower water activity of the concentrated juice relative to the natural juice protects former against microbial growth and provides extended shelf-life. In addition, concentrated juice has higher tinctorial strength and lower storage and transportation costs.

Concentrated Cactus Pear Juice Production Using Membrane Technology

The basic properties of membrane operations make them ideal for production of fruit juice concentrate of additive-free high quality natural fresh taste. The membrane processes are generally non-thermal and do not involve phase changes and addition of chemical additives. They are simple in

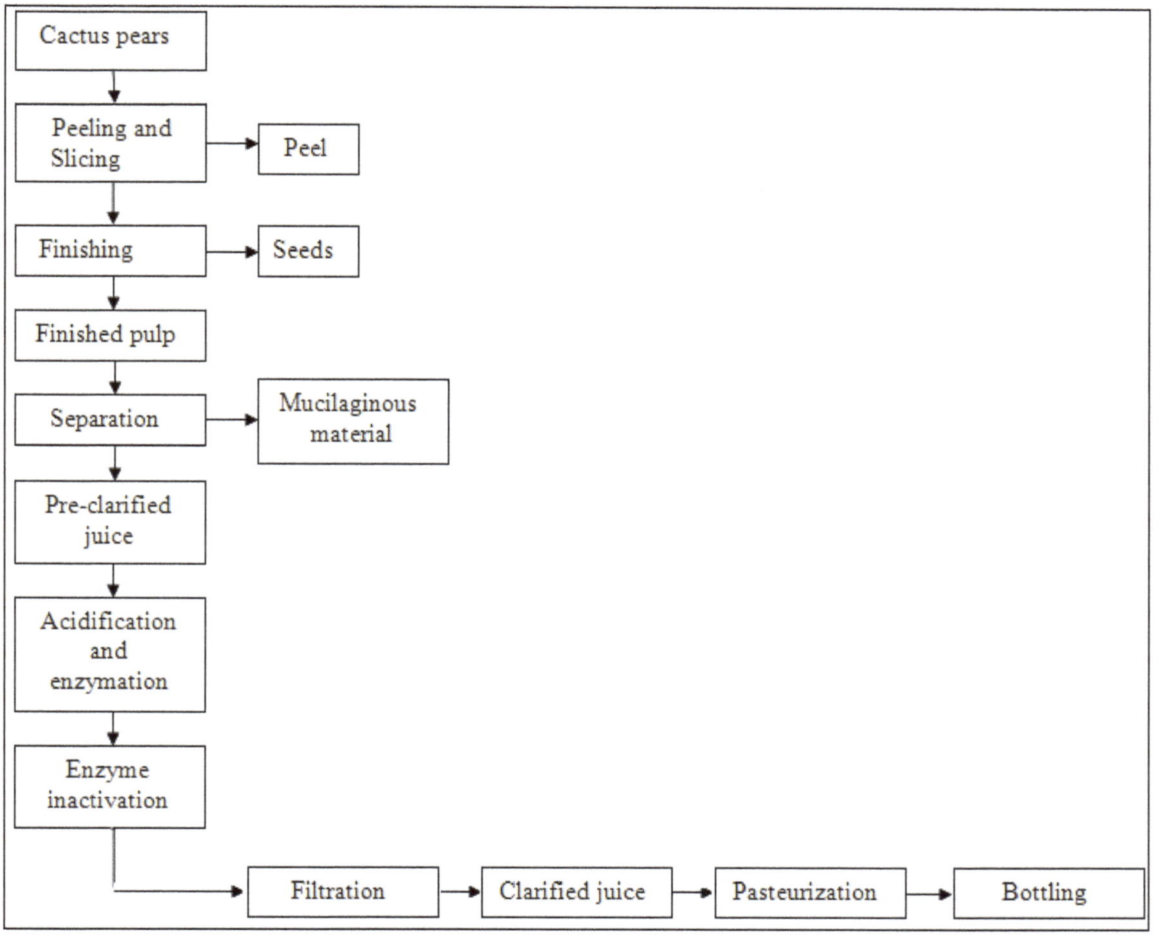

Figure 6.10: A flow diagram of cactus pear juice production.
Source: Moßhammer *et al.* (2005).

concept and operation and characterized by low energy consumption and reduced pigment decay (thus maintaining nutritional and visual appeal). Membrane based processes protect highly reactive molecules like free radical-scavengers and antioxidants in juice concentrates which may be otherwise damaged during conventional operations (such as thermal sterilization or evaporation) for juice preparation.

Membrane based processes like microfiltration (MF) and ultrafiltration (UF) represent a valid approach for recovery of nutritional and pharmaceutical compounds from cactus pear juice. The clarified juice presents physicochemical and nutritional properties comparable with those of fresh cactus pear juice except for absence of suspended solids and betalains which remain concentrated in the retentate. The permeates of both processes are a clear solution with a low protein content enriched with antioxidant compounds (polyphenols and vitamin C), sugars and minerals (Table 5.6). A remarkable reduction in protein content was observed in the permeate fractions of both processes-76.9 per cent for the MF process and 78.3 per cent for the UF process-, respectively (Sáenz *et al.*, 1993).

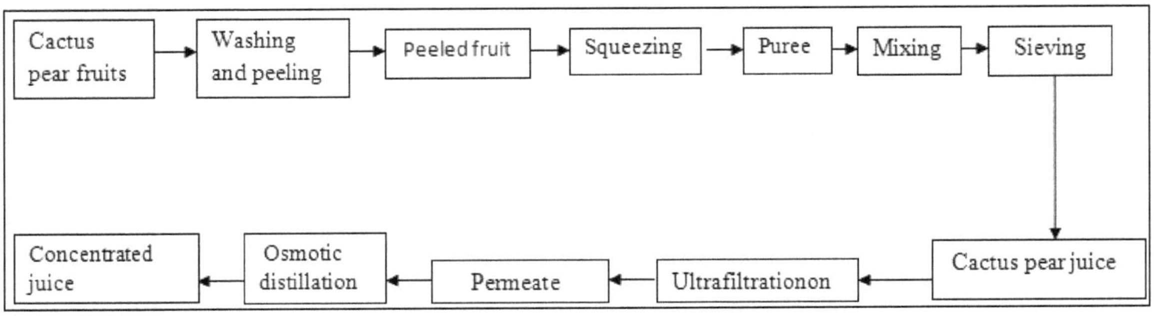

Figure 6.11: A flowchart for the production of concentrated juice from cactus pear.
*Source: Cassano *et al.* (2007).

The membrane based processes are advantageous since they produce juice concentrates without damaging the color and the herbaceous aroma (Cassano *et al.*, 2007).

The cactus pear juice concentrate is processed as shown in Figure 6.11. The fruit is washed, peeled and seeds and mesocarp fibers removed with a squeezer. The juice thus obtained is mixed with acid pectinases and filtered through hollow fiber membrane module (membrane material polysulphone at inlet pressure of 2.1 bar at 25°C, molecular weight cut off 10 kDa and feed rate: 18 L/h). The permeate is collected separately and retentate recycled to feed. Ultrafiltration is carried out to clarify the juice upto a desirable 'volume reduction factor' (ratio between initial feed volume and volume of the resulting retentate), producing a clarified juice (permeate) and fibrous concentrated pulp (retentate). The permeate stream is subjected to concentration by osmotic distillation equipped with a membrane contactor (polypropylene hollow fiber membrane, maximum pressure 4.2 Kg/cm^2, maximum temperature 40°C) and reject recycled. The juice (initial concentration 11°Brix) is concentrated in

Table 6.5: Physicochemical characteristics of cactus pear juice by membrane treatment.

Parameters	*Feed*	*Permeate*	*Retentate*	*Feed*	*Permeate*	*Retentate*
		MF			*UF*	
Proteins (g/L)	5.07	1.17	4.94	4.9	1.06	1.92
Total phenolics (as GAE) (mg/L)	546.75	546.92	545.42	545.00	511.67	552.17
Ascorbic acid (mg/L)	115.5	12.0	14.0	14.0	10.0	14.5
Flavonoids (mg/L)	212.00	134.33	237.00	231.67	135.00	260.66
TAA (mM Trolox)	3.93	3.43	4.33	4.27	3.67	3.93
Betaxanthins (mg/L)	41.12	11.38	53.57	39.29	9.02	46.93
Betacyanins (mg/L)	26.64	–	40.12	28.54	–	34.93
Suspended solids (per cent)	6.55	–	4.79	5.13	–	3.73
pH	5.71	5.58	5.54	5.57	5.59	5.50
TSS (°Brix)	9.5	9.50	10.0	9.00	9.0	10.0
Total acidity (per cent citric acid)	0.037	0.024	0.029	0.031	0.022	0.030

MF: Microfiltration; UF: Ultrafiltration; GAE: Gallic acid equivalent; TSS: Total soluble solids.

Source: Cassano *et al.* (2010).

countercurrent mode (juice on shell side and calcium chloride on the tube side), with recycle at 28°C. Osmotic distillation was conducted with slightly higher pressure on the shell side than tube side to avoid leakage of the product (Cassano *et al.*, 2007).

The type of membrane process employed for concentration determines the physicochemical characteristics of permeate and retentate streams. The constituents such as proteins, total phenolics, ascorbic acid and other flavonoids are present in various proportions depending on the type of stream (permeate or retentate) as shown in Table 6.5 (Cassano *et al.*, 2010).

The yield of each concentration step is calculated as:

$$\text{Apparent yield (per cent)} = (100/m_{JC}) \times m_{concentrate} \quad (1)$$

$$\text{Yield of respective processing step (per cent)} = (100/m_{JC}) \times (m_{concentrate} \times (TSS_{concentrate}/TSS_{JC})) \quad (2)$$

where,

m_{JC} is the mass prior to concentration, $m_{concentrate}$ is the resulting mass after concentration, TSS_{JC} total soluble solids prior to concentration, $TSS_{concentrate}$ is total soluble solids after concentration (Moßhammer *et al.*, 2006). Finally, further relevant quality parameters like haze, glucose and fructose together with organic acids, proline, formol numbers, and 5-hydroxymethylfurfural (HMF) should also be determined.

Commercial Importance

There are multiple domains of use of cactus pear and cacti plant, from food to medicines which have diverse applications and economic potential.

Healthcare and Medicine

General Medicine

The dried cladode powder of *O. ficus-indica var. Saboten* is useful in controlling gastric lesion and ulcer in rats with significant inhibition in indomethacin induced gastric lesion (Lee *et al.*, 2002). The cladodes of *O. ficus-indica* (L.) Mill are used in wound-healing and in Sicilian folk medicine as a cicatrizant (Galati *et al.*, 2003b). Cactus pear has hypoglycemic effect on type-II diabetes and hypolipidemic effect (lowers low density lipoprotein (LDL) and cholesterol levels) and checks high blood pressure and obesity. Cactus pear also confers anti-inflammatory effects and heals superficial wounds like cuts and scrapes (Fernandez *et al.*, 1992). In Chinese medicine, cactus fruit is considered a weak poison and used as medicine for treatment of inflammation and pain and used as a detoxifying agent for snake bite (Wang, 1988).

The ends of the hanging roots are given for obstinate vomiting. The roots are heated and applied as a poultice to abscesses to promote suppuration and discharge of pus. An infusion of the small branches is useful in haemoptysis. Research has shown that cacti can help to reduce the effects of excessive alcohol consumption, if used prior to consuming alcoholic beverages (Wiese *et al.*, 2004).

The root fibers in the form of decoction with or without addition of honey, resemble sarsaparilla in action and are useful in treating gonorrhea. Concentrated cactus pear fruit juice in combination with the fruit is reported to be an aphrodisiac and is also of value in spermatorrhoea and gonorrhea.

The young buds, like the milky juice, are astringent and are useful for treating dysentery and diarrhoea. The milky juice of cladode may be used in the treatment of rheumatism and lumbago. A decoction of the bark, containing about 10 per cent tannin is used as an astringent lotion in leucorrhoea (Sahelian, 2001).

Antioxidants in cactus pear interfere with oxidative stress by scavenging free radicals, chelating free catalytic metals and by acting as oxygen scavengers, contributing to health maintenance by acting against oxidative damages (Gurbachan and Felker, 1998).

Currently, extensive medical research using prickly pear cacti is focused on the management and treatment of type II diabetes, hyperlipidemias, obesity, investigation of antiviral properties and in treatment of benign prostate hypertrophy.

Cancer Treatment

Cactus pear is reported to inhibit cancer cell growth at concentrations as low as 5 per cent and also affects cell cycle with significant effect in G1 phase. In animal tumors, cactus pear extracts are found to increase annexin IV. Annexin IV is a Ca^{2+}-dependent membrane binding protein, expressed in many epithelial cancers and plays pivotal role in the early phases of apoptosis. It is observed that Native Americans have lower cancer rate in comparison to white and Afro-Americans since the former population has been incorporating cactus pear as dietary supplement for years (Zou *et al.*, 2005).

Although the mechanism of anti-cancer effect of cactus pear extracts is not yet completely understood, studies show that cactus pear affects apoptosis by altering expression of certain genes. The potential of cactus pear in inducing apoptosis in cancerous cells was significantly observed in cervical cell lines, where 25 per cent cactus extracts increased apoptosis by more than 50 per cent compared to untreated cells. In ovarian cancer cells, apoptosis induction was reported to increase by 10 per cent, while in T 24 bladder cancer cells apoptosis was reportedly 30 per cent. Arizona prickly pear is demonstrated effective in inhibiting cell growth in immortalized and cancer cell cultures *in vitro* and suppress tumor growth in nude mouse with ovarian cancer (Zou *et al.*, 2005).

Food Processing

Cactus pear fruit and juice can be consumed directly. The juice is a good source of natural sweeteners and used in salads, syrups, squashes, alcoholic drinks, jams, marmalades and in cheese production. Cactus cladodes are a new source of fiber in human diet and are incorporated into bakery products. Fruit pulp of the red cactus pear (*O. ficus-indica*) is used in ice creams, cosmetics, lotions, soaps and shampoo production. Cactus pears affect lipid peroxidation in oils and emulsion model systems and are heat stable and stabilize oils (Siriwardhana and Jeon, 2004).

The retentate in membrane extraction of cactus pear juice and/or concentrate is a fraction enriched in fibers, sugars and betalains. This can be used as a colorant, since the betalains retain their color from pH 3 to 7 where the more common anthocyanins lose their performance and degrade (Stintzing and Carle, 2004). Juice concentrates can be used for coloring foodstuff for low acid products and as an ingredient in baby foods (Cassano *et al.*, 2010). The concentrated juice may also be used to produce dehydrated cactus pulp sheet (Moreno-Álvarez *et al.*, 2009; El-Samahy *et al.*, 2007).

Environment Management

Cacti plants are employed for phytoremediation since they are efficient scavengers of pollutants, as they have very shallow and massive roots which can spread widely close to the soil surface. Cactus pear may also be used in soil fixation, soil mulching and in hedges (Sáenz, 1998). *In vitro* cultures of cacti are used to test the effect of degradation and/or biotransformation of dye precursors in plants (Golan-Goldhirsh *et al.*, 2004).

Other Uses

Mortar made with cladodes as an additive has a greater compressive strength than without additive. Cactus pear acts as a retardant of slowing the setting time in mortar manufacturing processes. Cactus pear juice is used as a binding agent in plaster and whitewash (Hernández-Zaragoza *et al.*, 2008). Cactus pear is also used to produce dehydrated pulp sheet (Moreno-Álvarez *et al.*, 2009).

Conclusions

The plant *Opuntia ficus-indica* (L.) Miller, is a potential future crop owing to its diverse useful characteristics and a fewer growth requirements. The fruit (cactus pear) has potential nutritive value and can be consumed and used in various forms. The growing demand and production of the same in Mexican countries and parts of Europe are indicators of its quality and value. The cladodes of the plant also offer multiple usages. Cactus pear can act as a boon for the underdeveloped countries around the globe providing nutrition at an affordable cost. Low resource requirements for growth would allow its plantation across the geographical terrain, especially in the arid areas which are generally unproductive. Thus, cactus pear is expected to be a fast growing commercial crop around the world in the years to come.

References

Angulo-Bejarano PI and Paredés-Lopez O. 2011. Development of a regeneration protocol through indirect organogenesis in prickly pear cactus (*Opuntia ficus-indica* (L.) Mill), Scientia Horticulturae 128: 283–288.

Anorve Morga J, Aquino EN and Mercado Silva E. 2006. Effect of controlled atmosphere on the preservation of minimally processed cactus pears. Acta Horticulturae 728: 211–216.

Ayadi MA, Abdelmaksoud W, Ennouri M and Attia H. 2009. Cladodes from *Opuntia ficus-indica* as a source of dietary fiber: effect on dough characteristics and cake making. Industrial Crops and Products 30: 40-47.

Barbera G, Inglese P and Pimienta-Barrios E (Eds). 1995. Agro-ecology, cultivation and uses of cactus pear. FAO Plant Production and Protection Paper 132.

Brutsch MO. 1988. The role of prickly pear in less developed agriculture. Report of the First National Symposium on Fruit Production from Spineless Prickly Pear, University of Pretoria, South Africa.

Cassano A, Conidi C, Timpone R, D'Avella M. and Drioli E. 2007. A membrane based process for the clarification and the concentration of cactus pear juice. Journal of Food Engineering 80: 914-921.

Cassano A, Conidi C and Drioli E. 2010. Physico-chemical parameters of cactus pear (*Opuntia ficus-indica*) juice clarified by microfiltration and ultrafiltration processes. Desalination 250: 1101-1104.

Cruz ARR, Soares EL, Campos FAP and Aragao FJL. 2009. Biolistic-mediated genetic *transformation* of prickly-pear cactus (*Opuntia ficus-indica* Mill.). Acta Horticulture 811: 255–257.

El-Samahy SK, Abd El-Hady EA, Habiba RA and Moussa-Ayoub TE. 2007. Cactus pear sheet and pasteurized and sterilized cactus pear juices. Journal of the Professional Association for Cactus Development 9: 148–164.

Fernandez ML, Lin EC, Trejo A and Mcnamara DJ. 1992. Prickly pear (*Opuntia sp.*) pectin reverses low-density lipoprotein receptor suppression induced by a hypercholesterolemic diet in guinea pigs. Journal of Nutrition 122: 2330–2340.

Galati E, Mondello M, Giuffrida D, Dugo G, Miceli N and Pergolizzi S. 2003a. Chemical characterization and biological effects of Sicilian *Opuntia ficus-indica* (L.) Mill. Fruit juice: Antioxidant and antiulcerogenic activity. Journal of Agricultural and Food Chemistry 51(17): 4903–4908.

Galati E, Mondello M, Monforte MT, Galluzzo M, Miceli N and Tripodo MM. 2003b. Effect of *Opuntia ficus-indica* (L.) Mill. cladodes in the wound-healing process. Journal of the Professional Association for Cactus Development 5, 1–16.

Golan-Goldhirsh A, Barazani O, Nepovim A, Soudek P, Smrcek S, Dufkova L, Krenkova S, Yrjala K, Schröder P and Vanek T. 2004. Plant response to heavy metals and organic pollutants in cell culture and at whole plant level. Journal of soils and sediments 4: 130–140.

Gurbachan S and Felker P. 1998. Cactus: new world foods. Indian Horticulture 43(1): 29–31.

Hernández-Zaragoza JB, Coronado-Márquez A, López-Lara T and Horta- Rangel J. 2008. Mortar improvement using nopal additive. Journal of the Professional Association for Cactus Development 10: 120–125.

Joubert E. 1993. Processing of the fruit of five prickly pear cultivars grown in South Africa. International Journal of Food Science and Technology 28: 377–387.

Kossori RL, Villaume C, Boustani EE, Sauvaire Y and Méjean L. 1998. Composition of pulp, skin and seeds of prickly pears fruit (*Opuntia ficus-indica* sp.) Plant Foods for Human Nutrition 52: 263–270.

Kuti JO. 2004. Antioxidant compounds from four *Opuntia* cactus pear fruit varities. Food Chemistry 85: 527-533.

Livrea MA and Tesoriere L. 2006. Health benefits and bioactive components of fruits from *Opuntia ficus-indica* [L.] Mill. Journal of the Professional Association for Cactus Development 8: 73-90.

Lee EB, Hyun JE, Li DW and Moon YI. 2002. Effect of *Opuntia ficusindica var. Saboten* stem on gastric damages in rats. Archives of Pharmacal Research 25: 67–70.

Magloire JF, Konarski P, Zou D, Stintzing FC and Zou C. 2006. Nutritional and medicinal uses of cactus pear (*Opuntia* spp.) cladodes and fruits. Frontiers in Bioscience 11: 2574"2589.

Moßhammer MR, Stintzing FC and Carle R. 2005. Development of a process for the production of a betalain based colouring food stuff from cactus pear. Innovative Food Science and Emerging Technologies 6: 221-231.

Moßhammer MR, Stintzing FC and Carle R. 2006. Evaluation of different methods for the production of juice concentrates and fruit powders from cactus pear. Innovative Food Science and Emerging Technologies 7: 275-287.

Moreno-Álvarez MJ, Hernández R, Belén-Camacho DR, Medina-Martínez CA, Ojeda-Escalona CE and García-Pantaleón DM. 2009. Making of bakery products using composite flours: wheat and cactus pear *(Opuntia boldinghii)* stems (cladodes). Journal of the Professional Association for Cactus Development 11: 78–87.

Moussa-Ayoub TE, El-Samahy SK, Rohn S and Kroh LW. 2011. Flavonols, betacyanins content and antioxidant activity of cactus *Opuntia macorhiza* fruits. Food Research International. doi: 10.1016/ j. foodres.2011.02.014

Osorio-Esquivel O., Moreno AO, Alvarez VB, Alvarez LD and Guisti MM. 2011. Phenolics, betacyanins and antioxidant activity in *Opuntia joconstle* fruits. Food Research International. doi: 10.1016/j. foodres.2011.02.011.

Ramdan MF and Mörsel J. 2003. Oil cactus pear (*Opuntia ficus-indica* L.). Food Chemistry 82: 339-345.

Rodríguez G, De Lira C, Hernández B, Cornejo V, Palacios F and Rojas M. 2007. Physicochemical characterization of nopal pads (*Opuntia ficus-indica*) and dry vacuum nopal powders as a function of the maturation. Plant Foods for Human Nutrition 62: 107-112.

Sáenz C, Sepúlveda E, Araya E and Calvo C. 1993. Colour changes in concentrated juices of prickly pear (*Opuntia ficus-indica*) during storage at different temperatures. Lebensmittel-Wissenschaft und Technologie 26: 417–421.

Sáenz C. 1998. Cladodes: A source of dietary fiber. Journal of the Professional Association for Cactus Development 2: 117–123.

Sáenz C. 2000. Processing technologies: an alternative for cactus pear (*Opuntia* spp.) fruits and cladodes. Journal of Arid Environments 46: 209–225.

Sahelian R. 2001. An anti-inflammatory principle from prickly pear cactus. Fitoterapia 72(3): 288–290.

Salim N, Abdelwahed C, Rabah C and Ahcene B. 2009. Chemical composition of *Ficus indica* (L.) fruit-short communication. African Journal of Biotechnology 8(8): 1623-1624.

Siriwardhana N and Jeon Y. 2004. Antioxidative effect of cactus pear fruit (*Opuntia ficus-indica*) extract on lipid peroxidation inhibition in oils and emulsion model systems. European Food Research and Technology 219: 369–376.

Stintzing FC, Schieber A and Carle R. 2000. Cactus pear, a promising component of functional food. Obst, Gemu¨se und Kartoffelverarbeitung 85(1): 40–47.

Stintzing FC, Schieber A and Carle R. 2001. Phytochemical and nutritional significance of cactus pear. European Food Research and Technology 212: 396–407.

Stintzing FC and Carle R. 2004. Functional properties of anthocyanins and betalains in plants, food, and in human nutrition. Trends in Food Science and Technology 15: 19–38.

Sudzuki F. 1995. Anatomy and morphology. Page 213. (In) Agroecology, Cultivation and Uses of Cactus Pear. (Barbera G, Inglese P, Pimienta-Barrios E, Eds.). Plant Production and Protection Paper 132.

Wang PZ. 1988. *Chinese Medicine Surgery*. Ancient Chinese Medicine Press, 164-183.

Walker RP, Famiani F, Baldicchi A, Cruz-Castillo JG and Inglese P. 2011. Changes in enzymes involved in photosynthesis and other metabolic processes in the fruit of *Opuntia ficus-indica* during growth and ripening. Scientia Horticulturae 128: 213–219.

Wiese J, McPherson S, Odden MC and Shlipak MG. 2004. Effect of *Opuntia ficus-indica* on symptoms of the alcohol hangover. Archives of Internal Medicine 164: 1334–1340.

Yahya EM and Mondragon-Jacobo, C. 2011. Nutritional components and anti-oxidant capacity of ten cultivars and lines of cactus pear fruit (*Opuntia* spp.). Food Research International doi: 10.1016/ j. foodres.2011.02.042.

Zou D, Brewer M, Garcia F, Feugang JM, Wang J, Zang R, Liu H and Zou C. 2005. Cactus pear: a natural product in cancer chemoprevention, Nutritional Journal : 10.1186/1475-2891-4-25.

2014, Future Crops Vol. 2 *Pages 153–177*
Editor: Professor K.V. Peter
Published by: DAYA PUBLISHING HOUSE, NEW DELHI

Chapter 7
Camelina

Ankur Agarwal*, Anuja Kumari, M C Arya and Zakwan Ahmed
Defence Institute of Bio-Energy Research (DIBER)
DRDO, Ministry of Defence, Field Station, Pithorgarh, Uttarakhand – 262 501, India
**E-mail: darl_ankur@rediffmail.com*

Camelina [*Camelina sativa* (L.) Crtz.] belongs to family Cruciferae (Brassicaceae), which includes mustard, rapeseed, turnip, broccoli, cabbage, rutabaga, cauliflower, brussels sprout and several other crops, lesser known (Al-Shaehbaz, 1987). Camelina is grown in Europe for centuries and was an important crop during Iron and Bronze ages (Schultze-Motel,1979). It is also known as German sesame, Siberian oilseed, False flax, Dutch flax, Linseed dodder and Gold-of-pleasure. Other names are Oljedodre (Norway), Saed-Dodder (Denmark), Ruistankio (Finland), Saat-leindotter (Germany), *Cameline cultivee* (France) and *Camelina pilosa* (Spain).

It is an under-exploited low input oilseed crop (Putnam *et al.*, 1993) and was the favourite one used in oil lamps and edible purpose from the Roman empire to the discovery of gas and electricity. The crop was widely grown in Europe and Russia up to early 1940's with some production lasting up to the 1950's (Crowley and Frohlich, 1998) and later replaced by modern oilseed rape due to their widespread use (Hebard, 1998). It is reported that camelina with high content of unsaturated fatty acids (approx. 90 per cent) was more difficult and expensive to hydrogenate than rapeseed oil and this led to its decline (Crowley and Frohlich, 1998).

Recent interest in this crop is mainly due to search of new crops with high oil content for industrial application (Bramm, 1993; Schuster and Friedt, 1995). Although linolenic acid content of camelina (35-40 per cent) is less than that of linseed (60-70 per cent) but it places this oil crop in the category of drying oils and too high to penetrate the edible oil market (Downeey, 1971). The most unusual characteristic of camelina oil is its relatively high content of cholesterol and relatively large amount of another sterol *i.e.* brassicasterol. Camelina oil has also exhibited a high eicosenoic acid content than other vegetable oil. The usefulness or disadvantage of eicosenoic acid is not clear; however it may pose a hurdle to obtain food approval (Leonard, 1998). Being a low input with frost and drought tolerance and being an early maturity crop with high linolenic acid (35-40 per cent) and cholesterol (188ppm), camelina serves as a good raw material for biofuel (Putnam *et al.*, 1993; Abramovic and Abram, 2005;

Frohlich and Rice, 2005; Pilgeram *et al.*, 2007; Lu, 2008; Perkins, 2010; Agarwal *et al.*, 2010). Camelina oil being rich in Omega-3 fatty acids (38 per cent of the total fatty acid content) also offers an opportunity to meet the growing demand for high quality fatty acids (Zubr, 1997) after removal of non-edible fractions.

Biology, Taxonomy and Physiology

Scientific Classification

Kingdom: Plantae
Division: Magnoliophyta
Class: Magnoliopsida
Order: Brassicales
Family: Brassicaceae
Genus: Camelina
Species: 1. *C. alyssum* (Mill.) Thell.
2. *C. microcarpa* Andrz. ex DC. (Known as little pod false flax)
3. *C. rumelica* Velen (Known as graceful false flax: n=13)
4. *C. sativa* Crantz.(Known as graceful false flax: 2n=40)

Camelina occurs in two different forms, the summer annual and winter hardy annual or biennial form. The major morphological differences between these two forms are the shape of leaves, pods, and seeds and some other specific traits (Zubr, 1997). *Camelina microcarpa* and *C. sativa* are two major species of camelina. Two subspecies of *C. sativa*, winter annual *C. sativa* ssp. *pilosa* and summer annual *Camelina sativa* are the most important species grown in agricultural systems (Plessers *et al.*, 1962). Shu (2001) reported predominance of these two species of camelina in China also.

C. sativa is an annual herbaceous oil-seed species that exhibits perfect sexual reproduction. It is a short season crop that matures in 85-100 days. The plant grows 0.3 to 1.0 m tall and have branched stems which become woody as they mature (Ehrensing and Guy, 2008). Initially, camelina grows as a rosette (Zubr, 1997), then bolts to heights of up to 90 cm by the end of the growing season (McVay and Lamb 2008; Putnam *et al.*, 1993). The leaves are lanceolate, arrow shaped, 5-8 cm long with slight pubescence and light green in colour. As the plant matures, it produces a flowering stalk and the lower leaves desiccate. *C. sativa* produces numerous four- petaled yellow flowers aggregated on a raceme, terminally (USFS-ILPIN, 2002). The pods are 6-14 mm long and 0.7-2.5 mm in diameter, borne in pear shaped capsules superficially resembling the bolls of flax (Putnam *et al.*, 1993) and contains small seed (0.7mm × 1.5mm), pale yellow-brown, oblong and rough with a ridged surface. Average 15 seeds per pod are formed from self pollination with a 1000-seed-weight of 0.8-1.8 gm (Zubr, 1997). Fruiting pedicels are slender and divaricate. Sepals oblong or ovate, erect, base of lateral pair not saccate. Petals are yellow or rarely white, spatulate, longer than sepals with obtuse apex. Stamens are 6, in 3 pairs of unequal length; filaments not dilated at base. Anthers are ovate or oblong, obtuse at apex. Nectar glands are 4; 1 on each side of lateral stamen; median glands absent. Ovules 8–25 per ovary. Fruit dehiscent silicles or rarely siliques, obpyriform, obovoid, or rarely linear, shortly stipitate; valves leathery, often with a prominent midvein, smooth, margins of 2 valves flattened and connate, apex abruptly caudate and extending 1–2 mm into style; replum rounded, concealed by connate margins of valves; septum complete; style 1.5–4 mm; stigma capitate, entire. Seeds biseriate or rarely uniseriate, wingless or narrowly margined, oblong, plump or slightly flattened; seed coat minutely reticulate, copiously mucilaginous when wetted; cotyledons incumbent or rarely accumbent (Shu, 2001).

In general, *C. microcarpa* grows shorter (80cm) than *C. sativa* (70-100cm). While *C. microcarpa* is densely hairy on the lower leaves, *C. sativa* is glaborous or only slightly hairy on the lower leaves. Also, the pedicels of *C. microcarpa* are rarely greater than 17 mm long, while the pedicels of *C. sativa* usually longer than 17 mm. The seeds of *C. microcarpa* are about half the size of *C. sativa*, and the pods are 2-4 mm smaller (Dorn 1984; Gilkey 1957).

Environmental Adaptation

Camelina is one of the oldest oil crops in the temperate regions (Agegnehu and Honermeier, 1997). In the late 1930's some varieties of this crop were grown on poor sandy soil in most parts of European countries like Russia, Poland, The Netherlands, Belgium etc. mainly to replace winter-killed rape (Plessers *et al.*, 1962). This crop is adapted to Eastern Europe and south west Asia where wild weedy forms survive and appears very adaptable to climate and soil type. It grows well in cool and semi-arid regions which have a relatively short growing season (it matures 21 days earlier than flaxseed). Camelina grows well on relatively poor and saline soils and can survive frost and freeze-thaw cycle after emergence (Putnam *et al.*, 1993).

Camelina is a short duration crop (85-100 days) and germinates at low temperature. No seedling damage has been seen at temperature as low as -11°C in Montana (Ehrensing and Guy, 2008). According to Russo *et al.* (2010) in a germination trial, 100 per cent germination was observed in camelina line NEB C-1 in the temperature range of 4 to 27°C. At 32°C, germination was less than 80 per cent. Seed germination at 16°C was the maximum and quick (by 2 days). Camelina is often grown on marginal lands and responds well under drought stress conditions and better suited to low rainfall regions than other oilseeds. The crop has low fertility requirements and exhibits allelopathic properties (Lovet and Duffield, 1981).

Camelina is listed as being adapted to the flax-growing regions of the Northern Midwest (Minnesota, North Dakota, South Dakota) and considered as minor weed in flax with no seed dormancy (Robinson, 1987). It is adapted to a wide range of climatic conditions, as demonstrated by its long history of cultivation across the European continent and its more recent favourable performance in North America, Australia and Canada (Gugel and Falk, 2006). Today Camelina is being grown successfully in Slovenia, Ukraine, China, Finland, Germany, Austria, Montana and Canada.

Origin and History

Camelina sativa originated in the Mediterranean to Central Asia (Putnam *et al.*, 1993) and is native to South and South-West Asian countries (Agegnehu and Honermerier, 1997). According to Budin *et al.* (1995) camelina crop originated in Germany about 600 B. C. and later spread to Central Europe and is native from Finland to Romania and east to the Ural Mountains (Ehrensing and Guy, 2008)

Camelina seed was found in the stomach of Tollund man, a 4th century BC mummy recovered from peat blog in Denmark (Glob, 1969). Anthropologist postulates that the man's last meal had been a soup made from vegetables and seeds including barley, linseed, camelina, knotweed, bristle grass and chamomile (Pilgeram *et al.*, 2007). Archeological evidence suggests that it was grown in Europe for atleast 3000 years. The earliest sites include the Neolithic levels at Auvernier, Switzerland (dated to the second millennium BC), the Chalcolithic level at Pefkafia (dated to the third millennium BC), and Sucidava- Celei *i.e.* circa 2200 BC (Jones and Valamoti, 2005). *C. sativa*, *C. microcarpa* and *C. linicola* existence dates back to Bronze (1500-400 BC) and Iron (400 BC–500 AD) ages of Scandinavia and western Europe where it was cultivated for human consumption (Zubr, 1997 and 2003). As evident from carbonized seed, its cultivation was shown to occur in regions surrounding the North Sea during the Bronze Age. Camelina monocultures occurred in the Rhine River Valley as early as 600 BC.

It is suggested that camelina probably spread in mixtures with flax and as monocultures, similarly to small grains, which also often spread as crop mixtures. It was cultivated in antiquity from Rome to Southeastern Europe and the Southwestern Asian steppes (Knorzer, 1978). From the beginning of 20th century up to the 1930's *C. sativa* was grown sporadically in France, Belgium, Holland, the Balkans and Russia (Frohlich and Rice, 2005; Zubr, 1997, 2003). After World War II (mid 20th century) its production declined, but it co-evolved as weed with flax thus, known as false flax (Zubr, 1997). According to reports camelina production lasted up to 1950's and it was replaced with the introduction and widespread use of oilseed rape (Hebard, 1998).

Although camelina has a long history of cultivation, it is currently under-exploited. The recent focus on this crop was inspired by search for new sources of oil and essential fatty acids, especially vegetable sources of omega-3 fatty acids (Karvonen *et al.*, 2002) and its use as a source of biofuel (Cardone *et al.*, 2003; Francis and Campbell 2003).

Nutritional Value

Triglyceride fats and oils fulfilled the needs of people for centuries. Many oilseed crops have been in cultivation for their nutritional and industrial use. Camelina has the reputation of the oldest oil crop in the temperate region as evident by its history of domestication. Its oil was used for food, medicinal use and lamp oil as early as in Bronze Age (Ehrensing and Guy, 2008). Recent interest in this crop is mainly due to search of new crops with high oil content for industrial application (Bramm, 1993; Schuster and Friedt, 1995). Although linolenic acid content of camelina (35-40 per cent) is less than that of linseed (60-70 per cent) but it places this oil crop in the category of drying oils and too high to penetrate the edible oil market (Downeey, 1971).

Camelina has seed oil content ranging from 290 to 430 g/kg (Putnam *et al.*, 1993, Vollmann *et al.*, 1996 and Zubr, 1997) comprising important fatty acids like oleic (C18: I), linoleic (C18: 2), alpha linolenic (C18: 3) and eicosenoic (C20: 1) (Gilbertson *et al.*, 2007). Camelina oil has a unique fatty acid pattern and is characterized by a linolenic acid content of 30-40 per cent, eicosenic acid around 15 per cent and erucic acid less than 4 per cent (Seehuber 1984, Marquard and Kuhlmann 1986, Budin *et al.*, 1995), which suggests utilization of camelina oil as drying oil with environmental safe painting and coating applications (Luehs and Friedt, 1993). Lange *et al.* (1995) reported low content of glucosinolates in camelina as compared to other brassica crop making its meal more palatable to cattle. Feed augmented with camelina meal increased Omega-3 fatty acid content in meat and eggs (Peiretti *et al.*, 2007; Rokka *et al.*, 2002).

Plant oils generally contain significant amount of plant sterols, particularly campesterol, stigmasterol and/or sitosterol, usually in the range of 0.2 to 2 per cent (ww) of total sterols in the oil with very little cholesterol (a few ppm) (Itoh *et al.*, 1973; Kochhar, 1983; Gunstone *et al.*, 1986). A few of the field crops oil (peanut, cotton seed, linseed and rapeseed) and a few tropical tree crops (palm, coconut and cocoa) have modest amounts of cholesterol (upto ~ 60ppm), although concentration varies depending on the source and method of oil purification. The most unusual characteristic of camelina oil is its relatively high content of cholesterol and relatively large amount of another sterol *i.e.* brassicasterol. The major sterols identified in camelina oil include cholesterol (188 ppm), brassicasterol

(133 ppm), campesterol (893ppm), stigmasterol (103 ppm), sitosterol (1,884 ppm), and Δ^5- avenasterol (393 ppm) (Shukla *et al.*, 2002). Unlike many other crucifers, camelina contains low amount of erucic acid (2.5 per cent) almost equal to that found in low-erucic acid rapeseed oil (~2 per cent) (Gunstone, 1996). Camelina oil has also exhibited a high eicosenoic acid content than other vegetable oil. The usefulness or disadvantage of eicosenoic acid is not clear; however it may pose a hurdle to obtain food approval (Leonard, 1998). Recently, Schwartz *et al.* (2008) reported a cholesterol level of 350 mg/kg in camelina oil and a total sterol level of 5110 mg/kg, which is predominantly ß-sitosterol. Polyunsaturated to monounsaturated fatty acid ratio of camelina oil is similar to soybean oil, higher than canola oil and lower than flax seed and traditional sunflower oil exhibiting its less susceptibility to auto-oxidation as monounsaturated fatty acids are much less susceptible to oxidation than the PUFA (Shukla *et al.*, 2002). Fatty acids in camelina oil are primarily unsaturated with only about 12 per cent being saturated. About 54 per cent of fatty acids are polyunsaturated and 34 per cent are monosaturated as presented in Table 7.1 (Putnam *et al.*, 1993).

Table 7.1: Fatty acid composition of camelina compared with other oilseeds (Putnam *et al.*, 1993).

Fatty Acid	*Fatty Acid Content (Per cent of oil)*					
	Camelina	*Flax*	*Canola*	*Soybean*	*Sunflower*	*Crambe*
Palmitic (16: 0)	7.80	5.12	6.19	10.44	6.05	2.41
Stearic (18: 0)	2.96	4.56	0.0	3.95	3.83	0.40
Oleic (18: 1)	16.77	24.27	61.33	27.17	17.36	18.36
Linoleic (18: 2)	23.08	16.25	21.55	45.49	69.26	10.67
Linolenic (18: 3)	31.20	45.12	6.55	7.16	0.0	5.09
Arachidic (20: 0)	0.0	0.0	0.0	0.0	0.0	0.50
Eicosenoic (20: 1)	11.99	0.0	0.0	0.0	0.0	2.56
Erucic (22: 1)	2.80	0.88	0.0	0.0	0.0	54.00
Other FA	3.40	3.80	4.38	5.79	3.5	6.01

Marquard and Kuhlmann (1986) reported oil content in the range of 37-41 per cent and seed protein content 23-30 per cent. Protein content and elemental composition of camelina are similar to flax (*Linum usitatissimum* L.) with the exception of higher sulphur content (Robinson, 1987). Camelina meal is comparable to soybean meal, containing 45-47 per cent crude protein and 10-11 per cent fiber (Korsrud *et al.*, 1978). Zero to trace levels of volatile isothiocyanates was found in camelina meal (Korsrud *et al.*, 1978; Sang and Salisbury, 1987) compared to rapeseed meal, which contains higher level of glucosinolates (Lange *et al.*, 1995).

The camelina oil is also very rich in natural antioxidants like tocopherols making it stable, resistant to oxidation and rancidity. Vitamin E content of oil is approximately 100mg/100g. Schwartz *et al.* (2008) found tocopherol content of oil 780 mg/kg. These compounds have favourable nutritional implications and beneficial physiological effects. Vitamin E is a fat soluble vitamin and exists in eight different forms. Each form has its own biological activity in the body. Alpha-tocopherol is the most active form of vitamin E in humans and is a powerful biological antioxidant to protect cells against the effects of free radicals, which are potentially damaging by products of the body's metabolism. Free radicals can cause cell damage that may contribute to development of cardiovascular disease and cancer. Free radicals are produced by normal human metabolism and increase by smoking, alcohol, exposure to heavy metals and radiations.

Omega-3 fatty acids are important for proper growth, brain development and immune function in children, improving heart health by raising HDL cholesterol, helps in reduction of inflammation and prevention of certain chronic diseases like heart diseases and arthritis (Ruxton *et al.*, 2007). Omega-3 fatty acids are highly concentrated in the brain and appear to be particularly important for cognitive and behavioral function, which otherwise lead to risk for developing vision and nerve problems in infants. Francis and Campbell (2003) also reported comparative oil content among a few oil seed crops (Table 7.2), which revealed that camelina is the only crop having eicosenoic acid (16 per cent of the total fatty acid composition).

Table 7.2: Percentage oil content and quality of a few oil seed crops.

Fatty Acid	*Camelina*	*Canola*	*Borage*	*Evening Primrose*	*Crambe*	*Guizotia*	*Matthiola*	*Linseed*
Saturated	8	7	17	10	5	13	11	9
Oleic	16.5	60	19	10.5	14.5	6	14	18
Linolenic	16.5	21	36	65	8	80	13	18.5
γ-linolenic	0.1	–	22	9	–	–	–	–
α-linolenic	36	10	–	–	7	0.5	60	54
Eicosenoic	16	–	–	–	–	–	–	–
Erucic	3	1	3	–	62	0.5	–	0.5
Oil content	38	40	35	22	32	38	22	38

Cold pressed oil extractions of camelina (from 1 kg seed) have revealed 81.1 per cent recovery of oil during first press yielding 300ml oil from 1 kg seed (Francis and Campbell, 2003) as presented in Table 7.3.

Table 7.3: Cold pressed oil extractions- first press from 1 kg of seed.

Species	*Oil Extracted (ml)*	*Percent Extracted (First Press)*	*Oil Colour**
Canola	350	83.3	7A yellow
Niger	310	86.0	6D pale yellow
Camelina	300	81.1	5D pale strain
Linseed	275	88.0	8A yellow
Crambe (in hull)	225	72.5	7B yellow

* New Gardens Horticultural flower colour standard.

Applications and Uses

Camelina oil is the main product from camelina seeds and the average oil yield from seeds is reported 30-40 per cent (Budin *et al.*, 1995, Rode 2002, Zubr 2003) although oil content up to 480 g/kg was reported (Vollmann *et al.*, 2007). Camelina oil is golden yellow in colour with mild nutty and characteristic mustard aroma. Some physical properties of camelina oil are as under (Table 7.4) (Abramovic and Abram, 2005; Anon, 2002).

Recent research work is focused on the application of camelina oil as biodiesel. Rice *et al.* (1998) revealed that ester yields for camelina oil were higher than that of waste cooking oil. The fuel properties of the methyl ester were also with in specification with the exception of cold filter plug point (CFPP). This group further suggested that a blend with roughly equal amount of camelina ester and mineral diesel would have adequate low temperature properties for Irish winter conditions (Table 7.5).

Table 7.4: Some physical properties of camelina oil.

Characteristics	*Value*
Refractive index (at 25ºC)	1.4756
Density (at 25ºC) (g/cc)	0.92
Iodine number (g/100 g oil)	105
Saponification Value (mg KOH/g oil)	187.8
Peroxidase value (meq/kg oil)	2.0_{max}
Acid value (mg KOH/g oil)	0.2_{max}

Table 7.5: Low temperature properties of blends of camelina ester and mineral diesel oil (Rice *et al.*, 1998).

Camelina Esters (per cent)	*Mineral Diesel (per cent)*	*Cloud Point (ºC)*	*CFPP (ºC)*	*Pour Point (ºC)*
100	0	+3	–3	–4
80	20	+3	–7	–6
60	40	+3	–9	–9
40	60	+3	–11	–12
20	80	+3	–13	<-18
0	100	+3	–15	<-21

Camelina oil can be used in edible products after removal of non-edible fractions. Being rich in Omega-3 fatty acids and low in saturated fatty acids, its demand is high in some European countries especially for baking and confectionary industry. Camelina oil is being marketed in Europe in salad dressing and also used in cosmetics, skin care products, soaps and soft detergents. The oil is used successfully as an adjuvant in agricultural spraying applications. It has a range of possible industrial application from environmentally safe paintings, coatings varnishes, pharmaceuticals and low emission biodiesel fuels (Bonjean and Le Goffic, 1998; Bernardo *et al.*, 2003; Ehrensing and Guy, 2008).

Extraction of oil from seeds through cold pressing results in recovery of main by-products *i.e.* oil cake/meal. Oil cake consists of 5-10 per cent residual oil, 45 per cent crude protein, 13 per cent fiber, 5 per cent of minerals and some minor levels of vitamins (Zubr, 1997) including ash (6.6 per cent) and non-nitrogenous matter (27.2 per cent) as reported by Mikersch (1952). Oil cakes being rich in protein and minerals are important to be used as nutritive supplement in animal feed formulations. Total 18 amino acids including 9 essential amino acids are found in cake (Zubr, 2003). High Trypsin Inhibitor Activity (TIA) (12-28 mg/g) is reported in camelina but TIA can be reduced by heating as it was reduced in soybean meal from 22.1 to 2.1 mg/g by heating for 40 minutes (Zubr and Matthaus, 2002). A similar cooking process like heating, baking etc could help in reducing the TIA in camelina seeds and meals also. Camelina meal is low in glucosinolates and therefore good as animal feed. Use of camelina oil and meal for food and feed purpose is not approved in USA and currently undergoing trials to receive FDA approval for animal feeding (Ehrensing and Guy, 2008). Biodiesel from camelina has strong advantage over soybean and canola biodiesel because of its low cloud point and pour point (Johnson, 2007). Recent success in use of camelina biofuel as jet fuel by US Navy in 2010 demonstrated its potential for aviation industry also (Perkins, 2010). As a biofuel, camelina methyl esters had similar properties and expected yields as canola methyl ester (Frohlich and Rice, 2005). Other uses are as bird feed, manufacturing of linoleum and use in dry flower arrangements.

Crop Improvement

Camelina breeding has not been carried out intensively in the past and availability of germplasm is limited. Recent interest in this under-exploited crop mainly due to demand for alternative low-input oil crop has led to crop improvement studies in European countries. Genetic improvement in this crop was initiated in Germany during 1980's including collection and evaluation of germplasm (Seehuber, 1984) which led to the development of advanced lines using single seed descent method for further improvement (Seehuber *et al.*, 1987; Seehuber and Dambroth, 1987). Vollmann *et al.* (1996) reported two populations of camelina CA13x and CA2x derived from genotypes from FAL Institute of Agronomy (Braunschweig, Germany) and M. I. Vavilov Institute (St. Petersburg, Russia).

Seed quality characters are important parameters for improvement of this crop. Most of the breeding programme as reported by various workers are devoted to selection for 1000 seed weight, oil content or modified concentrations of particular fatty acids. Camelina is a crop with small seeds (1000 seed weight ranges 0.6 to 2.0 g). Breeding programmes in several European countries have not changed this aspect to greater extent possibly because there is an inverse correlation between seed size and oil content. Moreover, large seeded genotypes exhibited reduced number of seeds per pod, malformation of pods and higher lodging (Vollmann *et al.*, 1996). The positive correlation between seed yield and oil content was reported which indicated the simultaneous improvement of both the traits was possible in this crop. An improvement in seed oil content will make camelina more competitive to other oil seed crop. In breeding approach, selection can be done for oil properties to favour biodiesel use, lubricant use (hydroxyl acids), nutrition (high omega-3 fatty acids), antioxidants to avoid oil oxidation and rancidity (high gamma tocoperol content) and lower erucic acid and gondoic acid content and low glucosinolate content. China has also developed a new oil crop, camelina No. 1 by hybridizing *C. sativa* (France) and *C. macrocarpa f. longistipata* (China). The content of unsaturated fatty acids in this crop is as high as 90 per cent comprising about 71.1 per cent polyunsaturated fatty acids including linolenic acid content of 34.5 per cent (Huang *et al.*, 2005).

Camelina sativa is a strict self pollinating crop (Vollmann *et al.*, 2005) and probably the closest crop plant to completely sequenced cousin *Arabidopsis thaliana* (Flannery *et al.*, 2006) with 2n=40 (Shu, 2001) and allohexaploid (Hutcheon *et al.*, 2010). This genetic proximity is of great interest in marker assisted breeding and genetic transformation for crop improvement.

Vollmann *et al.* (2005) evaluated a set of 130 accessions of camelina collected from all across the world. RAPD polymorphism in a representative set of 41 accessories revealed high degree of variability. Ghamkhar *et al.* (2010) carried out AFLP among the 53 diverse accessions of camelina and found high degree of variation. They also concluded that oil quality characteristics are strongly influenced by environmental factors. Vollmann *et al.* (2007) evaluated 60 genotypes across 3 environments and reported significant effect of temperature and precipitation on fatty acid content. Highly significant negative correlation was found between the amount of precipitation during seed filling and fatty acid content. Higher temperature during seed filling stage promoted saturated fatty acid content while low temperature promoted unsaturated fatty acids. Lu (2008) screened over 50,000 M_2 lines and identified a number of mutant lines containing altered contents of different fatty acids. A mutant with significantly increased oleic acid (C18: 1) content (>30 per cent) was identified exhibiting its suitability for food and non-food uses.

Figure 7.1: Camelina crop in field.

Figure 7.2: Camelina plant with pods.

Figure 7.3: Harvesting of mature crop.

Figure 7.4: Camelina seeds.

Figure 7.5: Camelina pod and its component.

Figure 7.6: Camelina oil.

Variety

Most of the work on this crop was done in European countries and hence some known varieties are from these countries only. In France, winter variety 'Epona' and spring variety 'Celine' were released. Varieties 'Calena', 'Lindo' and 'Iwan' are from Austria. Germany has released varieties like 'Ligena', 'Lindo' and 'Soledo'. Montana State University released two varieties 'Blaine Creek' and 'Suneson'. Sustainable Oils (Pvt company) during 2009 sold 3 high yielding genotypes SO-10, SO-20, SO-30 in US and Canada.

North Western Agricultural Research Centre of Montana State University (Montana, USA) started camelina breeding programme to develop camelina cultivars suitable for Montana and Northern Great plains. Over 50 accessions from world collection have been evaluated and three lines (MT-1, MT-3 and MT-5) were selected for further development as cultivar (Pilgeram *et al.*, 2007).

Intergeneric Hybridization

Hybridization attempts between cultivated *Brassica* species and *Camelina sativa* were unsuccessful due to occurrence of strong reproductive barriers. Intergeneric hybridization through protoplast fusion was attempted to introgress alternaria blight resistance genes from camelina to *Brassica carianata* by Narasimhulu *et al.* (1994). Somatic hybrids of *B. oleracea* and *C. sativa* were produced to improve *B. oleracea* resistance to alternaria blight (Sigareva and Earle, 1999). Protoplast electrofusion between *B. napus* and *C. sativa* was attempted by Jiang *et al.* (2009) to produce somatic hybrids with high linolenic acid content.

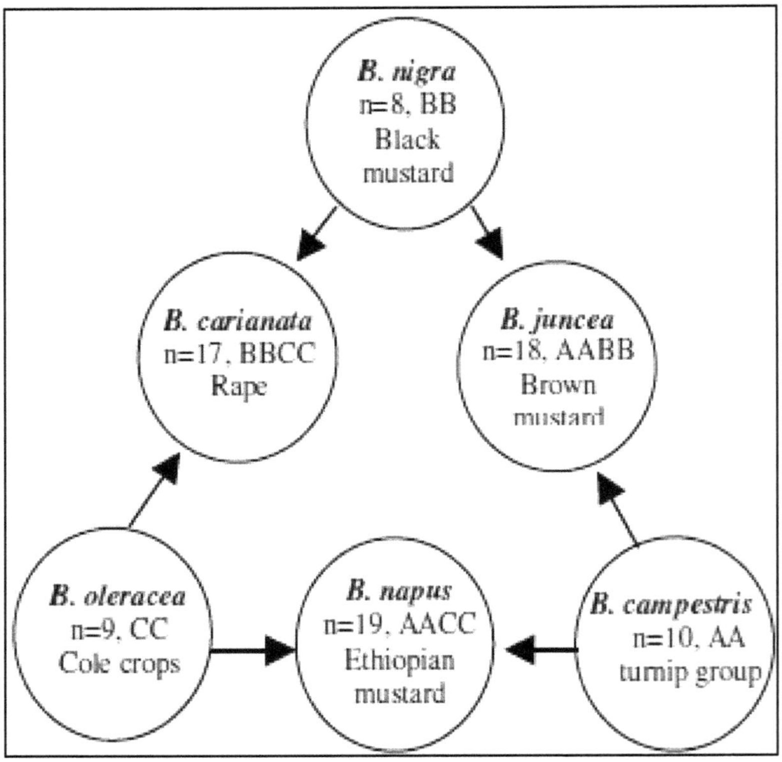

Camelina with its unique number of haploid chromosome (n=20) does not show affinity to any brassica family crops as revealed in U-triangle proposed by Korean scientist Nagaharu in 1935 (Ram, 1998). Although a few reports have established unsuccessful hybridization attempts between camelina and *Brassica* sp. but more studies are required in this direction especially in subtropical climate to check cross contaminating nature of this crop to maintain genetic purity of *Brassica* sp. in field.

Because of a short breeding history, the availability of germplasm is limited. Most camelina genotypes accessible to date originated from germplasm collections and botanical gardens (Vollmann *et al.*, 2005). Genetic variability and promising types are not available in India as the crop is not native to this area, although crop has recently been introduced in India successfully (Agarwal *et al.*, 2010). Two lines of camelina from BOKU University of Natural Resources and Applied Life Sciences, Vienna (Austria) were introduced with accession number EC 643909 (cv. Iwan) and EC 643910 (cv. Calena) through NBPGR, New Delhi

Climate and Soil

Camelina is adapted to a wide range of climatic conditions, as demonstrated by its long history of cultivation across the European continent (Gugel and Falk, 2006). It is well suited to cool arid regions and is a cool season plant which germinates at a soil temp of 3°C. Seedlings are frost tolerant, making early planting possible in areas of late frosts. Optimal growth occurs at temperatures of 15.6-18°C (60-65°F). Camelina also grows well in regions with low rainfall, and exhibits drought tolerance (Ehrensing and Guy, 2008). Similar to other cruciferous species air temperature may be important in setting flower and fruits and excessive heat during flowering is not suitable.

Camelina can be cultivated in a variety of soil types including marginal soils. It is the best suited to light to medium soils with good drainage and responds well under drought stress conditions and may be better suited to low rainfall regions than most other oilseed (Ehrensing and Guy, 2008). Optimum pH for soils is 6.5 but can be grown in more alkaline soils up to pH 8.0. Acidic soils can have a negative effect on Camelina's root development, thus resulting in a poor stand, or uneven development (Meakin, 2007).

Season of Planting

Camelina has two forms: spring and winter (Foller and Paul, 2002) Camelina can be sown as a winter crop in areas where winters are mild, but sowing must be delayed to avoid plants becoming reproductive in early spring with subsequent frost damage. Spring sown crops may be more appropriate when compared to winter for yield (IENICA, 2004).

Table 7.6: Performance of *Camelina sativa* cv. Calena under polyhouse at Pithoragarh.

Date of Sowing	*Plant Height (cm)*	*Days to Harvest*	*Branches/ Plant*	*Seed Pods/ Plant*	*Seeds/ Pod*	*Seed Yield/ Plant (g)*	*Seed Yield (g/m²)*
Jun 09	47.25	82	11.4	185.4	9.40	0.865	120.2
Jul 09	47.50	84	11.6	190.8	9.80	0.970	123.5
Aug 09	50.75	86	12.4	201.0	11.20	1.245	136.4
Sep 09	51.50	86	12.8	203.0	13.75	1.400	150.1

Most of the agronomic standardization reports are from European countries and U. S. A. Some recommend planting in late March to late April, while recommendations from Montana and Wyoming agronomists, state that planting from mid-March to mid-April is better (Crowley and Frohlich, 1998; Meakin, 2007 and Lafferty *et al.*, 2009). *Camelina* can be seeded earlier than other spring crops due to its frost tolerance. Early seeding favours increased seed yields and higher oil content. Ehrensing and Guy (2008) found 25 per cent yield reduction when sowing was delayed by 30 days from March to April. Late winter planting does not exhibit a significant yield increase, and thus is not preferred (Meakin, 2007). The suggested sowing date for camelina is early spring; dry conditions in the flowering stage lower the produce, whereas enough moisture at seed-filling stage increases the share of oil. Agro-technology standardization of this crop in India is in progress as the crop has recently been introduced in the country. The results of the experiments revealed earliness of the crop (82-86 days for seed maturity) with a seed yield (120-150 g/m^2) as shown in Table 7.6 (Agarwal *et al.*, 2010). The results of this preliminary study revealed that the crop showed better adaptability to cooler climate of hilly areas in Indian condition. Further studies are in progress for its utilization as bio-fuel crop in degraded land at DIBER project sites at Ahmednagar, Mhow and Secunderabad in central India as intercrop with jatropha and at Pithoragarh as large scale cultivation in mid hills of central Himalaya.

Method of Sowing

Camelina is relatively easy to grow and requires a few agricultural inputs compared to other crops except successful stand establishment being small seeded crop. Minimal seed bed preparation is needed to establish camelina, which has the ability to germinate with minimum rainfall and established quickly (Ehrensing and Guy, 2008). Good crops can be produced by broadcasting followed by packing the seed to assure seed-soil contact. Seed can be drilled very shallow with at least some seed visually on the soil surface. Seedbed preparation is critical to even emergence of camelina from the soil and competitiveness against weeds.

One of the biggest challenges for camelina small seed to achieving optimum yield is successful stand establishment. Zubr (1997) reported that careful seedbed preparation before sowing, such as repeated harrowing, is necessary to increase emergence percentage. Seedbed should be composed of fine well drained soil that has been firmed. If clay soil is present, soil should be worked several times until soil is light (Meakin, 2007). Drills should be set to plant at a shallow depth of 1-2 cm. Recommended inter row spacing is 20-25 cm but can be adjusted depending upon weed pressure. Experiments conducted at Defence Institute of Bio-Energy Research, field station Pithoragarh (1830m amsl) revealed significant effect of sowing time, environment and spacing on the seed yield of camelina. Line spacing exhibited significantly higher seed yield (12.05-12.64 q/ha) over broadcasting (10.23 q/ha) with 30cm line spacing showing superiority over 40 cm (Table 7.7) (Agarwal *et al.*, 2011).

Table 7.7: Effect of line sowing vs. broadcasting (Dec 2009 sowing in open condition) on Camelina seed yield.

Treatments	*Maturity (Days)*	*1000 Seed Wt (g)*	*Seed Yield/ Plant (g)*	*Seed Yield (q/ha)*	*Oil Content (per cent)*
Broadcasting	117.6	1.184	2.86	10.23	38.37
Line sowing (30 cm)	119.3	1.188	3.40	12.64	38.77
Line sowing (40 cm)	119.3	1.185	3.55	12.05	39.07
CD (5 per cent)	NS	NS	NS	0.703	NS

Seed Rate

Seed rates for camelina depends on a number of factors, like spatial arrangement of the plants, varieties, available moisture, soil fertility, germination, cultural practices, suitability of the growing season, disease and insect considerations, emergence and crop establishment, seedbed preparation and weed pressure (Granberry *et al.*, 2008, Zubr, 1997). Seeding rates of approximately 7-9 kg/ha will produce a stand 220-250 plants/m^2 and 6-10 kg/ha are recommended for sowing (Meakin, 2007). However, in field trials performed in Montana, a seeding rate of 4-6 kg/ha was adequate. In areas where establishment is difficult, the higher seeding rate should be used (Ehrensing and Guy, 2008). Vollmann *et al.* (1996 and 2007) found a seed rate of 3.0 kg/ha as optimum for growing good crop of camelina.

Agegnehu and Honermeier (1997) evaluated the effect of seeding rates on yield and its components of summer *C. sativa* in Germany and reported that increasing the higher seed rate of 800 seeds/m^2 reduced the number of branches/plant, pods/plant, seeds/pod and seed weight/plant. They also found that the seeding rate of 400 seeds/m^2 was the most efficient rate to obtain the highest yield. However, Crowley and Frohlich (1998) found that the seeding rate had no significant effect on growth and yield of camelina and recommended that 500 seeds/m^2 would be the most suitable. Urbaniak *et al.* (2008) suggested that the optimum seed yield could be achieved at 400-600 seeds/m^2. Johnson *et al.* (2008) noticed that higher plant densities hastened camelina maturity by as much as 7 days; optimum camelina plant densities varied between 125 and 200 plants/m^2 which would correspond to a seeding rate of 500 seeds/m^2 or higher. Therefore, it could be concluded that seeding rate recommendations are varying as per locations and further research needs to be carried out for its standardization.

Nutrition and Fertilization

Camelina is a low input crop requiring low amounts of fertilizers for cultivation. Fertilization depends on many factors including soil type, rainfall or irrigation and yield goals. Moudry and Strasil (1999) studied consumption of nutrients by the camelina crop (Table 7.8) and opinioned that camelina is not a nutrient demanding crop.

Table 7.8: Consumption of nutrients by camelina (Moudry and Strasil, 1999).

Components	*Consumption of Nutrients (kg/tone of harvested dry mass)*				
	N	*P*	*K*	*Ca*	*Mg*
Seeds	40.7	7.1	9.6	3.5	3.2
Straw	8.0	1.5	10.4	8.3	1.4

Studies indicated that nutrient requirement for camelina is moderate to low but it is necessary to provide approximately 30 kg/ha phosphorous and 50 kg/ha potassium to the field before sowing (Zubr, 1997). The basic nitrogen requirement for camelina is relatively low compared with other oilseed crops (Putnam *et al.*, 1993; Vollmann *et al.*, 1996). Generally, the maximum seed yield can be achieved when nitrogen application is approximately 75 kg/ha. Other studies showed a linear response of seed yield to N rates up to 120 kg/ha (Urbaniak *et al.*, 2008). Agegnehu and Honermeier (1997) found that the amount of nitrogen supplied greatly influenced the yield components. The number of branches, pods, seeds per pod and seed weight have all increased as the nitrogen increased.

For dry land farmers where water is a limiting factor, it is important to apply slightly less nitrogen than the recommended quantity. Fertilizer application should take place in two stages. Half the total fertilization should take place at the time of planting either at time of seeding or incorporated into the seedbed prior to sowing. The second portion should be applied to the crop at the four-leaf stage (Meakin, 2007; Zubr, 1997). Seed quality, oil yields and composition of fatty acids of camelina are influenced by balanced nutrition and fertilization (Zadernowski *et al.*, 1999; Honermeier and Agegnehu, 1996). Sczebiot (2002) reported that seed yields were the highest after split application of total dose of 80 kg N/ha. On the basis of research carried out in UK and France, the N dose recommended for camelina sown out in late March is 100 kg/ha provided seeding amount is 350 seeds/m^2 (Pearson *et al.*, 1999; Merrien and Chatenet, 1996). Agegnehu and Honermeier (1997) reported an increase in yield attributing traits and seed yields with an increasing dose of 120 kg/ha of nitrogen but no response to increased levels of phosphorus or potassium. Sulfur also seems to have no effect on the yield of camelina (Pearson *et al.*, 1999), although it does have an effect on the oil content of seed (Ehrensing and Guy, 2008). Therefore, sulfur should also be applied at a rate of 25kg/ha for seed oil content and to maintain a balanced ratio between nitrogen and sulfur (Meakin, 2007).

Irrigation

Plant response to water stress varies with developmental stages. For production of camelina available moisture should be between 152- 381mm and should be higher during seeding and seed emergence (Lafferty *et al.*, 2009). Once plants have established they exhibit positive drought tolerance (Meakin, 2007). The drought avoidance characteristic might make camelina better suited to drier regions than other oilseeds (Putnam *et al.*, 1993). Vollmann *et al.* (1996) reported that water stress during the flowering stage decreased plant yield while sufficient irrigation during the seed filling period increased oil content. According to French *et al.* (2009) irrigation requirements for the crop in the central Arizona environment ranged between 333 and 423mm. They also observed the weak correlation between seed yield and total evapo-traspiration up to a water stress of 75 per cent soil water depletion but further depletion (85 per cent) led to drastic yield decline.

Weed Control

Camelina is naturally tolerant to weed competition. Weed control may or may not be required depending on cultivation technique and location of production. In many situations, weed control is not necessary due to the crops rapid growth (Meakin, 2007). Winter sown camelina is especially competitive against many weeds due to its low germination temperature of 3°C which is lower than many weed seeds. Camelina also has allelopathic properties (ability to inhibit growth of other plants), and many weeds are suppressed until leaf drop occurs in the crop (Ehrensing and Guy, 2008). Perennial or biennial weeds are likely to be more difficult to control in Camelina. However, the competitiveness of camelina with annual weeds presents the possibility that camelina could be grown both without tillage and without pre- emergence weed control (Putnam *et al.*, 1993). Camelina is similar to canola in its sensitivity to residual soil herbicides (Ehrensing and Guy, 2008).

Harvesting

Harvesting time for crop depends on maturity and maturity varies with precipitation, temperature and sowing time. However, typical harvesting takes place from late June to mid September in places with mild winter (McVay and Lamb, 2008). In studies conducted in Central and Northern Europe, harvesting takes place prior to late July (Zubr, 1997). Harvesting takes place when seeds reached full maturity and seed moisture content is 8-16 percent (Meakin, 2007). Mature pods are dark tan or brown. Unlike other members of the mustard family, Camelina pods hold their seeds tightly. Due to their resilient seed hulls and low risk of shattering, Camelina crop stands well up to six weeks after maturity, without any damage; allowing for greater flexibility in harvesting time. This flexibility also allows farmers to harvest during favorable weather conditions, which can result in a lower post harvest seed moisture content (Meakin, 2007). Mature crop can be harvested manually using the sickle like other crops without too much shattering problem. Harvesting during early morning hours result in reduced shattering loss.

Camelina can be harvested with unmodified combines to make the process less labour intensive (Ehrensing and Guy, 2008). Harvesting with standard equipment posed no problems, they are less labour intensive but can be swathed (Meakin, 2007). Barring high wind damage, the crop can be left to stand until pods are a straw to tan color. This occurs within few days of the first yellow coloration of pods in the field. Swathing is done prior to maturity and can be used for quickly dry drop of combining and is recommended if lodging occurs or there is a significant amount of green weed material in the field. Swathing can occur when about two-thirds of the pods changed color from green to golden-brown and seed is orange in color, it is recommended that the entire crop be at least 50 per cent mature prior to swathing (Ehrensing and Guy, 2008; Meakin, 2007). After swathing, the crop can be harvested with the use of a combine 7-14 days later, which is dependent on favorable weather conditions (Meakin, 2007).

An ordinary combine harvester adjusted for rapeseed, canola or alfalfa seed (Zubr, 1997; Ehrensing and Guy, 2008) could be used for camelina. Although combine fan's speed (air flow) must be reduced (Bonjean and Le Goffic, 1999) and opening screen must be adjusted to minimize seed loss (Ehrensing and Guy, 2008). Camelina seeds are quite small; any leak in equipment should be sealed properly to reduce seed loss during harvest and transport. According to recommendations released for Montana, header height can be set fairly high to minimize green material through the combine. A screen of 9/64 must be installed over the lower sieve to allow seed separation from the capsule. Generally, the water content of seeds should be less than 11 per cent during harvesting. Avoiding harvest in humid weather is an advisable management practice for protecting the seeds in capsules. The optimum moisture content for seed storage is below 8 per cent; therefore, post harvest drying is a common practice for maintaining seed viability (Zubr, 1997).

Yield

Camelina has yield potential similar to many other members of the Brassica family. While the yield of other Brassicas have significantly increased in recent decades through plant breeding and agronomic improvements, camelina's potential remains unexplored. Putnam *et al.* (1993) reported camelina yield between 600-1700 kg/ha. Vollmann *et al.* (2007) found camelina grain yields up to 2800 kg/ha. Pilgeram *et al.* (2007) revealed that camelina could be increased at a rate of 500-1000 fold in a single cropping year. In Montana under dryland condition *camelina* yields are 1800-2000 kg seeds/ha, under rainfed condition 900-1700 kg/ha and under irrigated condition up to 2400 kg/ha. At Moscow, camelina showed yield potential of 2100-2400 kg/ha under rainfed condition (Ehrensing and Guy, 2008). Camelina possesses 35-40 per cent oil, if grown twice in a year with an average seed recovery of 1700 kg/ha will yield 1400-1500 litre/ha/annum. Camelina has a seed oil content ranging from 290-480 g/kg (Putnam *et al.*, 1993, Vollamann *et al.*, 1996, Zubr, 1997 and Vollamann *et al.*, 2007).

Experiments conducted at Defence Institute of Bio-Energy Research, field station Pithoragarh (1833 m amsl) revealed the significant effect of sowing time, environment and spacing on the seed yield of camelina. Results of this experiment revealed that Oct-Nov is the best month for sowing the crop with a seed yield of 1.37 –1.48 tones/ha irrespective of the growing environment (Table 7.9). Line spacing exhibited significantly higher seed yield (1.20-1.26 tones/ha) over broadcasting (1.02 tones/ha) with 30cm line spacing showing superiority over 40 cm (Agarwal *et al.*, 2011).

Storage

Camelina seeds are surrounded by a light mucilage layer that easily absorbs moisture which aids in seed germination (like a super slurper), but is detrimental to seed storage. Seed stored at high moisture can sweat and seeds can stack in a bin if seed moisture exceeds 10 per cent. Storage is recommended at 8 per cent (McVay and Lamb, 2008; Meakin, 2007). Seed deteriorates during storage due to high moisture content, high relative humidity, high temperature that can result in fungal infection (Weiss, 2000). Low moisture content reduces deterioration of oil and avoid clumping (McVay and Lamb, 2008). If moisture content is above the desired 8 per cent, seed can be dried using floor drying, continuous flow system, or small batch dryers with a maximum temperature of 40°C. Mechanical drying of seed is not always necessary and floor drying at shallow depths is often sufficient (Meakin, 2007). Although there have been no recorded insect problems with Camelina during storage; insects can be a problem when storing seed for over three weeks. It is advisable to pre-treat storage containers (Weiss, 2000).

Table 7.9: Effect of sowing time and environment on camelina seed yield and yield attributing traits.

Treatments		Plant Height (cm)	Seed Pods/ Plant	Maturity (days)	1000 Seed Wt (g)	Seed Yield/ Plant (g)	Seed Yield (q/ha)	Oil Content (per cent)
Sowing Time	Environment							
Oct 2009	Polyhouse	104.7	481.0	128.6	1.211	3.17	15.59	38.9
	Open	94.7	438.7	138.6	1.166	3.02	14.04	38.9
Nov 2009	Polyhouse	104.2	450.0	120.0	1.188	3.12	14.77	38.5
	Open	94.2	422.7	123.6	1.195	2.85	12.60	38.8
Dec 2009	Polyhouse	98.7	444.3	110.3	1.193	3.14	13.68	38.5
	Open	95.2	404.3	115.0	1.190	2.77	11.56	38.7
Jan 2010	Polyhouse	100.8	410.7	110.0	1.106	2.75	12.79	37.8
	Open	89.8	388.3	111.6	1.120	2.64	10.62	38.3
CD sowing time		NS	17.41	1.88	0.009	NS	0.81	0.438
CD Environment		NS	NS	NS	0.006	0.242	NS	0.310
CD interaction		NS	NS	2.67	0.012	NS	NS	NS

Diseases and Pests

An important feature of camelina is its high level of resistance against insect pests and plant pathogens, which maybe partly due to production of anti-microbial phytoalexins. Downy mildew (*Peronospora parasitica*), the seed-borne fungal disease was observed in many commercial trials (McVay and Lamb, 2008). Crowley and Frohlich (1998) also found disease infection in Ireland like *Sclerotinia, Botrytris* and *Ustilago*. Conn *et al.* (1988) and Browne *et al.* (1991) reported that camelina with two special phytoalexins which are camalexin and methoxycamalexin, has high resistance to some plant diseases - *Alternaria brassicae, Leptosphaeria maculans* (Black leg) and *Rhizoctonia solani* as well as insects like *Phyllotreta nemorum*.

While conducting trials at Pithoragarh, disease Aster Yellows (*Candidatus* Phytoplasma asteris) was observed. The characteristics symptoms are greenish-yellow and reddish-purple leaves, pods and stems along with stunting and distorted inflorescence.

Seed Certification

Originally camelina spread from Europe as a contaminant of flax and became a weed throughout North America and Canada but recent research interest in this crop lifted its status from a weedy annual to under-exploited crop with enormous potential for food, feed and industrial utilization. Recently Canadian Seed Growers' Association issued regulations vide circular No. 6/Rev.01-5-2010 (Section 14.10) for seed certification of camelina as presented in Table 7.10 (CSGA, 2010).

Table 7.10: Foundation, registered and certified seed production of all varieties of *Camelina sativa* as specified by Section 14.10 of CGSA.

14.10.1 Seed classes and generations	a) The number of official pedigreed classes is determined by the breeder of the variety and are normally Foundation, Registered and Certified.
14.10.2 Land requirements	a) Camelina crops must not be grown on land which in the previous year grew a non-pedigreed crop of camelina or a different variety of camelina.
	b) Camelina crops must not be grown on land which in the previous year grew a crop of canola, mustard, oilseed radish or rapeseed.
14.10.3 Crop inspection	a) It is the grower's responsibility to ensure that crops are inspected by an authorized inspector prior to swathing or harvesting.
	b) A crop that is cut, swathed or harvested prior to crop inspection is not eligible for pedigree.
	c) The crop must be inspected at a stage of growth when varietal purity is the best determined. Crops not inspected at the proper stage for best determining varietal purity may be the cause for declining pedigreed status.
	d) A field inspection should be made during the bloom stage but not before at least 50 per cent of the plants are showing one or more blossoms.
14.10.4 Crop standards	
14.10.4.1 Isolation	a) Camelina crops for foundation or registered status must be isolated by a distance of 100 meters from other varieties of camelina or from a non-pedigreed crop of camelina.
	b) Camelina crops for certified status must be isolated by a distance of 16 meters from other varieties of camelina or from a non-pedigreed crop of camelina.
	c) The required isolation must be provided prior to the time of flowering and crop inspection.
14.10.4.2 Weeds	a) All crops for pedigree must be free of prohibited noxious weeds.
	b) Very weedy crops may be declined pedigreed status.
	c) Prickly Lettuce (*Lactuca serriola*), Stinkweed (*Thlaspi arvensis*) and Shepherds Purse (*Capsella bursa-pastoris*) plants produce seeds that are difficult to separate from camelina and seed crops with excessive number of these weeds may be declined pedigreed status.
14.10.4.3 Maximum Impurity Standards	The maximum impurity levels outlined below apply, unless variants are specified by the breeder.

Impurity	*Maximum Permitted in each Class*		
	Foundation	*Registered*	*Certified*
Other varieties of camelina	2 per 10,000 plants	5 per 10,000 plants	10 per 10,000 plants

Conclusion

Camelina is an under-exploited crop species with high multiplication ratio, drought and frost tolerance, low seed and input requirement and short maturity. It has enormous potential and great economic significance not only for industrial utilization but also for nutrition specially being rich in omega 3 fatty acids after removal of non-edible fractions. Research reports across the globe reported useful genetic variation in agronomic and seed quality characters in breeding lines, which offers opportunity for further improvement in various traits for its food and non-food utilization. Significant environmental effects on various quality traits have necessitated the requirement of multilocation data before recommending any package of practices for this crop in a new area. Preliminary studies showed promising results and its usefulness as a potential biofuel crop. An improvement in the seed oil content using modern biotechnological tools would be of great significance to make it a more attractive crop for industrial purpose.

References

Abramovic H. and Abram V. 2005. Properties of *Camelina sativa* oil. *Food Technol. Biotechnol.* 43 (1) 63–70.

Agarwal A., Arya M. C. and Ahmed Z. 2011. Influence of sowing time, environment and spacing on seed yield and oil recovery in *Camelina sativa*. *Indian J. Agril. Sci.* (submitted)

Agarwal A., Pant T. and Ahmed Z. 2010. *Camelina sativa*: A new crop with bio-fuel potential introduced in India. *Current Science*, 99 (9): 1194-95.

Agegnehu M. and Honermeier, B. 1997. Effects of seeding rates and nitrogen fertilization on seed yield, seed quality and yield components of false flax. *Die Bodenkultur*.48: 15-21

Al-Shaehbaz I. A. 1987. The genera of Alysseae (Cruciferae; Brassicaceae) in the Southeastern United States. *Journal of the Arnold Arboretum*. 68: 185-240.

Anonymous. 2002. Gold-of-Pleasure. http://www. ienica. net/crops/goldofpleasure. pdf.

Bernardo, A., Howard-Hildige, R., O'Connell, A., Nichol, R., Ryan, J., Rice, B., Roche, E. and Leahy, J. J., 2003. Camelina oil as a fuel for diesel transport engines. *Ind. Crops Prod.* 17, 191–197.

Bonjean A. and Le Goffic F. 1999. La cameline – *Camelina sativa* (L.) Crantz : une opportunité pour l'agriculture et l'industrie européennes. OCL 6: 28–34.

Bonjean A., Le Goffic F. 1998. La cameline - *Camelina sativa* (L.) Crantz: une opportunit´e pour l'agriculture et l'industrie europ´eenne. OCL-Oleagineux Crops Gras Lipides 5 (5), 20–25.

Bramm, A. 1993. Produktionsmöglichkeiten pflanzlicher Öle und ihrer Verwendung im technischen Bereich. Landbauforschung Völkenrode, 43: 12-120.

Browne L. M., Conn K. L., Ayer W. A. and Tewari J. P. 1991. The camalexins: new phytoalexins produced in the leaves of *Camelina sativa* (Cruciferae). *Tetrahedron*. 47(24): 3909-3914.

Budin J. T., Breene W. M. and Putnam D. H. 1995. Some compositional properties of camelina (*Camelina sativa L.* Crantz) seeds and oils. *J. Amer. Oil Chem. Soc.* 72(3): 309-315.

Cardone M., Mazzoncini M., Menini S., Rocco V., Senatore A., Seggiani M. and Vitolo S. 2003. *Brassica carinata* as an alternative oil crop for the production of biodiesel in Italy: agronomic evaluation, fuel production by transesterication and characterization. *Biomass and Bioenergy*. 25: 623-636.

CGSA. 2010. Foundation, Registered and Certified seed of camelina. Canadian Seed Growers' Association Circular 6/Rev.01.5-2010 Section 14.10.

Conn K. L., Tewari J. P. and Dahiya J. S. 1988. Resistance to *Alternaria brassicae* and phytoalexin-elicitation in rapeseed and other crucifers. *Plant Science*. 56(1): 21- 25.

Crowley J. G. 1999. Evaluation of *Camelina sativa* as an alternative oilseed crop. TEAGASC Irish Agriculture and Food Development Authority. Project Report 4320.

Crowley J. G., and Frohlich A. 1998. Factors affecting the composition and use of *camelina*. Teagasc Project Report 4319, Crops Research Centre, Teagasc, Dublin, Ireland.

Dorn R. D. 1984. Vascular Plants of Montana. Mountain West Publishing, Cheyenne.

Downey R. K. 1971. Agricultural and genetic potential of cruciferous oilseed crops. *J. Amer. Oil Chem. Soc.* 48: 718-722.

Ehrensing D. and Guy S. 2008. *Camelina.* Available at: http://extension. oregonstate. edu/catalog/pdf/em/em8953-e. pdf [accessed on27 Jun. 2009].

Flannery M. L., Mitchell F. J. G, Coyne S., T. A. Kavanagh, J. I. Burke, N. Salamin, P. Dowding, and Hodkinson T. R. 2006. Plastid genome characterisation in *Brassica* and Brassicaceae using a new set of nine SSRs. *Theor. Applied Genet.* 113(7): 1221–1231

Foller I. and Paul V. H. 2002. Investigations on diseases of false flax (*Camelina sativa* (L.) Crtz.) with special regard to downy mildew (*Peronospora parasitica* (pers.) Fr.). IQBC/wprs Bulletin 25: 43-52.

Francis C. M. and Campbell M. C. 2003. New high quality oil seed crops for temperate biodiesel production. *Industrial Crops and Products.* 21: 25-31.

French A. N., Hunsaker D., Thorp K. and T. Clarke. 2009. Evapotranspiration over a camelina crop at Maricopa, Arizona. *Industrial Crops and Products,* 29 : 289–300

Friedt W and Luhs W. 1998. Recent development and perspectives of industrial rapeseed breeding. *Fetty Lipid,*100: 219–226.

Frohlich, A. and Rice, B. 2005. Evaluation of Camelina sativa oil as a feedstock for biodiesel production. *Industrial Crops and Products* 21: 25-31.

Ghamkhar K., Croser J., Aryamanesh N., Campbell M., Kon'kova N., and Francis C. 2010. Camelina (*Camelina sativa* (L.) Crantz) as an alternative oilseed: molecular and ecogeographic analyses. *Genome,* 53 (7): 558-567.

Gilbertson P. K., Johnson B. L., Berti M. T. and Halvorson M. A. 2007. Seeding Date and Performance of Specialty Oilseeds in North Dakota. Issues in new crops and new uses. Janick, J and Whipkey, A (Eds.). ASHS Press, Alexandria, VA, pp. 105-10.

Gilkey H. M. 1957. Weeds of the Pacific Northwest. Oregon State College.

Glob P. V.1969. The bog people: Iron –Age man preserved. Cornell Univ Press, USA.

Granberry D. M., Kelley T. and Boyhan G. 2008. Seeding rates for vegetable crops. The University of Georgia and Ft. Valley State University, the U. S. Department of Agriculture and counties of the state cooperating. Available from: http://pubs. caes. uga. edu/caespubs/pubcd/b1128-w. html. [Assessed by July 28th, 2008]

Gugel R. K. and Falk K. C. 2006. Agronomic and seed quality evaluation of *Camelina sativa* in western Canada. *Can. J. Plant Sci.,* 86: 1047-1058.

Gunstone F. D. 1996. Fatty Acids and Lipid Chemistry, Chapmann and Hall, New York, 1996, pp. 7-69.

Gunstone F. D., Harwood J. L. and Padley F. B. 1986. Occurrence and characteristics of oils and fats. In: Lipid Handbook (Eds) Gunstone F. D., Harwood J. L. and Padley F. B., Chapmann and Hall, New York, 1986, pp. 104-105.

Hebard A. 1998. *Camelina sativa* –a pleasurable experience or another false hope? *Lipid Technol.* 10: 81–83.

Honermeier B. and Agegnehu M. 1996. Camelina has a future as a non-food crop. Neue-Landwirtschaft., No. 12, 44-46.

Huang F. H., Xie B. J. and Liu C. S. 2005. A new oil crop: *Camelina sativa*. *China Oils and Fats*. 10.

Hutcheon C., Ditt R. F., Beilstein M., Comai L., Schroeder J., Goldstein E., Shewmaker C. K., Nguyen T. Rocher J. de and Kiser J. 2010. Polyploid genome of Camelina sativa revealed by isolation of fatty acid synthesis genes. *BMC Plant Biology*, 10: 233

IENICA 2004. Agronomic Guide: *Camelina sativa*, [Online]. Available at: http://www. ienica. net/ agronomyguide/agronomyguide05. pdf [accessed on 15 Jul 2009].

Itoh T., Tamura T. and Matsumoto T. 1973. Sterol composition of nineteen vegetable oils. *J. Ame. Oil. Chem. Soc.*, 50: 122-25.

Jiang J. J., Zhao X. X., Tian W., T. B. Li and Wang, Y P. 2009. Intertribal somatic hybrids between *Brassica napus* and *Camelina sativa* with high linolenic acid content. *Plant Cell Tiss Organ Cult*, 99: 91–95.

Johnson D. 2007. Camelina: An Emerging Crop for Bioenergy, *In vitro* biology meeting abstract issue, *In Vitro* Cell. Dev. Biol. (DOI 10.1007/s11626-007-9026-4)

Johnson EN, Falk K, Klein-Gebbinck H, Lewis L, Vera C, Malhi S, Shirtliffe S, Gan Y, Hall L, Topinka K, Nybo B, Sluth D, Bauche C and Phelps, S. 2008. Agronomy of *Camelina sativa* and *Brassica carinata*, Western Applied Research Corporation Annual Report 2008. p 17.

Jones G. and Valamoti S. M. 2005, Lallemantia, an imported or introduced oil plant in Bronze Age northern Greece. *Vegetation History and Archaeobotany* 14 (4): 571–577.

Karvonen H. M., Aro A., Tapola N. S., Salminen I., Uusitupa M. I. J. and Sarkkinen E. S. 2002. Effect of -Linolenic Acid-Rich Camelina sativa Oil on Serum Fatty Acid Composition and Serum Lipids in Hypercholesterolemic Subjects. *Metabolism*. 51(10): 1253-1260

Knorzer K. H. 1978. Entwicklung und Ausbreitung des Leindotters (*Camelina sativa*.). Ber. Dtsch. Bot. Ges. 91: 187–195.

Kochhar S. P. 1983. Influence of processing of sterols of vegetable oils. *Prog. Lipid Res.*, 22: 161-188.

Korsrud G. O., M. O. Keith, and Bell, J M. 1978. A comparison of the nutritional value of crambe and camelina seed meals with egg and casein. *Can. J. Anim. Sci*. 58: 493-499.

Lafferty R., Rife C. and Foster G. 2009. Spring Camelina Production Guide for the Central High Plains. Golden: Blue Sun Agriculture Research and Development.

Lange R., Schumann W., Petrzika M., Busch H. and Marquard R. 1995. Glucosinolate in Leindottersamen. *Fat. Sci. Technol*. 97: 146–152.

Leonard C.1998. Camelina oil: A linolenic source. *INFORM*, 1998, 9: 830-838.

Lovet J. V. and Duffield A. M. 1981. Allelochemicals of *Camelina sativa*. *J. Appl. Ecol.* 18 (I): 283-290.

Lu Chaofu. 2008. Camelina sativa: A potential oilseed crop for biofuels and genetically engineered products. Information systems for biotechnology, 2008 p 7-9. www. isb. vt. edu

Luehs W. and Friedt W. 1993. Non-food uses of vegetable oils and fatty acids. p. 73-130. (In) Murphy, D. J. (Ed.), Designer oil crops, breeding, processing and biotechnology. VCH Verlagsgesellschaft, Weinheim, Germany.

Marquard R. and Kuhlmann H. 1986. Untersuchungen ¨uber Ertragsleistung und Samenqualit¨at von Leindotter (*Camelina sativa* Crtz.). *Fette-Seifen-Anstrichmittel* 88, 245–249.

McVay, K. A. and Lamb, P. F. 2008. Camelina production in Montana. Montana State University. Available from: www. msuextension. org. [Assessed by May 12th, 2008]

Meakin. S. 2007. Crops for Industry: A Practical Guide to Non-food and Oilseed Agriculture. United Kingdom: The Crowood Press Ltd.

Merrien A. and Chatenet F. 1996. *Camelina sativa*: yield development. *Oleoscope*, 35: 24-27.

Mikersch Von, J. D. F. and Thoris Verharburger Olfabriken A. G. 1952. The composition of Camelina Oil. Farbe Lack (Abstract in English) 8,402-406.

Moudry J. and Strasil Z. 1999. Pµstování alternativních plodin. ZF JU v Èeských Budµjovicích, 1999, 165 s.

Narasimhulu S. B, Kirti P. B and Bhatt S. R. 1994. Intergeneric protoplast fusion between *Brassica carinata* and *Camelina sativa*. *Plant Cell Rep.*, 13: 657–660

Pearson N., Walker K. C., Barrow A., Biddle A., Cook S., Knott C. M., Lainsbury M. A. and Walker, K. 1999. The performance of *Camelina sativa* in the UK. Protection and production of combinable break crops, Royal Agricultural College, Cirencester, UK, 14-15 Dec. 1999. *Aspects of Applied Biology*, 56: 249-255

Peiretti P. G.; Mussa P. P., Prola L. and Meiner G. 2007. Use of different levels of false flax (*Camelina sativa* L.) seed in diets for fattening rabbits. *Livestock Science*,107: 192-198.

Perkins J. 2010. Defence Biofuels- U. S. Navy tests biofuel blend in the 'Green Hornet' jet. *Biofuel Journal*, second quarter, p 74.

Pilgeram A. L., Sands D. C., Boss D., Dale N., Wichman D., Lamb P., Lu Chaofu, Rick B., Kirkpatrick M., Thompson B. and Johnson D. L. 2007. *Camelina sativa*, A Montana omega-3 and fuel crop, P. 129-131. In: J. Janick and Whipkey, A. (eds.) Issues in New Crops and New Uses, ASHS Press, Alexandria, V. A., 2007.

Plessers A. G., McGregor W. G., Carson R. B. and Nakonesshny W. 1962. Species trials with oilseed plants. Camelina. *Canadian Journal of Plant Science*. 42: 452-459.

Putnam D. H., Budin J. T., Field L. A. and Breene W. M. 1993. Camelina: A promising low input oilseed, p 314-322. In: J. Janick and J. E. Simon (Eds.), New Crops, Wiley, New York.

Ram H. H. 1998. Vegetable Breeding: Principles and Practices. Kalyani Publishers, New Delhi. p 421.

Rice B., Frohlich A. and R. Leonard. 1998. Bio-diesel production from camelina oil, waste cooking oil and tallow. Final report by Teagasc (ISBN 1 901138 67 4).

Robinson R. G. 1987. *Camelina: A useful research crop and a potential oilseed crop*. Minnesota Agricultural Experiment Station, University of Minnesota, Bulletin 579-1987 (Item No. AD-SB-3275).

Rode J. 2002. Study of Autochthon *Camelina sativa* (L.) Crantz in Slovenia. *Journal of Herbs, Spices and Medicinal Plants*, 9 (4): 313-318.

Rokka T., Alen K., Valaja J. and Ryhanen E. L. 2002. The effect of a *Camelina sativa* enriched diet on the composition and sensory quality of hen eggs. *Food Research International*, 35,253-256.

Russo V. M., Bruton B. D. and Sams C. E. 2010. Classification of temperature response in germination of Brassicas. *Industrial Crops and Products*, 31: 48–51.

Ruxton C. H. S., Reed S. C., Simpson M. J. A. and Millington K. J. 2007. The health benefits of omega-3 polyunsaturated fatty acids: a review of the evidence. *J. Hum. Nutr. Diet.*, 20: 275–285.

Sang J. P. and P. A. Salisbury. 1987. Wild Crucifer species and 4-hydroxyglucobrassicin. *Cruciferae Newsl.* 12: 113.

Schultz-Motel J. 1979. Die Anbaugeschichte des leindotters, *Camelina sativa* (L.) Crantz. *Archaeo-physica* 8: 267-281.

Schuster A. and Friedt W. 1995. *Camelina sativa*: old face - new prospects. *Eurcarpia Cruciferae Newsletter*, 17,6-7.

Schwartz H., Velimatti O., Vieno P. and Anna-Maija L. 2008. Tocopherol, tocotrienol and plan sterol contents of vegetable oils and industrial fats. *Journal of Food Composition and Analysis*, 21: 152-161.

Sczebiot M. 2002. Effect of mineral fertilization on yielding of spring false flax and crambe. *Rosliny Oleiste*. 2002, 23: 1, 141-150.

Seehuber R. 1984. Genotypische Variabilit¨at in Ertrags-und Qualit¨atsmerkmalen bei Mohn und Leindotter. *Fette-Seifen-Anstrichmittel* 86, 177–180.

Seehuber R. and Dambroth M. 1987. Development of basic populations of plant species suitable for the production of fatty acids, especially considering linseed, false flax and poppy. Landbauforsch. Voelkenrode (Germany) 37: 219-223.

Seehuber R., J. Vollmann, and Dambroth M. 1987. Application of the single-seed-descent method in false flax to increase the yield level. *Landbauforsch.* Voelkenrode (Germany) 37: 132-136.

Shu Ya Ma Qui. 2001. Camelina. *Flora of China* 8: 189.

Shukla V. K. S., Dutta P. C. and Artz W. E. 2002. Camelina oil and its unusual cholesterol content. *J. Ame. Oil Chem. Soc.*, 79 (10): 965-969.

Sigareva M. A. and Earle E. D.1999. Camalexin induction in intertribal somatic hybrids between *Camelina sativa* and rapid-cycling *Brassica oleracea*. *Theo Appl Genet* 98: 164–170.

Urbaniak S. D., Caldwell C. D., Zheljazkov V. D., Lada R. and Luan L. 2008. The effect of seeding rate, seeding date and seeder type on the performance of *Camelina sativa* L. in the Maritime Provinces of Canada. Canadian Journal of Plant Science. 88(3): 501-508.

USFS-ILPIN. 2002. "Illinois Plant Information Network." Retrieved 10-8-2008, from http: / /www. fs. fed. us/ne/delaware/ilpin/492. co.

Vollmann J., Damboeck A. Eckl Schrems H. and Ruckenbauer P.1996. Improvement of *Camelina sativa*, an under-exploited oilseed, P. 357-362. *In*: J. Janick (ed.), Progress in New Crops, ASHS Press, Alexandria, V. A.

Vollmann J., Grausgruber H., Stift G., Dryzhyruk V and Lelley T. 2005. Genetic diversity in camelina germplasm as revealed by seed quality characterstics and RAPD polymorphism. *Plant Breeding*,124: 446-453.

Vollmann J., Moritz T., Kargl C., Baumgartner S. and Wagentristl H. 2007. Agronomic evaluation of camelina genotypes selected for seed quality characteristics. *Industrial Crop and Products*, 2007, 26: 270-277.

Weiss E. 2000. Oilseed Crops. Oxford: Blackwell Science Ltd.

Zadernowski R., Budzynski W., Nowak-Polakowska H., Rashed A. A. and Jankowski, K. 1999. Effect of fertilization on the composition of lipids from false flax (*Camelina sativa*) and crambe (*Crambe abissinica*). *Rosliny-Oleiste*, 20: 2, 503-510.

Zubr J. 1997. Oil-seed crop: *Camelina sativa*. *Industrial Crops and Products*, 6: 113-119.

Zubr J. 2003. Qualitative variation of *Camelina sativa* seed from different locations. *Industrial Crops and Products*, 17: 161-169.

Zubr J. and Matthaus B., 2002. Effects of growth conditions on fatty acids and tocopherols in *Camelina sativa* oil. *Industrial Crops and Products*, 15: 155–162.

2014, Future Crops Vol. 2 Pages 179–198
Editor: Professor K.V. Peter
Published by: DAYA PUBLISHING HOUSE, NEW DELHI

Chapter 8
Castor Bean

Swati Sinha[1], C. Lavanya[2], R.K. Chaudhary[3] and Rajib Bandopadhyay[1]*
[1]Department of Biotechnology, Birla Institute of Technology, Mesra, Ranchi – 835 215, Jharkhand
[2]Directorate of Oilseeds Research Rajendranagar, Hyderabad – 500 030
[3]Department of Botany, University of Calcutta, Kolkata – 700 019
*E-mail: rajib_bandopadhyay@bitmesra.ac.in

Since ancient times the use of castor bean (*Ricinus communis* L.) was documented in several parts of the world. In Egyptian tombs, which date back to 4000 B. C., oil from castor beans was used in slow burning oil lamps. The use of castor oil was also known since 2000 B. C. in Ayurvedic and Unani medicines. But today in the 21st century we find that not only *Ricinus communis* is an unexploited agricultural crop, it also remains largely unused despite of its various known advantages. In the present study, we would like to enumerate about castor bean plant and how it can aptly fit into the category of future plant.

Ricinus communis L is a non-food crop mainly utilised as a natural source of hydroxylated fatty acids (Castor oil). Castor bean is an oilseed crop that belongs to the spurge (Euphorbiaceae) family, which comprises ~6,300 species which include cassava (*Manihot esculenta*), rubber tree (*Hevea brasiliensis*) and physic nut (*Jatropha curcas*). The genus *Ricinus* is considered monotypic; *Ricinus communis* is the only species and includes many polymorphic types (Weiss, 1983). Nine oil seed crops are grown in India. They include seven edible oil crops, *viz.*, groundnut, rapeseed-mustard, soybean, sunflower, sesame, safflower and Niger and two non-edible ones, *viz.*, castor and linseed. India is the first country in the world to produce hybrids in castor. Castor seed is the source of castor oil. The seeds contain between 40 per cent and 60 per cent oil that is rich in triglycerides, mainly ricinoleic acid. It also has tremendous economic use beside castor oil.

Origin of the Castor Plant

It is known as 'earand' in Hindi, 'rehri' in Bengali and Hindi and 'eri' in Assamese. Due to the resemblance of castor bean seeds with the cannie pest '*Ixodes ricinus*' (Figure 8.1), Linnaeus coined the generic name *Ricinus* to this plant. In latin, *communis* means common.

Figure 8.1: (a) *Ixodes ricinus* (source: www. discoverlife. org) (b) Castor seeds.

According to Hooker (1890) and De Candolle (1860) the origin of castor plant is in Africa. But, according to Fluckiger and Hambury (1892), castor is native to India. Watt (1892) believed that castor may have originated in India and as well as in Africa. Perhaps, the plant originated in Gondowana land and then diversified independently in India and Africa. Kulkarni and Ramamurthy (1977) summarized the literature on castor bean distribution.

Botanical Description

R. communis is a hardy plant. It thrives well in diverse climates especially in dry waste lands of tropics and semi tropics, but can also adjust to semi temperate climate. The botanical description of the monotypic genus, *Ricinus*, was given by Hooker (1890) as an evergreen bush or small tree, tall, annual or perennial, sometimes shrubby or sub-arborous, shoot and panicle glabrous, leaves alternate, green or reddish, 30-60 cm in diameter, membranous, broad, palmately 7 to many-lobed (Figure 8.2).

Figure 8.2: Castor plant (*Ricinus communis*) in cultivated field (Source: www. ecolibrary. org).

Male Flowers

Calyx membranous, splitting into 3-5 valvate segments; stamens vary many, filaments crowded, variously connate or in branching clusters, anther cells distinct, distant, sub-globose, divergent; pistillode 0.

Female Flowers

Calyx spathaceous, caducous; ovary 3-celled; styles short or long, spreading, often very large, coloured, entire 2-fid or partite, feathery or papillose; cells 1-ovuled. Capsule 1.5-2.5 cm long, with or without spine, 3, 2-valved cocci. Seeds oblong, testa coruscates, albumen fleshy, cotyledons broad, flat. The flowers of castor bean plant are incomplete and no colored petals are found, despite the fact that the leaves, stems, stamens, stigma and even seed coat are brilliantly colored (Figure 8.3).

Figure 8.3: *Ricinus communis* inflorescence showing red stigma.

Pollination in castor plant bears a mixed breeding system favoring selfing by geitonogamy. As it lacks petals it relies mainly on wind, although the wind promotes mostly out crossing. Although it has also been observed that honey bee can also transfer pollen from male to female flowers within the same raceme and boost the release of airborne pollen by male flowers (Figure 8.4).

Figure 8.4: Bee polinator on *Ricinus communis* inflorescence.

Stem Colour

White (1918) reported five categories of stem colour, *viz.* (i) bright green, (ii) green with reddish-bluish, (iii) carmine or rose red, (iv) mahogany red, and (v) purple. Experiments done by White (1918) and Harland (1922, 1928) indicate that stem colours are determined by a single factor-pair difference. At Rajendranagar Farm (DOR, Hyderabad, India), it was observed that, among the cultivated varieties, there are two main colors, namely (i) red and (ii) green (Figure 8.5). The color shows different intensity, depending on the presence or absence of bloom. Besides, pink and sulphur-white stems were also found in ornamental varieties grown in gardens.

Waxy Coating on Plant

Waxy coatings known as bloom are found on the stem and certain other part of the castor plant, is a natural protection to the plant from extremes of weather and some insect pests. The color of the bloom is ashy-white. According to White (1918) and Harland (1922), bloom is partially dominant and in F_2, a ratio of 3 bloom: 1 no bloom was obtained. Patwardhan (1931) showed that bloom is intermediate in the F_1 and that triple bloom is dominant over double bloom.

Figure 8.5: Stem color (a) mahogany red and (b) green.

Leaves

Normal plant has leaves with lobes and is lacinate (Figure 8.6), which is not very deep. In F_2 the ratio fitted a monogenic difference (Peat, 1926).

Node Number to First Inflorescence

Node number is greatly influenced by environment factors (Peat, 1926). Masur (1933) found that first inflorescence was fairly constant within cultivars and hereditary character.

Femaleness of Spike

The spikes are usually moneocious, the pistillate flowers develop at the top and staminate flowers develop at the base as shown in Figure 8.7. But there are types where the entire length of the spike is full of pistillate flowers, occasionally with only a few staminate flowers at the base. This is a very desirable character for the (increased) seed production. In a cross between such 100 per cent pistillate spike and local castor, which has 35 per cent pistillate flowers, the F_1 progenies have spike with 100 per cent pistillate flower. In F_2 all intermediate stages were found. Katayama (1952) found that the female condition is determined by the gene '*a*'; plants carrying dominant gene '*A*' is normal moneocious.

Figure 8.6: Red coloured castor plant.

Classen and Hoffman (1950) found that 90 to 100 per cent female was controlled by a single recessive gene.

Sex in Castor Plant

Shifriss (1960) had done extensive work on castor plant from Israel and some parts of Africa. His study on sex variation reveals the existence of two genetic systems, which are designated, as 'conventional' and 'unconventional' types.

Conventional or Mendelian forms include moneocious variants and a rare recessive female mutant. The moneocious variants differ from each other in sex tendency in proneness the apical region of their racemes, which bears normally, pistillate to interspersed staminate flowers. Sex tendency is probably governed by polygenes; proneness to differentiate into interspersed staminate flowers is controlled by recessive genes that may be environmentally resistant or sensitive.

Unconventional forms evolved from dominant female mutants under conditions favouring strong female tendency. These spontaneous mutants are genetically unstable. Actual mechanism has not yet

Figure 8.7: Castor plant showing pistillate flowers at top and staminate flowers at the base.

been understood clearly. Phenotypically they are either sex revertants or non-reverted females. Hybrid seeds can be produced, without much labour, by the use of unconventional dominant femaleness and certain environmentally sensitive gene for interspersed staminate flowers. They give better percentage of pistillate plants than those obtained by conventional breeding.

Cytology

The 2n number in *R. communis* was reported 20 by Nemec and Hagerup (Kulkarni *et al.,* 1977). *Ricinus communis* is a secondary balanced polyploid (Richharia, 1937). Based on the maximum number of secondary associations, the haploid constitution may be: AAA; BB; CC; DD and E. The maximum number of secondary association observed at metaphase I and metaphase II is five. Secondary association between two pairs of bivalents at metaphase was observed by Kurita (1952). Narain (1953) succeeded in inducing polyploidy in apical meristem and germinating seeds in *Ricinus communis* by application of 0.3 per cent colchicine solution.

Genetics

Castor plants have a number of contrasting characters such as green and red stem, spiny and non-spiny fruits (as in Figure 8.8), size of fruit and seed, mottling on seeds and oil content. Therefore it is a good material for genetic studies. However, not much work has been done so far on the genetics of this crop. A few cases of genetic transformation, to a high oil yielding variety (VP-1), by particle bombardment and also by *Agrobacterium tumefaciens*, were reported by Ravi Kumar and Sujatha (1998).

Earliness of Flowering

A progeny of pure strain do not all flowers at one time, flowering of the main spike extend over weeks. In a cross between early and late types, White (1918) obtained intermediates in F_1 and with all the three types in F_2.

Figure 8.8: Phenotypic variation in the capsule spines in castor (a) Spineless capsule, (b) semi-spine, (c) normal spines.

Ricin and Toxicity

Castor seed contains two lectins called ricin and RCA (*Ricinus communis* agglutinin) (Hartley and Lord, 2004). Both ricin and RCA are synthesized in the endosperm cells of maturing seeds, and are stored in a 'protein body'. The toxicity of raw castor beans is mainly due to the presence of ricin. Many cytotoxic proteins are found to be identical to ricin, both in structure and function. They inhibit protein synthesis by binding irreversibly to eukaryotic ribosomes.

Ricin is a byproduct of castor oil production: when castor beans are crushed, they form a pulp from which castor oil is extracted, and ricin is what remains. Ricin is one of the strongest poisons of biological origin. Seventy microgram of ricin can kill an adult person, and its peptide sequence resembles to cobra venom protein sequence. Ricin is more potent than cobra venom, and so, the cultivation of castor plant is banned in USA.

Ricin is a polypeptide. It belongs to the Type 2 ribosome-inactivating proteins (RIPs). It consists of two glycoprotein chains – A-chain (alanine chain) and B-chain (isoleucine chain), which are covalently bound by disulfide bridges (Figure 8.9). The A-chain is an enzyme (N-glycosidase) and is considered the active toxic component, which can enzymatically inactivate a large number of ribosomes; while the B chain is necessary for binding the toxin to the cell surface (Audi *et al.*, 2005). Pure, crystalline ricin is odourless and tasteless, and practically only soluble in water. Ricin is inactivated by boiling water.

Ricin possesses antitumour properties; the most promising use of ricin is the production of immunotoxins. RCA is a weak cytotoxin but is a potent hemagglutinin. Poisoning by castor bean seed is due to ricin only, since RCA does not penetrate the intestinal wall. Ricin is therefore a potential weapon for biological warfare.

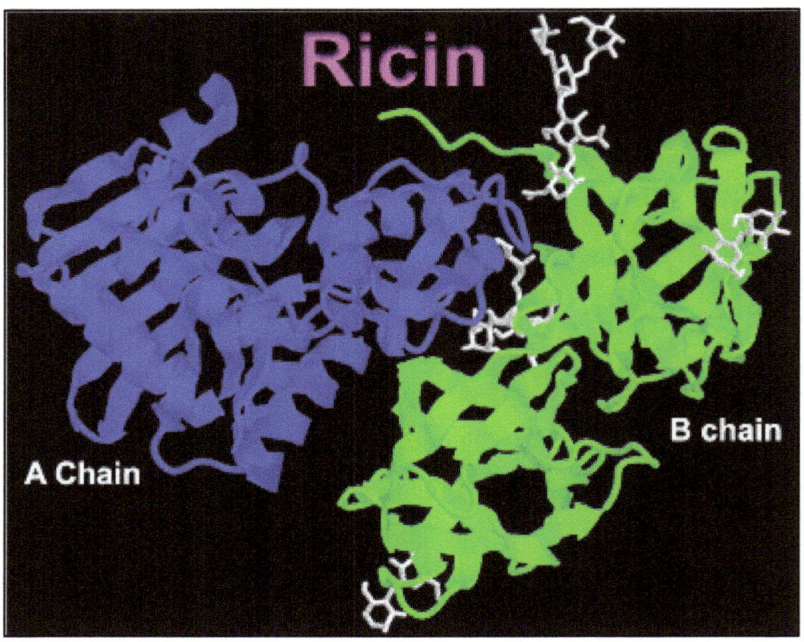

Figure 8.9: 3 D structure of Ricin protein (Source: www. wiley. com).

Composition of Castor Oil

Castor oil is one of the few naturally occurring glycerides with high purity (Akpan *et al.*, 2006), since the fatty acid portion is nearly 90 per cent of ricinoleic acid (Figure 8.10). The average composition of castor seed oil is mentioned in Table 8.1.

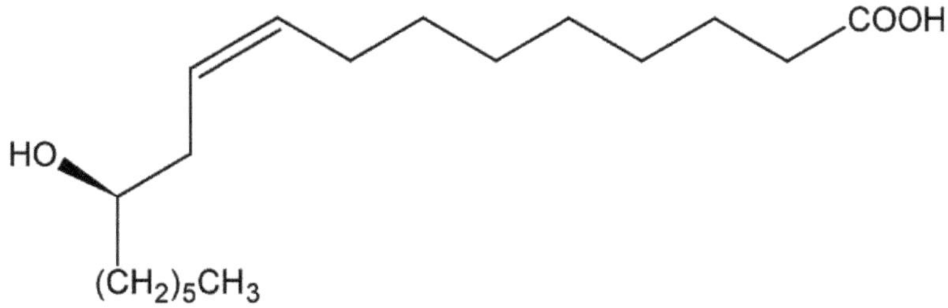

Figure 8.10: Chemical structure for ricinoleic acid.

Genes Determining Floral Organ Development

The principles of *evodevotics* (Evolutionary developmental genetics) are exemplified by outlining the role of MADS-box genes in the evolution of plant reproductive structures, such as sepals, petals, stamens and pistils. MADS-box gene products act as transcription factors of homeotic selector genes that determine floral formation.

Table 8.1: Average composition of Castor seed oil.

Acid Name	*Average Percentage Range*
Ricinoleic acid	85 to 95 per cent
Oleic acid	6 to 2 per cent
Linoleic acid	5 to 1 per cent
Linolenic acid	1 to 0.5 per cent
Stearic acid	1 to 0.5 per cent
Palmitic acid	1 to 0.5 per cent
Dihydroxystearic acid	0.5 to 0.3 per cent
Others	0.5 to 0.2 per cent

MADS-box is highly conversed and has a 180 bp long DNA sequence. MADS is an acronym for the four founder proteins *MCM1* (form baker's yeast), *AGAMOUS* (from *Arabidopsis*), *DEFICIENS* (from *Antirrhinum*) and *SRF* (a human protein). These four proteins have conserved domains that bind to DNA and act as transcription factors. The MADS-box genes of plants are scattered throughout the entire plant genomes (Fischer *et al.*, 1998; Liljegren *et al.*, 1998).

The *R. communis* shows a very interesting phenomenon of sex reversion as shown in Figure 8.11. It has been proposed that MADS-box genes along with transposon activity induced in the castor bean plant due to extreme environment changes is responsible for this phenomenon (Chaudhari *et al.*, 2004).

Castor bean plant is a monoecious plant and has pistillate (female) flowers on the upper portion and the staminate (male) flowers on the lower portion of the raceme. Inflorescence develops either at the apex (terminal raceme) or at axils of (axillary racemes). By the action of a single recessive gene (*f*) which exists in the pistillate line, the sex tendency moves either towards femaleness or towards maleness on a raceme. When pistillate and staminate flowers follow an interspersed situation along the entire raceme axis, the situation is known as 'interspersed monoecious'. Seed coat ornamentation and sex reversal phenomenon are unique for castor bean plant.

Figure 8.11: Sex reversion in castor plant.

Breeding and Development of Hybrid Castor Plants

As mentioned earlier, *Ricinus communis* L. shows interesting sex variation that may or may not be influenced by environmental changes. There is a considerable variation both within and among genotypes in proportion of pistillate and staminate flowers on racemes; this proportion can be strongly influenced by environment. This sensitivity of the sex genes to environment was utilized for the production of F_1 hybrid seed (Zimmerman, 1966).

In normal-monoecious varieties, the percentage of pistillate flowers on racemes is usually highest on the first, decreasing on subsequent racemes. Works from Indian agricultural institutions showed that there is a high heritability value for number of pistillate flowers onto a raceme and subsequently number of capsules on the main raceme (Sindagi, 1965).

Pistillate lines are of three types- N, S, and NES. In N-type pistillate line, a single switching recessive gene (*f*) governs the pistillate condition similar to genetic male sterility in other crops. The S pistillate type was originally obtained by selection within sex reversal variants at Israel. The S type behaves like polygenic complex with dominance and epistasis (Shifris, 1960). This system operates through dominant pistillate line, selected from sex revertants and maintained through unidentified environmentally sensitive genes for interspersed staminate flower expression. The NES-system is homozygous for sex-switching gene (*f*), and contains environmentally sensitive genes that control staminate flower development. (Ankineedu and Rao, 1973). The development of an S-type pistillate line (VP-1) from a cross between two complex genotypes had marked a significant breakthrough in hybrid castor development in India. VP-1 is a pistillate line from Virjapur pistillate line, whose description is given in Table 8.2. In this, TSP-10R is one of the S type exotic pistillate plants from USA and JP-5 is another pistillate line. Mostly used females in commercial hybrid castor bean were derived from VP-1.

Table 8.2: General description of VP-1.

1.	Growth habit	Dwarf to Medium
2.	Stem color	Green
3.	Bloom	Triple
4.	Branching	Convergent
5.	Nodes up to primary raceme	
	Range	12 to 2
	Mean	16
6.	Nature of internodes	Condensed
7.	Leaf shape	Cup
8.	Sex expression	Stability of femaleness depends on method of maintenance and multiplication employed
9.	Spike	Semi compact
10.	Capsule	Spiny
11.	Shattering	Partial shattering
12.	Seed size	Length 1.62 cmBreadth 2.32 cm
13.	Seed Color	Light Chocolate
14.	Seed Molting	Conspicuous
15.	Thousand Seed Weight (gm)	260
16.	Days to flowering	40-55

A cross was made at the field of D. O. R., Rajendranagar, Hyderabad, from seeds of castor plants obtained from one female plant (VP-1), one male parent (48-1), to form their hybrid their hybrid (GCH-4). This breeding line has sharp morphological differences as described in Table 8.3.

Table 8.3: Important morphological characteristics of male, female parents and its hybrid.

Characters	*Male Parent (48-1)*	*Female Parent (VP-1)*	*Hybrid (GCH-4)*
Plant height	Tall	Dwarf	Medium
Nature of internodes	Normal	Condensed	Normal
Stem color	Red	Green	Light red
Bloom	Double	Triple	Triple
Leaf shape	Flat	Cup	Flat
Branching	Divergent	Convergent	Divergent
Days of flowering	55-65	40-55	45-50
Nodes upto pr. Raceme	16-22(17)	12-20(16)	14-18(16)

Contd...

Table 8.3–Contd...

Characters	*Male Parent (48-1)*	*Female Parent (VP-1)*	*Hybrid (GCH-4)*
Sex expression	Mostly female	Irregular	Female
Spike nature	Loose	Semi-compact	Loose
Capsule nature	Non spiny	Spiny	Spiny (partially)
Seed colour	Deep chocolate	Light chocolate	Light chocolate
Seed size	Medium	Small	Large
Seed length	1.64 cm	1.62 cm	1.79 cm
Seed width	2.36 cm	2.32 cm	2.36 cm
Seed weight	0.233 gm	0.257 gm	0.276 gm
Seedling weight	2.40 gm	1.55 gm	2.50 gm

Jyoti DDS-9, Kranti PAK-4, GCH-5, RHC-1, TMVCH, NPH-1, GAUCH-4, DCH-177, 48-1 are a few improved varieties of castor grown in various parts of India. These varieties have different growth requirements and they vary in their stress response.

Production

According to an estimate done by FAO in 2011 the total area in the world harvested for castor oil is 1923597 Ha, with a total seeds produced accounts to 20598.76 tonnes. The total annual yield of castor oil seed throughout the world is 14387.36 kg/ha, with a total production summing up to 2767548 tonnes. The largest producers of castor oil seed are India, China, Brazil and Mozambique (Figure 8.12). These countries account for approximately 94 per cent of total castor produced in the world.

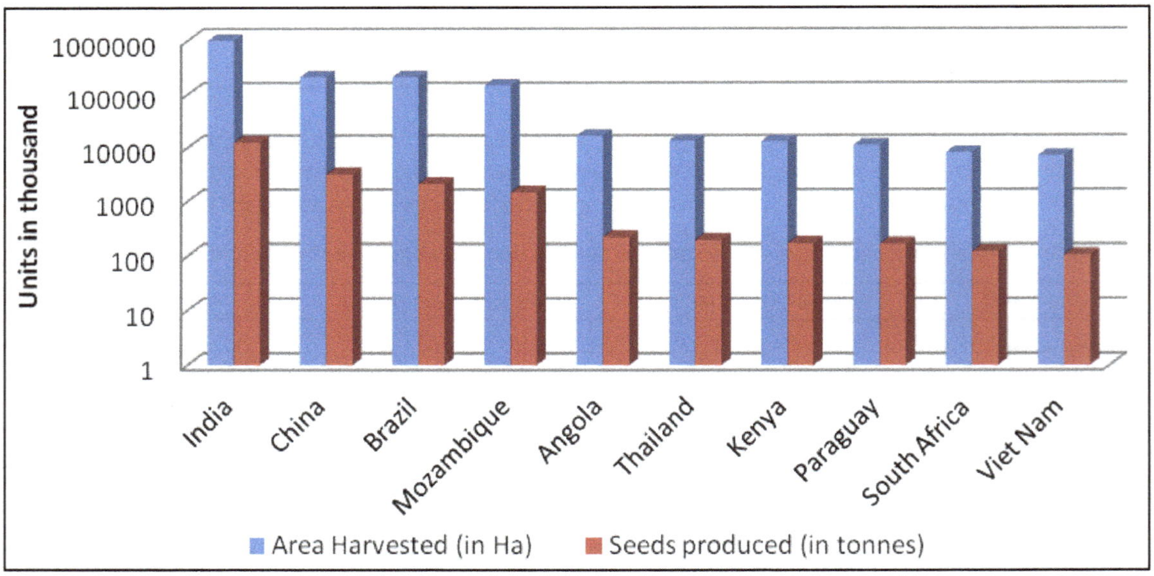

Figure 8.12: Largest Castor-producing countries and their production volume (*Source*: FAOSTAT, http://faostat.fao.org).

India is the largest producer of castor, contributing to almost 76 per cent of the world share. Gujarat is the leading state in castor seeds production in India followed by Rajasthan and Andhra Pradesh. Total area under Castor crop in India for the year 2011-12 is 11.50 lakh hectares. Estimated total production of Castor Seeds in India for the year 2011-12 is 16.19 lakh tones. Average yield for the year 2011-12 is 1417 kg/hectare as opposite to 1453 kg/Ha in the year 2010-2011 experiencing a decrease of 3 per cent (Table 8.4).

According to a latest crop survey report done by Solvent Extractors' Association (SEA) of India the estimated total production of castor seed is expected to go further down by 30 per cent in the year 2012-13.

Table 8.4: State-wise Area, Yield and Production of Castor Seeds in India (2011-12)*.

District	*Estimated Area Under Crop ('000 Ha)*			*Estimated Production ('000 tonnes)*			*Estimated Yield (kg/Ha)*		
	2010-11	*2011-12*	*Per cent Change*	*2010-11*	*2011-12*	*Per cent Change*	*2010-11*	*2011-12*	*Per cent Change*
Gujarat	483	697	44 per cent	900	1226	36 per cent	1863	1760	–6 per cent
Rajasthan	127	169	33 per cent	180	207	15 per cent	1417	1222	–14 per cent
Andhra Pradesh	198	230	16 per cent	140	156	11 per cent	707	677	–4 per cent
Other States	52	54	3 per cent	30	30	1 per cent	576	564	–2 per cent
Total	860	1150	34 per cent	1250	1619	30 per cent	1453	1417	–3 per cent

Source: SEA of India.

Draft Genome Sequence and Annotation of *Ricinus communis*

Castor bean is primarily of economic interest as a source of castor oil, used for the production of high-quality lubricants because of the presence of high proportion of the unusual fatty acid ricinoleic acid in it. However the castor bean seed has high content of ricin, which is an extremely toxic protein. The easy access and convenient purification of ricin has raised concerns about its possible use in bioterrorism. Therefore, knowledge of the genetics and enzy-mology of fatty acid metabolism in castor bean seeds is important in efforts to ensure a sustained supply of hydroxy fatty acids without the complications posed by the toxicity of ricin (Chan *et al.*, 2010).

The whole genome sequencing of *R. communis* was carried out in J. Craig Venter Institute, TIGR, USA. Castor bean seeds from inbred cultivar Hale49 (NSL 4773) obtained from the National Center for Genetic Resources Preservation (NCGRP) at Ft. Collins, Colorado (Germplasm Resources Information Network) were used. Using flow cytometry it was estimated that castor bean has 10 chromosomes ($2n$ =20), with a genome size of ~320 Mb. Due to the presence of limited genomic information and no available genetic map, the draft sequence of the castor bean genome was generated by producing ~2.1 million high-quality sequence reads from plasmid and fosmid libraries. Then Celera assembler was used to build contigs that were linked to form 25,800 scaffolds. The assembly covered the genome ~4.6x, spanning 350 Mb. On the whole >50 per cent of the genome was repetitive DNA, out of which one-third were retrotransposons and, <2 per cent were DNA transposons.

Protein-coding genes were annotated using multiple gene-prediction programs, homology searches against sequence databases and the cDNA spliced-alignment tool PASA (program to assemble spliced alignments). For genome annotation 52,165 expressed sequence tags (ESTs) from five cDNA non-normalized libraries were generated. Using PASA, these and other castor bean cDNA sequences from GenBank were aligned and predicted 5,491 genes and another 688 genomic regions where no gene had been predicted.

The annotation of the genomic sequence mainly emphasized on three main aspects: (a) Ricin gene family; (b) oil metabolism genes; (c) disease resistant genes.

The Ricin Gene Family

The ricin gene encodes three domains: an N-terminal RIP domain and two C-terminal lectin domains. The draft genome revealed the presence of 28 putative genes in the family, including potential pseudo genes or gene fragments. Now, with the whole genome sequence available, it is possible to produce ricin-free castor plants by silencing or knocking out the genes responsible for the production of ricin (Ramprasad *et al.).*

Oil Metabolism Genes

Due to the importance of castor bean as an oil seed crop, upon annotation 71 genes were found that were involved in the biosynthesis of fatty acid and triacyglycerols. It was observed that most of these pathways corresponded mainly to ricinoleic acid and tri-ricinolein. Castor bean has evolved an oleic acid hyroxylase which is utilized in the synthesis of ricinoleic acid, and also has developed capacity to build up high amount of ricinoleic acid in its seed oil.

The better understanding of the biosynthetic pathway of ricinoleic acid and the genes associated with it would help to transfer and engineer these genes into other conventional oil seed crops which unlike castor bean do not contain toxic proteins like ricin.

Disease Resistance Genes

One hundred and twenty-one disease-resistance proteins were identified in castor bead which can help in understanding and improving biotic stress resistant species of Euphorbiaceae, especially for cassava. The most of these proteins belonged to the nucleotide bind-ing–leucine-rich repeat class, followed by the less common extracellular leucine-rich repeat–containing proteins.

Conventional and Commercial Uses of Castor Oil and Wax

Castor bean is cultivated for the seeds which yield viscous, pale yellow non-volatile and non-drying oil. The oil extracted from castor seeds is inexpensive, environment friend and has a good shelf-life. It is somewhat difficult to imagine that, barely a few decades ago, castor oil was valued chiefly as a purgative. Through concerted research and development efforts castor oil has been elevated from the humble status of domestic medicinal oil to versatile hi-tech oil with a string of sophisticated applications. Food grade castor oil is used in food as additive, flavouring, candy and as preservative. It is also used as food grain preservative.

The chemical structure of castor oil is of great interest because of the wide range of reactions it affords to the oleochemical industry and unique chemicals that can be derived from it. These derivatives are at par with petrochemical products for use in several industrial applications. In fact, they are considerably superior since they are from renewable sources, biodegradable and eco-friendly.

Castor isostearate succinate a derivative of castor oil, is a polymeric mixture of esters with isostearic acid and succinic acid used for skin conditioning, such as in shampoo, lipstick and lip balm. Due to the presence of ricinoleic acid in castor oil, it exerts anti-inflammatory effects. In traditional medicines it is used to treat skin disorders, burns, sunburns, cuts and abrasions. The oil is also used as a rub or pack for various ailments, including abdominal complaints, headaches, muscle pains, inflammatory conditions, skin eruptions, lesions, and sinusitis.

On industrial level castor also finds its use in transportation, cosmetics, pharmaceutical and manufacturing industries. The oil is extracted from castor seeds. Due to its water resistant quality castor oil is used for coating fabrics, protective coverings of food containers and guns. Castor oil has low temperature viscosity and high temperature lubrication properties, making it useful as a lubricant in jet, diesel, and race car engines.

It is the primary raw materials for the production of sebacic acid, which is the basic ingredient for the production of nylon. It is also used for electrical condensers, carbon paper manufacturing, and it is the source of synthetic scents and fruit flavors (esters) like jasmine, apricot, peach, palm, rose, banana, and lemon flavors. The oil is also used for making soaps, inks, plastics, preserving leathers and burning lamps.

Castor oil cake (pomace) makes an excellent fertilizer. Another application of castor plant is in silk industry. Eri-silk caterpillars (*Philosamia ricini*) feed on castor leaves. As 25 per cent leaf plucking does not affect the oil-seed production, castor leaves can therefore be utilized for silkworm rearing, which would give additional income to farmers.

Castor wax is a hard, brittle, vegetable wax. It is produced by the hydrogenation of pure castor oil, in the presence of a nickel catalyst. It is odourless and insoluble in water. Castor wax is used in polishes, cosmetics, electrical capacitors, carbon paper, lubrication, coatings and greases, oils and petrochemical products. Castor wax is also useful in polyurethane coating formulation which is used as a top coat varnish for leather, wood and rubber.

Future Prospects for the Use of *Ricinus communis*

A bioenergy crop *Ricinus communis* has potential of phytoremediation of DDT and Cd co-contaminated soils. It has been shown that *R. communis* has great potential for removing DDTs and Cd from contaminated soils owing to its fast growth, high biomass, strong absorption and accumulation for both DDTs and Cd (Huanga,2011).

It has been reported that the root extracts of *R. Communis* contains flavonoids and saponins which possess mast cell stabilizing and antianaphylactic activity hence shows antiasthamatic activity (Taru, 2011).

Ricinus communis concentrated ethanolic leaf extract had shown effective antibacterial activity in comparison with the standard Gentamycin (Kota C. S., 2011). Antimicrobial activity of various extracts of roots of *Ricinus communis* were screened against pathogenic microorganisms such as *Escherichia coli, Staphylococcus aureus, Pseudomonas aeruginosa, Salmonella typhimurium, Proteus vulgaris, Bacillus subtilis, Candida albicans* and *Aspergillus niger.* It also inhibits the mitochondrial respiratory chain reactions (Verma *et al.*, 2011).

Castor oil is currently less expensive vegetable oils, and can be used in the production of biodiesel. Both cloud and pour points decline as more castor oil biodiesel is added to petroleum diesel. Hence it has a higher level of stability at lows temperatures, it an ideal combustible for those regions with extreme seasonal weather as it doesn't require any kind of additives to conserve its fluidity (Forero, 2004).

India being the largest producer of castor oil seeds, hence castor oil presents itself as a huge source of raw material for biodiesel production which would help in achieving energy security and sustainability. Also the castor crop has several advantages like high yielding varieties, short production cycle and consistent, superior yield which make it of a better-quality than other biodiesel crop (Anandan *et al.*, 2012).

Conclusion

Enormous scope exists for genetic improvement of castor through *in vitro* techniques adopting biotechnological tools. Genetic improvement in castor was initially based on selection followed by mutation breeding which led to transition of an economically unproductive semi-wild castor to a high yielding, dwarf and early maturing cultivated castor.

The major objective of castor genetic transformation is to develop ricin-free castor plant varieties with the potential to create new economic opportunities. This problem is tackled through selection and traditional breeding methods, advanced generation lines (F_6) with 70–75 per cent reduction in ricin and *Ricinus communis* agglutinin (RCA_{120}) toxins have been developed (Auld *et al.*, 2001 and Auld *et al.*, 2003). The second strategy is to develop transgenics using antisense genes for blocking ricin and allergen production that are highly expressed during seed development. Efforts are also being made for "knocking out" the genes responsible for ricin formation (Sujatha *et al.*, 2008).

In the search for biodegradable and environmentally friendly fuels, the use of castor oil has proven to have technical and ecological benefits, and stands as an opportunity for agricultural development in arid and impoverished areas (Gressel, 2008). From the aforesaid discussion it appears that castor seed may be used in global scenario for fossil fuel crisis, as it is a good candidate for biodiesel production especially by using genetically modified castor plants without ricin and RCA.

Hence the importance of castor bean plant (*Ricinus communis*) cannot be overlooked. It has been used for domestic purposes since ages and its commercial as well as market value cannot be under emphasized. With the draft genome sequence of castor bean plant along with its annotation will help us understand the oil metabolism and improve the castor oil yield. Also the knowledge of ricin gene family and disease resistant genes would help in the development of ricin free castor bean plant varieties and genetically improved plants with better disease resistance.

Acknowledgments

Authors are thankful to BTISNet SubDIC (BT/BI/065/2004) for providing internet facilities and the Government of Jharkhand, Department of Agriculture for providing infrastructure development fund (5/B. K. V/Misc/12/2001). RB is thankful to BIT for getting Cumulative Professional Development Grant (CPDG; Ref No. GO/PD/2011-12/269/3523; dated, August 04/2011) during preparation of this manuscript. We are thankful to Director, DOR, Rajendranagar for kind help and materials.

References

Akpan, U. G., Jimoh, A. and Mohammed, A. D. 2006. Extraction, Characterization and Modification of Castor Seed Oil. Leonardo Journal of Sciences (8): 43-52.

Anandan S., Gowda N. K. S. and Sampath K. T. 2012. Biofuel co-products as livestock feed: Opportunities and challenges. ISBN 978-92-5-107299-8 pp. 339-349.

Ankineedu, G. and G. P. Rao. 1973. Development of pistillate castor. Indian J. Genet. Plant Breed. 33: 416–422.

Audi, J., M. Belson, M. Patel, J. Schier, and Osterloh, J. 2005. Ricin poisoning: A comprehensive review. JAMA 294: 2342–2351. doi: 10.1001/jama.294.18.2342.

Auld, D. L., S. D. Pinkerton, K. A. Lombard, C. K. Murphy, K. E. Kenworthy, W. D. Becker, R. D. Rolfe, and Ghetie, V. 2003. Registration of TTU-LRC castor germplasm with reduced levels of ricin and RCA120. Crop Sci. 43: 746–747.

Auld, D. L., R. D. Rolfe, and McKeon, T. A. 2001. Development of castor with reduced toxicity. J. New Seeds 3: 61–69.

Chan A. P. *et al.,* 2010. Draft genome sequence of the oilseed species *Ricinus communis.* Nature Biotechnolgy, 28, 951–956.

Chaudhari R. K., Mukhopadhyay K., Chaudhari I. and Bandopadhyay R. 2004, Recent Advances in Life Sciences: 25-43 ISBN: 81-7736-230-5.

Classen C, and Hoffman A. 1950. The inheritance of pistillate character in castor and its possible utilization in the production of commercial hybrid seeds. J. Am. Soc. Agron. 42: 79-82.

Fischer WN, Kwart Andre B, Rentsch D, Krolkiewicz S, Tegeder M, Breitkreuz K and Frommer WB. 1998. Amino acid transport in plants. Trends. Plant Sci. 3: 188-195.

Forero C. L. B, 2004-10-12. Biodiesel from Castor Oil: A Promising Fuel for Cold Weather.

Flukiger and Hambury. 1892. Pharmacography 567. (c. f. Watts Dictionary of Economic Products of India).

Gressel J.2008. Transgenics are imperative for biofuel crops. Plant Sci, 174, pp. 246–263.

Harland S. C. 1922. Inheritance in *Ricinus communis* L. Part 1. J. Genet. 10: 207-218.

Hartley, M. R. and Lord, J. M. 2004. Cytotoxic ribosome-inactivating lectins from plants. Biochim. Biophys. Acta 1701: 1–14.

Hooker, J. D. 1890. Flora of British India. V. 5. AL. Reeve and Co, London.

Huanga H, Yua N, Wang L, Gupta D. K., *et al.* 2011. The phytoremediation potential of bioenergy crop *Ricinus communis* for DDTs and cadmium co-contaminated soil. Bioresource Technology, Vol. 102(23): 11034-11038.

Katayama Y. 1952. The progeny of the female individuals in the castor bean. Jap. J. Genet. 23: 19. 1958 [Pl. Breed. Abstr. 22930 : 215.

Kota C. S., Manthri S. *et al.*, 2011. Antibacterial activity of *Ricinus communis* leaf extract. IJPSR, Vol. 2(5): 1259-1261.

Kulkarni LG and Ramamurthy GV. 1977. Castor (2nd ed.) ICAR. New Delhi.

Kurita M. 1952. Secondary association of chromosome in castor oil plant. 1946. Jap. J. Genet. 21: 63.

Liljegren SJ, Ferrandiz C, Alvarez-Buylla ER, Pelaz S and Yanofsky MF. 1998. Arabidopsis MADS-box genes involved in fruit dehiscence. Flowering Newsl. 25 : 9-19.

Masur NG. 1933. Castor breeding in Bombay Presidency. Agric. Live-Stk. India. 3: 125-143.

Narain A. 1953. Bloom characters in castor oil plant (*Ricinus communis* L.). Curr. Sci. 21: 166-167.

Patwardhan GW. 1931. A preliminary note on inheritance of castors. J. of Indian Bot. Soc. 10: 100-109.

Peat, J. E. 1926. Genetic studies in *Ricinus communis* L. J. Genet. 19: 373-389.

Ramprasad R. and Bandopadhyay R. 2010. Future of *Ricinus communis* after completion of the draft genome sequence. Current Science, Vol 99 (10): 1316-1318.

Ravi Kumar G. and Sujatha M. 1998. Genetic transformation experiments in castor (*Ricinus communis* L.). National Seminar of Oilseeds and Oil Research and Development needs in the Millennium. Feb 2-4,2000, DOR, Hyderabad, India.

Richharia RH. 1937. A note on the Cytogenetics of *Ricinus communis*, L. Indian J. Agr. Sci. 7: 707-711

Shifriss O. 1960. Conventional and unconventional systems controlling sex variations in Ricinus. J. Genetics. 57: 361-388.

Sindagi S. S.1965. Genetic variability and correlation coefficient relating to yield a few other quantitative characters and use of selection indces in castor. Plant Breed. 36: 2466.

Sujatha M., Reddy, T. P. and Mahasi M. J. 2008. Role of biotechnological interventions in the improvement of castor (*Ricinus communis* L.) and *Jatropha curcas* L. Biotechnology Advances. Vol. 26 (5): 424–435.

Taur D. J. and Patil R. Y. 2011. Antiasthmatic activity of *Ricinus communis* L. roots. Asian Pacific Journal of Tropical Biomedicine. S13-S16.

Verma S. K., Yousuf S, Singh S. K., Prasad GBKS and Dua V. K. 2011. Antimicrobial potential of roots of *Riccinus communis* against pathogenic microorganisms. International Journal of Pharma and Bio Sciences, Vol. 2(1): 545-548.

Weiss EA. 1983. Tropical Agricultural Series: Oilseed Crops. Longman.

White OE. 1918. Breeding new castor beans. J. Hered. 9: 195-200.

Zimmerman L. H. and Smith J. D. 1966. Production of F_1 Seed in Castorbean by Use of Sex Genes Sensitive to Environment. Crop Science, Vol.6 : 406-409.

2014, Future Crops Vol. 2 Pages 199–211
Editor: Professor K.V. Peter
Published by: DAYA PUBLISHING HOUSE, NEW DELHI

Chapter 9
Citron

Lallan Ram*
Processing Laboratory
National Research Centre for Citrus, Amravati Road, Nagpur – 440 010, Maharashtra
**E-mail : lallanram09@gmail.com*

Citron (*C. medica* L.) is a fruit of *Citrus* species belonging to Rutaceae family. The citron fruits are characterized by thick rind impeded with very thick white albedo inner layer, adherent to segments which cannot be separated easily. Internationally, the fruits are used for fragrance or zest of its outer peel but the most important part is albedo, widely employed in the food industry at larger scale. Thus, from ancient through medieval, the citron was used mainly for medical purposes to combat sea sickness, pulmonary troubles, intestinal ailments, and other disorders. The essential oil of the flavedo (the outermost, pigmented layer of rind) was also regarded as an antibiotic. Citron juice with wine was considered an effective antidote to poison.

The fruit's name derives ultimately from Latin, citrus, also the origin of the genus name, and as a result it has many similar names in many European languages, *e.g. cederat, cédrat, cedro*, etc. Indeed, into the 16th century, the English name citron included the lemon and perhaps the lime as well. The botanical classification of fruits is described as under.

Kingdom :	Plantae
Division :	Magnoliophyta
Class :	Magnoliopsida
Subclass :	Rosidae
Order :	Sapindales
Family :	Rutaceae
Genus:	*Citrus*
Species :	*medica*

A fruit better known to most consumers in its preserved rather than in its natural form, the citron, *Citrus medica* Linn., is called in French, cedrat, cidratier, citronnier des Juifs; in Spanish, cidra, poncil, poncidre, cedro limón, limón cidra, limón Francés, though in Central America it is often referred to as toronja, the popular Spanish name for grapefruit. In Portuguese, it is cidraol, in Italian, cedro or cedrone; in German, cedratzitrone orcederappelen; in Dutch, citroen; in India, citron, beg-poora, or leemoo; in Malaya, limau susu, limau mata kerbau, limau kerat lingtang; in Thailand, som-mu, som manao or som ma-nguâ; in Laos, manao ripon, mak vo or mak nao; in Vietnam, thank-yen or chanh; in Samoa, tipolo or moli-apatupatu; in China, kou-yuan. Theophrastus wrote of it as the Persian, or Median, Apple, and it was later called the Citrus Apple.

Origin and Distribution

The citron's place of origin is not well defined but seeds were found in Mesopotamian excavations dating back to 4000 B. C. The armies of Alexander the Great are thought to have carried the citron to the Mediterranean region about 300 B. C. A Jewish coin struck in 136 B. C. bore a representation of the citron on one side. A Chinese writer in AD 300 spoke of a gift of "40 Chinese bushels of citrons from Ta-ch'in" in AD 284. Ta-ch'in is understood to mean the Roman Empire. The citron was a staple, commercial food item in Rome in AD 301. There are wild citron trees in Chittagong, Sitakund Hill, Khasi and Garo hills of northern India. Dioscorides mentioned citron in the 1st Century AD and Pliny called it malus medica, malus assyria and citrus in AD 177. The fruit was imported into Greece from Persia (now Iran). Greek colonists began growing the citron in Palestine about 200 B. C. The tree is assumed to have been successfully introduced into Italy in the 3rd Century. The trees were mostly destroyed by barbarians in the 4th Century but those in the "Kingdom of Naples" and in Sardinia and Sicily survived. By the year 1003, the citron was commonly cultivated at Salerno and fruits (called poma cedrina) were presented as a token of gratitude to Norman lords. For centuries, this area supplied citron to the Jews in Italy, France and Germany for their Feast of the Tabernacles (sukkot) ceremony. Moses had specified the cone of the cedar, hadar (kedros in Greek) and when it fell into disfavor, it was replaced by the citron, and the Palestine Greeks called the latter kedromelon (cedar apple). Kedros was Latinized as cedrus and this evolved into citrus, and subsequently into citron. For many years, most Citrus species were identified as botanical varieties of *Citrus medica*. Spaniards probably brought the citron with other Citrus species to St. Augustine, Florida, though it could have survived there only in greenhouses. The tree was introduced into Puerto Rico in 1640. Commercial citron culture and processing began in California in 1880. The trees suffered severe cold damage in 1913 and, within a few years, the project was abandoned. From 1926 to 1936, there were scattered small plantings of citron in Florida, and particularly one on Terra Ceia Island, supplying fruits to the Hills Brothers Canning Company. The groves eventually succumbed to cold and today the citron is grown in southern Florida only occasionally as a curiosity. The main producing areas of citron for food are Sicily, Corsica and Crete and other islands off the coasts of Italy, Greece and France, and the neighboring mainland. Citron is also grown commercially in the central, mountainous coffee regions of Puerto Rico. Some is candied locally but most is shipped in brine to the United States and Europe. Citron is casually grown in several other islands of the Caribbean and in Central and South America. It has been rather commonly grown in Brazil for many years. There have long been scattered citron trees in the Cauca Valley of Colombia. After 5 years of study, horticulturists decided in 1964 that commercial culture could be profitable. Citron trees are not uncommon in some of the Pacific Islands but are rare in the Philippines.

Taxonomy and Nomenclature

Although the Buddha's Hand tree bears fruits that are different from other citrons in shape, it is otherwise a typical member of the species. It is usually considered as one variety in the United States, but in China there are actually at least a dozen named Buddha's Hand varieties or subvarieties, differing in fruit shape, color and size, and the tree's growing habit, etc. These varieties are classed in several types grown for specific purposes, such as fruit production or bonsai tree sales.

In Yunnan, a variety called *"Muli"* or *"Xiangyanggo"* has fruit characteristics intermediate between the common and the Buddha's Hand citron. Along the same lines, Hodgson writes that in one clone "only parts of the fruits are fingered and the rest are corrugated, lacking in flesh, and contain seeds hanging free in the locules." There also exists a variegated form. Clearly the CVC accession is a standard Buddha's Hand (no pulp, no seeds, no variegation), but its particular variety or subvariety has not yet been determined.

"Buddha's Hand citron" and "Fingered citron" are alternative names for the same variety or group of varieties. In China, where numerous scientific studies of the Buddha's Hand citron have been published, its Chinese name is often mistranslated into English as "bergamot," which properly refers to *C. bergamia* [*C. aurantium x C. limetta*].

Important Attributes of Citron Fruit

The citron fruit is usually ovate or oblong, narrowing towards the stylar end. However, the citron's fruit shape is highly variable, due to the large quantity of albedo, which forms independently according to the fruits' position on the tree, twig orientation, and many other factors. The rind is leathery, furrowed and adherent. The inner portion is thick, white and fleshy; the outer is uniformly thin and very fragrant. The pulp is usually acidic, but also can be sweet, and even pulpless varieties are found.

Most citron varieties contain a large number of monoembryonic seeds. They are white, with dark innercoats and red-purplish chalazal spots for the acidic varieties, and colorless for the sweet ones. Some citron varieties are also distinct, having persistent styles which do not fall off after fecundation. Those are usually promoted for etrog use.

Some citrons have medium-sized oil bubbles at the outer surface, medially distant to each other. Some varieties are ribbed and faintly warted on the outer surface. There is also a fingered citron variety called Buddha's hand.

The color varies from green, when unripe, to a yellow-orange when overripe. The citron does not fall off the tree and can reach 4–5 kg, if not picked before fully mature. However, they should be picked

Figure 9.1: Citron fruit (Arunachal Pradesh accession).

Figure 9.2: Citron fruit (NEH accession).

before the winter, as the branches might bend or break to the ground, and may cause numerous fungal diseases for the tree.

Fruit accession from Arunachal Pradesh indicated that the fruiting season is midseason and dates from May to June. Each plant bears approximately 217 fruits. The shape of fruit is oblique (asymmetric) fruit base shape is convex and that of apex is mammiform weighs around 86.8 g, the diameter of each fruit is 48.6 mm and length is 79.36 mm. The colour of epicarp is green- yellow and surface texture is rough. Rind is 2.48 mm thick and the albedo (mesocarp) is strongly adhered to the pulp (endocarp). Oil glands are conspicuous and the density is medium. Albedo is greenish in colour and areola is absent. Fruit is strongly attached to the stalk; there are 11 segments in each fruit which are strongly adhered to each other. The segment walls are thick and fruit axis is solid with the diameter of 6.8 mm. Pulp is white in colour and the fruit contains medium juice. The acidity of fruit is 3.1 per cent and TSS is 6.56 ° Brix.

Accession of fruit from NEH region indicated that the fruiting season is midseason and dates July. Each plant bears approximately 89 fruits. The shape of fruit is oblong, fruit base shape is convex and that of apex is mammiform weighs around 170 g, the diameter of each fruit is 63.85 mm and length is 103.8 mm. The colour of epicarp is green- yellow and surface texture is rough. Rind is 6.87 mm thick and the albedo (mesocarp) is strongly adhered to the pulp (endocarp). Oil glands are strongly conspicuous and the density is low. Albedo is greenish in colour and areola is absent. Fruit is strongly attached to the stalk; there are 12.5 segments in each fruit which are strongly adhered to each other. The segment walls are thick and fruit axis is solid with the diameter of 10.5 mm. Pulp is yellow in colour and the fruit contains very less juice. The acidity of fruit is 3.84 per cent and TSS is 6.1 ° Brix.

Seed

Accession of seed from Arunachal Pradesh indicated that there are 28 seeds per fruit, length of the seed is 7.47 mm and width 5.06 mm. Weight of 20 seeds is approximately 1.56 g. seeds are semi-spheroid in shape and texture is smooth, colour is cream and the cotyledon colour is light yellow to cream and the seeds are monoembryonic.

Another accession of seed from Arunachal Pradesh indicated that there are 71 seeds per fruit, length of the seed is 9.52 mm and width is 5.24 mm seed thickness is 3.16mm. Weight of 20 seeds is approximately 1.33 g. seeds are semi-spheroid in shape and texture is smooth, colour is yellowish and the cotyledon colour is light yellow to cream and the seeds are monoembryonic.

Food/Nutritional Value of Fruits

About 87-88 per cent of available moisture is in fruits. It is a rich source of vitamin C, calcium, phosphorous and fibre. The details of the nutritional value of fruits are given in per hundred gram of edible portion of the fruits (Tables 9.1 and 9.2). Most of the citron fruits have very thick outer part of the peel which also contain large amount of moisture, acidity, sugars, vitamins like ascorbic acid and energy to the same extend indicating that it is highly useful for industrial implication point of view.

Table 9.1: Value per 100 g of edible portion.

Parameters	*Moisture*	*Protein*	*Fat*	*Fiber*	*Ash*			
Value (g)	87.1	0.081	0.04	1.1	0.41			
Parameters	*Calcium*	*Phosphorus*	*Iron*	*Carotene*	*Thiamine*	*Riboflavin*	*Niacin*	*Ascorbic Acid*
Value (mg)	36.5	13.0	0.55	0.009	0.052	0.029	0.125	368

Table 9.2: Chemical characteristics of citrus peel of Darunj (*Citrus medica*).

Parameter	*Moisture (Per cent)*	*Acidity (Per cent citric acid)*	*Total Sugars (Per cent)*	*Reducing Sugars (Per cent)*	*Ascorbic Acid (mg/100g)*	*Energy (kcal/g)*
Values	8946	*0.40*	12.17	9.28	87.49	0.03

Plant

Citrus medica is a slow-growing shrub or small tree that reaches a height of about 2 to 5 m. It has irregular straggling branches and stiff twigs and long spines at the leaf axils. The evergreen leaves are green and lemon-scented with slightly serrate edges, ovate-lanceolate or ovate elliptic 2.5 to 7.0 inches long. Petioles are usually wingless or with minor wings. The flowers are generally unisexual providing self-pollination, but some male individuals could be found due to pistil abortion. The clustered flowers of the acidic varieties are purplish tinted from outside, but the sweet ones are white-yellowish.

Figure 9.3: Arunachal Pradesh citron accession.

The acidic varieties include the Florentine and Diamante citron from Italy, the Greek citron and the Balady citron from Israel. The sweet varieties include the Corsican and Moroccan citrons. Between the pulpless are also some fingered varieties and the Yemenite citron.

The citron tree is very vigorous with almost no dormancy, blooming several times a year, and is therefore fragile and extremely sensitive to frost.

Figure 9.4: NEH region citron accession.

Accession of tree from Arunachal Pradesh indicated that shape of tree is obloid with drooping growth habit. Tree vigour and branch density are medium, the branch angle is wide and shoot tip colour is purple. An adult tree has medium spine density and the spine length is 13.6 mm.

The accession from NEH region indicated that the shape of tree is spheroid with drooping growth habit. Tree vigour is medium and branch density is sparse, the branch angle is medium and shoot tip colour purple. An adult tree has medium spine density and spine length is 14.7 mm.

Leaf

Accession from Arunachal Pradesh indicated that vegetative life cycle of leaf is evergreen with simple leaf division. The intensity of green colour of leaf blade is light. The attachment of leaf lamina is brevipetiolate and length is 79.66 mm, width is 45.51 mm and the length/width leaf lamina ratio is 1.75. Leaf lamina is elliptic in shape with dentate margin, the apex is emarginated. The length of petiole is 4.92 mm.

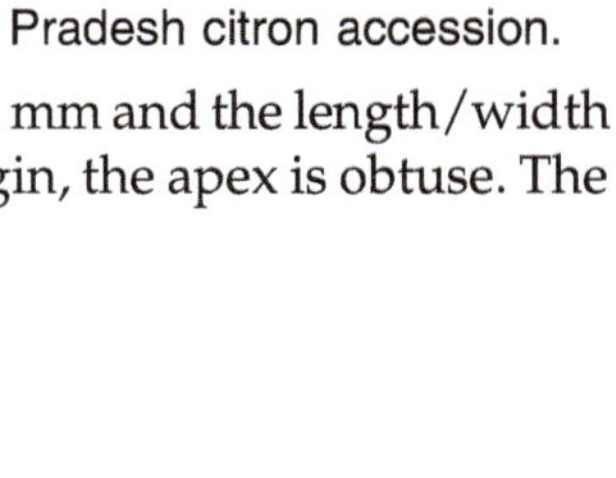

Figure 9.5: Arunachal Pradesh citron accession.

Vegetative life cycle of leaf from NEH region is evergreen with simple leaf division. The intensity of green colour of leaf blade is light. The attachment of leaf lamina is brevipetiolate and length is 85 mm, width is 41 mm and the length/width leaf lamina ratio is 2.07. Leaf lamina is elliptic in shape with dentate margin, the apex is obtuse. The length of petiole is 4.5 mm.

Flower

Figure 9.6: NEH region citron accession.

Flowers from different accessions indicated that the flowering month for Citron from Arunachal Pradesh is from Nov- Dec; flower bud is purple in colour 17.45 mm in length and 4.85 mm in width. The pedicel length is 3.72 mm; anthers are shorter in length as compared to stigma. The flowers are white in colour and Hermaphrodite type, made up of five petals. The length of petal is 19.45 mm and width is 4.22 mm. The colour of anthers is yellow and the stamens are 28.

The flower from citron of NEH region accession indicated that flowering month for Citron is December; flower bud is purple in colour 11.17 mm in length and 3.35 mm in width. The pedicel length is 2.68 mm; anthers are shorter in length as compared to stigma. The flowers are white in colour and hermaphrodite type, made up of five petals. The length of petal is 10.16 mm and width is 3.53 mm. The colour of anthers is yellow and the stamens are 33.

Growth Habit

The citron is borne by a slow-growing shrub or small tree reaching 2.4-4.5 m high with stiff branches and stiff twigs and short or long spines in the leaf axils. The leaflets are evergreen, lemon-scented, ovate-lanceolate or ovate elliptic, 6.25-18 cm long; leathery, with short, wingless or nearly wingless petioles; the flower buds are large and white or purplish; the fragrant flowers about 1 1/2 in (4 cm) wide, in short clusters, are mostly perfect but some male because of pistil abortion; 4- to 5-petalled, often pinkish or purplish on the outside, with 30 to 60 stamens. The fruit is fragrant, mostly oblong, obovoid or oval, occasionally pyriform, but highly variable; various shapes and smooth or rough fruits sometimes occurring on the same branch; one form is deeply divided from the apex into slender sections; frequently there is a protruding style; size also varies greatly from 9-22.8 or 30 cm long; peel is yellow when fully ripe; usually rough and bumpy but sometimes smooth; mostly very thick, fleshy, tightly clinging; pulp pale-yellow or greenish divided into as many as 14 or 15 segments, firm, not very juicy, acid or sweet; contains numerous mono embryonic seeds, ovoid, smooth, white within.

Quality Parameters

Climate

The citron tree is highly sensitive to frost; does not enter winter dormancy as early as other *Citrus* species. Foliage and fruit are easily damaged by very intense heat and drought. Best citron locations are those where there are no extremes of temperature.

Soil

The soils where the citron is grown vary considerably, but the tree requires good aeration. However, for better plant growth and yield, well drained and soil rich in nutrition is suitable for citron. Experimental orchard soil at The National Research Centre for Citrus, Nagpur is more suitable for citron which is performing well.

Cultivation of Citron

Citron trees are grown readily from cuttings from branches 2 to 4 years old and quickly buried deeply in soil without defoliation. For quicker growth, the citron may be budded onto rough lemon, grapefruit, sour orange or sweet orange but the fruits do not attain the size of those produced from cuttings, and the citron tends to overgrow the rootstock. Rough lemon is too susceptible to gummosis to be employed as a rootstock for citron in Colombia. The 'Etrog', to be acceptable for ritual use, must not be budded or grafted.

Culture

The citron tree tends to put out water sprouts which should be eliminated, and the grower should prune branches hanging so low that they touch the ground with the weight of fruit. Italian producers keep the tree low and stake the branches, and may even trim off the thorns, to avoid scarring of the fruits. The trees begin to bear when 3 years old and reach peak production in 15 years and die in about 25 years.

In 'Etrog' orchards, the Israeli growers are careful to take every precaution to protect the fruit, tying the fruiting branch securely in place and trimming away any twigs that might touch the fruit. To avoid moving irrigation equipment through the groves, the trees are manually watered and frequently sprayed to eliminate destructive insects.

If citrons are allowed to fully ripen on the tree they will be very aromatic and the peel yellow, the inner peel very tender. In India, a fruiting branch may be bent down and the immature fruit put into a jar shaped like a human head (or other form) so that the mature fruit will be of the same shape. These are sold as curiosities and are said to be intensely fragrant.

Genetic Diversity in Citron

It is suggested that the cultivated citrus are derived from the three true species, citron, pummelo, and mandarin. These three species reproduce sexually and if different cultivars within the species are intermated, the progenies are similar to their parents. The other important types (orange, grapefruit, lemon, and lime) originated from one or more generations of hybridization between these ancestral genera. Most of the cultivars of orange, grapefruit, and lemon are believed originated from nucellar seedlings or budsports. Currently citrus fruits have high level of morphologic variations and various fruit characteristics because of inter and intra specific interaction. Consequently, the amount of genetic diversity within these groups is relatively low, in spite of there being many named varieties. Conversely, mandarins, pummelos and citrons have higher levels of genetic diversity since many of the cultivars have arisen through sexual hybridization.

Citron, the major progenitor of some commercial Citrus cultivars like all true lemons and rough lemon was reported as one of the "basic" true Citrus species. Besides, lemon was notified as a hybrid of citron and sour orange in recent studies. Rough lemon was reported to be closely related with the citrons in previous studies and was also reported as a hybrid of mandarin and citron. *Citrus volkameriana* was reported as a hybrid between citron and sour orange.

In a recent study genetic diversity in citron, lemon, rough lemon and *C. volkameriana* group was carried out using SRAP and SSR markers. They evalutaed 56 accessions were evaluated. Similarity level of citrons to other accessions was ~0.70. Four citron accessions - 'Buddhas Hand', 'Etrog' and two Turkish selections- were distinguished clearly. Similarity level of citron and lemon-rough lemon group was 0.65 based on the ISSR data.

Varieties of Citron

This fruit has three types of varieties as given below:

Varieties	*Species*
Acidic-pulp varieties	Diamante citron, Greek citron, Balady citron, Florentine citron
Non-acidic varieties	Morocoan citron, Corsican citron
Pulpless varieties	Buddha's hand citron, Yemenite citron

Types of Citron Cultivars

Citron cultivars are mainly of two types, those with pinkish new growth, purple flower buds and purple-tinted petals, acid pulp and dark inner seed coat and chalazal spot, those with no pink or purple tint in the new growth nor the flowers, with non-acid pulp, colorless inner seed coat, and pale-yellow chalazal spot. Among the better-known cultivars are:

a) **'Diamante'** ('Cedro Liscio'; possibly the same as 'Italian' and 'Sicilian')–of unknown origin but the leading cultivar in Italy and preferred by processor's elsewhere; long-oval or ellipsoid, furrowed at base, broadly nippled at apex; peel yellow, smooth or faintly ribbed; very thick, fleshy; pulp crisp, non-juicy, acid; seedy. Tree small, spreading, thorny as 'Corsican'. Very similar is a cultivar called "Earle" in Cuba

b) **'Corsican'**–origin unknown but the leading citron of Corsica; introduced into the United States around 1891 and apparently the cultivar grown in California; ellipsoid or faintly obovate, furrowed at base; large; peel yellow, rough, lumpy, very thick, fleshy; pulp crisp, non-juicy, non-acid, seedy. Tree small, spreading, moderately thorny with some large spines.

c) **'Etrog'** ('Ethrog', 'Atrog'; *C. medica* var. *Ethrog* Engl.)– The leading cultivar in Israel; ellipsoid, spindle-shaped or lemon-like with moderate neck and often with persistent style at base; usually with prominent nipple at apex; medium-small as harvested; if not picked early, it will remain on the tree, continuing to enlarge for years until the branch cannot support it. For ritual use, the fruit should be about 142 g and not oblong in form. Peel is yellow, semi-rough and bumpy, faintly ribbed, thick, fleshy; flesh is crisp, firm, with little juice; acid; seedy. Tree is small, not vigorous; leaves rounded at apex and cupped. This cultivar is the official citron for use in the Feast of the Tabernacles ritual but if unavailable any yellow, unblemished, lemon-sized citron with adhering style can be substituted.

Figure 9.7: Fingered citron.

Figure 9.8: Madhankri citron.

Figure 9.9: Chhangura citron.

d) **Fingered Citron'** called *fu shou* in China, *bushukon* in Japan, *limau jari, jeruk tangan, limau kerat lingtang,* in Malaya; *djerook tangan* in Indonesia; *som-mu* in Thailand; *phât thu* in Vietnam. The fruit is corrugated, wholly or partly split into about 5 finger-like segments, with little or no flesh; seedless or with loose seeds. The fruit is highly fragrant and is placed as an offering on temple altars. It is commonly grown in China and Japan and is candied in China.

e) In India, there are several named types, in addition to the 'Fingered', in the northwest: **'Bajoura':** small, with thin peel, much acid juice

f) **'Chhangura':** believed to be the wild form and commonly found in a natural state; fruit rough, small, without pulp

g) **'Madhankri'** or 'Madhkunkur': fruit large with sweetish pulp

h) **'Turunj':** fruit large, with thick peel, white inner part sweet and edible; pulp scant, dry, and acidic. Leaves are oblong and distinctly notched at the apex.

Pests and Diseases

The citron tree is undoubtedly subjected to most of the pests which attack other *Citrus* species. The citrus bud mite (*Eriophyes sheldoni*), citrus rust mite (*Phyllocoptruta oleivora*), and snow scale (*Unaspis citri*) are among its major enemies.

Horticulturists in Florida report that citron trees are nearly always unthrifty, are subject to gummosis, and usually in a state of decline and dieback, and are accordingly poor bearers.

Branch knot, caused by the fungus *Sphaeropsis tumefaciens,* was first noticed on citron trees in Puerto Rico in 1977. By 1983, it became a serious threat to the local citron industry. The deformations become large and necrotic, led to witches' broom, dieback and breaking of branches.

Maturity and Harvesting

The citron tree blooms nearly all year, but mostly in spring and the spring blooms produce the major part of the crop. The fruit is dark-green when young, takes 3 months to turn yellow. To retain the green color, firmness and uniformity desired by the dealers in candied citron, the fruit must be picked when only 12.5-15 cm long and 7.5-10 cm wide. Mature trees yield an average of 30 kg/year but exceptional trees have borne as much as 68-100 kg. 'Etrog' fruits are wrapped in hemp fiber immediately after picking.

Processing and Value Addition

In India, citron cultivation has not been done on commercial scale as its availability is meager and unorganized. The National Research Centre for Citrus, Nagpur has a very large collection of citrus species but its utilization is limited and untapped. Being an important fruit all can be utilized for preparation of different value added products. An attempt has been made to utilize the whole citron fruits, juice, peel and leaf for developing the value-added products like juice beverage, juice powder and its shelf-life studies under ambient storage conditions. Extraction of peel oil, peel pectin and leaf oil use latest juice processing methods as applied successfully in other fruits (Jayaraman and Dasgupta, 1992), including spray juice drying technique for developing the juice powder and squashes, a well known juice beverages.

Citron Juice Powder Development

Figure 9.10: Juice powder.

Fruits with their attractive colour, pleasant taste and distinctive aroma were consumed since time immemorial. Fruits are seasonal commodities and are highly perishable in nature. The latest juice processing technique for product development is adopted as it offers products of high quality like high flavour and aroma retention, nutritional value, minimal change in colour, appearance and easily rehydrable of final product compared with other drying methods. The protocols for manufacturing the juice powder were standardized and accordingly the juices were fed @ 2.5ml/min in spray dryer under vacuum condition (-35 to -40 mm/WC) using 0.77 micron spray nozzle. Citron juice powder was manufactured and the flavour, colour, aroma, TSS, and limonin content were similar to the fresh juice (Tables 9.4 and 9.5). The shelf life of powder was better beyond one year in air tight poly glass container in ambient storage condition except Vit. C which was slightly altered. (Ram and Kumar, 2012)

Food Uses

The most important part of the citron is the peel which is a fairly important article in international trade. The fruits are halved, depulped, immersed in seawater or ordinary salt water to ferment for about 40 days, the brine being changed every 2 weeks; rinsed, put in denser brine in wooden barrels for storage and for export. After partial de-salting and boiling to soften the peel, it is candied in a strong sucrose/glucose solution. The candied peel is sun-dried or put up in jars for future use. Candying is done mainly in England, France and the United States. The candied peel is widely employed in the food industry, especially as an ingredient in fruit cake, plum pudding, buns, sweet rolls and candy.

Puerto Rican food technologists reported in 1970 that the desalted citron could be dehydrated in a hot air tray dryer at 42.22° C, reducing the weight by 95 per cent to lower costs of shipment, then stored in polyethylene bags and later reconstituted and candied. In 1979, after further experiments, it was announced that fresh citron cubes, blanched for 1/2 minute in water at 76.7° C can be candied and the product is equal in quality to the brined and candied peel, and this procedure saves the costs of salt, storage, and shipping of heavy barrels. If the citron lack flavor, a few orange or lemon leaves may be added to the syrup.

The fruit of the wild 'Chhangura' is pickled in India. In Indonesia, citron peel is eaten raw with rice. The entire fruit of the 'Fingered citron' is eaten.

If there is sufficient juice in better cultivars, it is utilized for beverages and to make desserts. In Guatemala, it is used as flavoring for carbonated soft-drinks. In Malaya, citron juice is used as a substitute for the juice of imported, expensive lemons. A product called "citron water" is made in Barbados and shipped to France for flavoring wine and vermouth. In order to expand the market for citron, Puerto Rican workers have established that the green-mature fruits can be peeled by immersing in a boiling lye solution to save the labor of hand-peeling and then the fruits can be made into marmalade, jelly, and fruit bars which are crusty on the outside, soft within. In Spain, syrup made from the peel is used to flavor unpalatable medical preparations.

Citron fruit juice and its squashes during storage at ambient conditions revealed that carotenoid content was stable upto 120 days. While limonin, naringin, hesperidin and non enzymatic browning value were recorded to be unchanged. Vit C was found reduced drastically after 120 days in storage. The limonin content was between the threshold limit (6 ppm). (Ram and Kumar, 2012)

Table 9.3: Biochemical composition of fruit juice and squash.

Parameters	*Fresh Juice*	*Squash (Initial)*	*Storage (30 ± 2°C Temp)*	
			360 (days)	*LSD (p=0.5)*
TSS (per cent)	6.00	41.8	46.0	NS
Acidity (per cent)	4.80	1.86	2.05	NS
Vit C (mg/100ml)	26.00	16.00	2.50	2.05
Limonin (ppm)	6.25	5.96	4.01	0.03
Naringin (per cent)	0.98	0.41	2.09	NS
Hesperidin (per cent)	1.52	0.14	1.66	NS
Carotene (mg/ml)	0.00	0.011	0.00	NS
NEB (OD440 nm)	0.28	1.813	0.00	0.05

Table 9.4: Chemical composition of citron juice powder.

Chemical Composition	*Fresh Juice*	*Juice Powder in Ambient Storage (30 ±2 C)*		
		Initial	*6 Months*	*12 Months*
TSS (per cent)	9.0	9.0	9.00	9.00
Acidity (per cent)	6.14	1.08	1.76	1.34
Vit C (mg/100ml)	32.0	10.0	8.00	9.00
Limonin (ppm)	5.57	3.41	3.61	4.02
Carotenoids (mg/ml)	–	0.05	0.05	0.04

Table 9.5: Physical composition of citron juice powder.

Physiological Characters	*Fresh Juice*	*Juice Powder in Ambient Storage (30 ±2 C)*		
		Initial	*6 Months*	*12 Months*
Colour Lightness/darkness	1.27	1.00	0.79	3.47
Greenness	–0.10	–0.12	0.11	0.17
Yellowness	0.35	0.04	–0.50	–0.72
Orange	–0.28	–3.00	–0.22	–0.23
Flavour	7.5	7.0	7.0	7.0
Aroma	8.1	7.4	7.5	7.5
Texture	8.0	7.2	7.1	7.0

Other Uses

Fruit

Chinese and Japanese people prize the citron for its fragrance and it is a common practice in central and northern China to carry a ripe fruit in the hand or place the fruit in a dish on a table to perfume the air of a room. The dried fruits are put with stored clothing to repel moths. In southern

China, the juice is used to wash fine linen. Formerly, the essential oil was distilled from the peel for use in perfumery.

Peel Leaves and Twigs

In some of the South Pacific islands, "Cedrat Petitgrain Oil" is distilled from the leaves and twigs of citron trees for the French perfume industry. The leaf and peel oil of citron grown in NRCC, Nagpur was extracted using hydro distillation technique. Table 9.6 showed that recovery of oil from leaf was more than that of peel, however quality of peel was better in respect of Aldehyde content (Ram and Kumar, 2012)

Table 9.6: Physico-chemical quality of peel and leaf oil in citron fruits.

Parameter	*Oil (per cent)*	*Aldehyde (per cent)*	*Optical rotation (+degree)*	*Refractive Index*	*pH*	*Temp (°C)*	*Colour*				*Visual Colour*
							*L**	*a**	*b**	*a/b**	
Leaf	0.14	1.87	49.15	1.28	6.0	27.0	4.99	–0.39	0.73	–0.53	Whitish yellow
Peel	0.052	2.18	60.20	1.36	6.0	27.2	2.11	–0.09	0.04	–2.25	yellowish

*L: Lightness/darkness, a: Greenness; b: Yellowness; a/b: Orange.

Flowers

The flowers are distilled for essential oil which has limited use in scent manufacturing.

Wood

Branches of the citron tree are used as walking-sticks in India. The wood is white, rather hard and heavy, and of fine grain. In India, it is used for agricultural implements.

Medicinal Uses

In ancient times and in middle Ages, the 'Etrog' was used as a remedy for seasickness, pulmonary troubles, intestinal ailments and other disorders. Citron juice with wine was an effective purgative to rid the system of poison. In India, the peel is a remedy for dysentery and is eaten to overcome halitosis. The distilled juice is given as a sedative. The candied peel is sold in China as a stomachic, stimulant, expectorant and tonic. In West Tropical Africa, the citron is used only as a medicine, particularly against rheumatism. The flowers are used medicinally by the Chinese. In Malaya, a decoction of the fruit is taken to drive off evil spirits. A decoction of the shoots of wild plants is administered to improve appetite, relieve stomachache and expel intestinal worms. The leaf juice, combined with that of *Polygonum* and *Indigofera* is taken after childbirth. A leaf infusion is given as an antispasmodic. In Southeast Asia, citron seeds are given as a vermifuge. In Panama, they are ground up and combined with other ingredients and given as an antidote for poison. The essential oil of the peel is regarded as an antibiotic.

References

Bonavia, E. 1888-90. The Cultivated Oranges and Lemons of India and Ceylon. London, W. H. Allen.

Jayaraman, K. S. and Dasgupta, D. K. 1992. Dehydration of fruits and vegetables-recent development in principle and technologies. *Drying Tech.* **10**: 1.

Lallan Ram L, and Kumar D, 2012. Underutilized citron (*Citrus medica L.*) fruit for development of value-added products and their ambient storage *Indian J. Hort.* 69(1): 140-143.

Morton, J. 1987. Citron. (In) Fruits of warm climates. Julia F. Morton(Ed), Miami, FL. p.179-182.

Ranganna, S. 1986. *Handbook of Analysis and Quality Control of Fruit and Vegetable Products* (2nd Edn.), Tata McGraw Hill Pub. Co. Ltd. New Delhi, India, p. 11-12.

Ram, L. 2008. Flavedo oil and leaf oil content in different citrus germplasm. *Ann. Rep. NRCC*, Nagpur, p. 67-68.

Ram, L. and Singh, S. 2006. Medicinal important of citrus products and by products- A review. *Agril. Rev.* **27**: 170-79.

Sharma, S. K., Sharma, P. C. and Lal Kaushal, B. B. 2004. Storage studies of foam mat dried hill lemon (*C. pseudolimon* Tan) juice powder. *J. Food Sci. Tech.* **41**: 9-13.

2014, Future Crops Vol. 2 Pages 213–224
Editor: Professor K.V. Peter
Published by: DAYA PUBLISHING HOUSE, NEW DELHI

Chapter 10
Delphiniums

*K.K. Dhatt**
Department of Floriculture and Landscaping,
Punjab Agricultural University, Ludhiana – 141 004, Punjab
**E-mail: dhattkiran@gmail.com*

Flowering annuals are grown usually for landscaping and commercially for seed production under northern Indian climatic conditions. The half-hardy annuals of temperate region are the best adapted as seasonal flowers during winter *viz.*, October–March. Under north - western plains of India, annuals pass their juvenile phase in Dec-Jan and bloom profusely in March–April and set seed in April–May. In India, commercial floriculture is also growing at 20-25 percent rate in the domestic and export market. The area under flower cultivation has doubled from 52,000 ha in 1993-94 to 103, 000 ha in 2001-02 (Satyanaryanan, 2006). Punjab is earning Rs. 70 million of foreign exchange by producing 90 tonnes of flower seed from 400 ha of land annually to Holland, France, England, Germany, USA and Japan (Nancy 2006). Emphasis of government on diversification, free trade policy under WTO, small land holding, assured irrigation and progressive nature of the farmers would enhance the area under floricultural crops.

Delphiniums are a genus of around 250 species in Ranunculaceae family and consist of annual, biennial and herbaceous perennial species. These species are mainly found in mountainous areas worldwide, except Australia and Polar Regions (Christopher Brickell, 1996). The modern hybrid *Delphinium* cultivars are selected for garden use and many species are cultivated as bedding plants in background, borders or in front of a tall fence or hedge and as cut flowers. The dwarf Chinese Delphinium is used as pot plant, window boxes and rock gardens. *Delphinium ajacis* is an annual species with 30-60 cm height and bears blue, violet, rose pink or white flowers grown for seed production in winter in North India. These annual delphiniums are generally called larkspur to distinguish them from the perennial kinds. The hardy perennial kinds fall into two main groups: the tall, spire flowered Elatum varieties and the shorter, branching Belladona varieties. The Elatum cultivar reaches up to 2.5m height and bears flowers in long spikes from late spring to mid summer. These types make the boldest display and are magnificent plants for the back of herbaceous border. Belladona cultivars are relatively small type growing 100-125 cm high with wiry stems and looser spikes. These are more graceful and are used to break up heavier masses of bloom in the garden and for as cut flower.

The perennial Pacific hybrids of *Delphinium elatum* are the most admired producing flower spikes of 1.75m. The perennial delphinium thrive best in the hills but it can be grown as an annual in northern plains and neighboring states which have sufficiently cool weather. The delphinium is propagated from seed, cuttings and divisions. In the hills the seeds are sown in February-March under heat to raise early seedlings to make them flower in August- September. In the Northern plains the seeds are sown during late September or early October and the plants flower during March-April. The vegetative propagation is only possible in the hills, where delphinium can be grown as perennial. Delphinium requires a deeply dug and rich soil with adequate drainage. A medium loam is preferable to a light sandy soil. Where the soil is light, dig in deeply plenty of compost or old farm yard manure before planting. Delphinium flowers are heavy feeders and require regular fertilization for healthier and superior blooms. The location of site should be such that it receives sun light at least for 8-hours. Larkspur grows the best in well drained soils with a pH of 6.0-7.0. The cultivation on raised bed improves drainage resulting in better plant growth and flowering. There are three basic colours of delphinium flowers, yellow, red and blue. This means that the beautiful violet delphinium of today is the result of hybridization. With the help of hybridization a few other colour have emerged like pale sky blue, indigo blue, purple, lavender, pink and pure white. Although modern hybrids include shades of red, pink, yellow and white but graceful delphinium will always be associated with the colour blue. Most species are easy to grow, though they do the best in cooler areas. A reciprocal relationship was found between brightness and lightness on a CIELAB (Commission Internationale de l'Eclairage) diagram in which flower colour ranges from pale to vivid. The types and concentration of anthocyanins are important factors in flower coloration; the composition of such pigments as non-acylated delphinium glycosides, viodelphin and cynodelphin produced in the calyx immediately reflect on perceived colour. (Hashimoto, *et al.*, 2000).

Origin and History

The name delphinium is derived from the Greek word "dolphin", because of the shape of the unopened flower bears definite resemblance to that of dolphin. Other names are, lark's heel (Shakespeare), lark's claw and knight's spur. The earliest references to delphinium cultivation are in England and the only species indigenous to the British Isles is *Delphinium ajacis*. About the middle of sixteenth century, the Italian species, *Delphinium consolida* was imported into England, while *Delphinium staphisagria* was brought from Italy during seventeenth century. Apparently no delphinium were cultivated in English gardens before the eighteenth century, when French nurseryman Vilmorin offered seeds of *Delphinium alatum* for sale. This species is the original parent of all our modern hybrid delphiniums. Shortly after the introduction of *Delphinium alatum*, *Delphinium formosum* and *Delphinium grandiflorum* arrived on the scene, although it was not until middle of nineteenth century that the popular delphinium that we know today appeared. In 1857, the grand old firm of Nancy offered its customers seventeen varieties, and continued to produce new hybrids for many years. The firms of Lemonie and Kelway produced 1091 varieties between 1878 and 1918. The delphinium flower has an interesting story in both real life and in myth. During early 20th century Frank Reinelt – who was head gardener for the Romanian Queen, Marie – relocated to the United States and started his own nursery. In his gardens, he began hybridizing delphiniums, and by the 1930s he fathered one of the more popular breeds, the Pacific Giant. In Greek mythology, the delphinium flower was believed blossomed from the blood of Aias – or Ajax – and the petals were inscribed with the word ai, ai - or alas, alas. The delphinium flower is said to represent big-heartedness and the essence of divine qualities. They are also meant to symbolize levity, fun and a general sense of joy. As a gift, these flowers brighten up any arrangement, be it fresh or dried. Given their general meaning of warmth and buoyancy, it is not surprising that these blossoms are also representative of summery July births (Ippolito and Henry, 1975).

Taxonomy

This genus *Delphinium* of Ranunculaceae family consists of annual and herbaceous perennial from northern temperate zone and mountains farther south and diffused throughout the cooler parts of Asia. The plants have erect stems and leaves are, alternate, or digitately deeply lobed with 3-7 toothed, pointed lobes in a palmate shape. The main flowering stem is erect, and varies greatly in size between the species from 10 cm in some alpine species, up to 2 meters tall in the larger meadowland species. It is topped with a raceme of many flowers, varying in color from purple and blue to red, yellow or white. The flower has five petal-like sepals which grow together to form a hollow pocket with a spur at the end, which gives the plant its name. Within the sepals are four true petals. The seeds are small and often shiny black and terminal racemes of spurred flowers with 5 petaloid sepals, contrasting coloured petals, 1 or 2 of them have nectiferous spurs inserted into large sepal spur. The fruit consists of erect clusters of 1-3, sometimes to 5 cylindrical follicles (Linda, 1984).

Species

There are about 300 species distributed in Europe and North America with a few isolated species originated in Tropical Africa and South America. In the United States, they thrive in such areas as the Atlantic coast of Boston, and the Pacific Northwest (Ippolito and Henry, 1975). A brief description of some important species is given below:

Delphinium ajacis

Delphinium ajacis is native to Mediterranean region. The plants grow up to 45 -100 cm high and are generally alike with small finely divided dark green linear leaves about 6 cm long. The inflorescence is loose spike of blue flowers. The flower colour varies from rose – pink, white, blue or violet and scarlet. Strains like Giant Imperial are of particular value for their habit of branching from the base and are ideal for cut flower. The popular larkspurs are hybrids of *Delphinium ajacis*.

Delphinium consolida

Delphinium consolida is commonly known as forking larkspur and this is one of the wild species of larkspur. It grows wild in cornfields, but has become very rare nowadays. The flowers are commonly purple, but a white variety exists as well. A plant with much branched stems and leaves, it prefers chalky soil.

Delphinium brunonianum

An interesting dwarf species which rarely exceeds 25 cm, found throughout the great Himalaya range and in Afghanistan and China. It is a tufted perennial with hairy, with 5 lobed kidney shaped basal leaves which are somewhat musk scented when bruised. The hooded flowers have purple tinted blue sepals and almost black petals.

Delphinium cardinale

Native of California and Mexico, this fleshly rooted perennial is surprisingly handy in a sheltered well drained site though tended to be short lived. This is one of the least hardy and unusual species, slow growing, compact habit with pale green with much divided palmate foliage. It has deeply 5-lobed leaves and branched racemes of scarlet, cupped blossoms of 2-3 cm diameter.

Delphinium grandiflorum

This is a short lived perennial; usually grown as annual. Plants grow 60-100 cm high and bear flowers in blue and lilac shade which are the best suited for planting in large groups. Leaves are 5-lobed with each lobe further divided into narrow segments

Delphinium hybrids

This group includes the Belladonnas, the Elatum and Pacific Giants. Growing up to 2 m high, and it flowers in early to mid summer in shades of blue, white, cream, pink, lilac and purple.

Most of the modern perennial delphinium was evolved from crosses between *Delphinium elatum* and other species like *Delphinium formosum, D. cardinale, D belladonna, D. nudicaule* (Herbert Aswith, 1961). *Delphinium elatum* is a natural tetraploid species but Dr Legro succeeded in overcoming sterility barrier when he made a number of interspecific crosses at diploid level and tetraploided the resulting plants and then successfully married them to hybrid *elatums*.

Cultivation and Uses

At present delphinium is rarely used in folk medicine. The seeds of *Delphinium staphisagria* are deadly poisonous and contain several alkaloids. This is helpful in curing diseases or irritations of the reproductive system, as well as aiding in mental and acute respiratory ailments. Earlier minute doses of extracts from these seeds were used to relieve pain, but the practice was abandoned. Poisoning from these sources is characterized by gastro enteric irritation accompanied by vomiting, diarrhoea followed by heart failure and death. All parts of the plant contain an alkaloid delphinine and are very poisonous. In small amounts, extracts of the plant are used in herbal medicine (Ippolito and Henry, 1975). Gerard's *Herbal* reports indicate that drinking the seed of larkspur helps against the stings of scorpions, and a tincture is used against asthma and dropsy. Larkspur, especially tall larkspur (*Delphinium barbeyi, D. occidentale, D. glaucescens, D. glaucum*) is a significant cause of cattle poisoning on rangelands in the Western United States. Larkspur is more common in high-elevation areas, and many ranchers will delay moving cattle onto such ranges until late summer when toxicity of the plants is reduced. Death is through cardio toxic and neuromuscular blocking effects, and occurs within a few hours of ingestion. Studies conducted on low larkspur (*Delphinium nuttalianum*) revealed that it is a toxic range plant and is often fatal to cattle when ingested during spring or early summer on foothills or mountain rangelands. Grazing materials near Price and Utah(USA) during 1996-1997 examined toxicity and consumption of low larkspur by cow-calf pairs. Toxic alkaloid concentrations were relatively stable although larkspur plants rapidly matured. Stage of plant growth also affected consumption as the cattle apparently avoided eating low larkspur before flowering and eat little if other forage is available. Species of tall larkspur- *Delphinium barbeyi, D. oxidentale, D. glaucum-* pose a serious poisoning threat to cattle on many summer ranges. Toxic alkaloid concentration in pods was x=2.9mg/g exceeded leaf alkaloid concentrations (x=1.5mg/g) in all species. Leaves showed a more rapid decrease in concentration of toxic alkaloids with plant maturity compared to pods (Gardner and Pfister, 2000).

Larkspur poisoning kills sporadically 5-15 per cent of the cattle on North American mountain rangelands. *Delphinium* spp. contains more than 40 different diterpenoid alkaloids with variable prevalence and concentration. Several of these alkaloids, including highly toxic methyllycaconitine, reversibly bind and block nicotinic acetylcholine receptors (Stegelmeier, 2007).

The antibacterial activity of the aqueous and alcoholic extracts of the root, stem, leaf, flower and fruit of *D. ajacis* [*Consolida ambigua*] against *Agrobacterium tumefaciens* and *Streptococcus mutans* was determined, alone or in combination with antibiotics. All the extracts inhibited the growth of

A. tumefaciens. Except for the alcoholic extract of the roots of *D. ajacis*, all the extracts inhibited the growth of *S. mutans*. The growth of *A. tumefaciens* and *S. mutans* were inhibited by the aqueous flower and leaf extracts of *D. ajacis*, respectively (Neelam and Bohra, 2006).

Delphinium is a medicinally important genus, several species of which are reported from Himalayan region. The survey conducted in Himachal Pradesh elucidated genetic diversity in *D. denudatum and studied phylogenetic relationship among Delphinium species inhabiting the area and the region inhabited three Delphinium* species *viz*., *D. denudatum*, *D. vestitum*, and *D. brunonianum*. From these, *D. vestitum*, and *D. brunonianum* had only small population of 150 and 16 plants respectively. The third species delphinium denduatum exhibited relatively large populations with 1-5 plants/m^2 as per the RAPD data, *D. denudatum* populations in the study area have evolved from two lineages (Sharma *et al*., 2007). The study further suggested *in situ* conservation such as botanical reserves of *D. vestitum*, and *D. brunonianum* due to low density and *ex situ* conservation of *D denudatum* like field banks and seed banks (Sharma *et al*., 2007).

Larkspurs (*Delphinium* species) grow wild on grazing meadows and stony slopes in the Himalaya region, with some species adapted to subtropical and others to sub temperate and temperate climatic conditions [Polunin and Stainton 1984; Chowdhery and Wadhwa 1984; Herbarium of Forest Research Institute (FRI), Dehradun, India; Herbarium of Botanical Survey of India (BSI), Dehradun, India]. Four species (*D. cashmerianum*, *D. kolzii*, *D. pyramidale* and *D. roylei*) might have been lost from previously reported locations, *i.e. D. cashmerianum* from Manglad Nala in Keylong and Bhaba Basin near the bridge on the way to Kardang in District Lahaul and Spiti (BSI, FRI); *D. kolzii* from Great Himalayan National Park, Kullu; *D. pyramidale* from Kangra and *D. roylei* from Kullu (FRI,), as no plants were sighted at these locations. Some of the alkaloids possess antifungal activities against human and plant pathogenic fungi and antifeedant properties for insects, including *Spodoptera littoralis* and *Leptinotarsa decemlineata* (Rahman *et al*., 1997; Gonzalez-Coloma *et al*., 1998). The bio molecules from *D. denudatum* are also being studied as potential cures for human diseases. Aqueous extracts from roots are shown to exhibit properties of antiepileptic drugs (Raza *et al*., 2001, 2004) and anti hepatotoxicity (Subramanium and Pushpangadan 1999). *D. denudatum* extracts are tried in morphine de-addiction (Zafar *et al*., 2001).

Delphinium species are known for their toxicity to livestock and excessive consumption can cause animal death (Pfister and Gardner, 1999; Pfister *et al*., 1999). Pastes and extracts of flowers and roots of many *Delphinium* species are used traditionally as insecticides, as ingredients of drugs for dysentery and diarrhoea, as tonics for toothache, as cardiac and respiratory depressants and also as a stimulant (Said 1970; CSIR 1986; Prajapati and Kumar 2003). Several alkaloids, which are derivatives of the norditerpenoid lycoctonine were isolated from roots and leaves of *Delphinium* species, including *D. brunonianum*, *D. vestitum* and *D. denudatum* (Bhandary *et al*., 1990; Rahman *et al*., 1997)

Propagation and Planting

Delphinium is propagated from seeds, cuttings or division. In the hills, seeds are sown in February –March under heat to raise early seedlings to make them flower in August -September. In the Northern plains, the seeds are sown during late September to early October and the plants bloom in March-April. Planting can be done directly *in situ* or the seedlings are transplanted after sowing nursery. The perennial types are planted through cuttings. Under sub-tropical condition of Punjab delphinium is being planted directly *in situ* or the seedling are transplanted in last week of October for landscaping and seed production. This annual is in full bloom during April - May and seed pods are picked manually in May. The optimum time for transplanting is end of October and the seed yield decreased as planting was delayed beyond first week of November. *Delphinium ajacus* transplanted on 20th October also produced the tallest plants (92.67cm) and gave the longest flowering duration of 47.97 days. The maximum seed yield (14.41g/m^2) and germination (85.67 per cent) were also recorded in this planting as reported by Dhatt(2008). Culture condition of plant growth regulators was examined for promoting efficient plant regeneration from petiole explants of eleven commercial cultivars of delphinium series and Pacific giant series. The optimum concentration of individual cytokine for shoot regeneration was 10 mg/L 6- benzylaminopurine, in 1mg/litre N phenyl-N' -1,2,3 –thiadiazol-5-yl urea and 5-10 mg/L zeatin, in combination with 1mg/litre2,4 dichlorophenoxy acetic acid (Hosokawa, *et al.*, 2001). Delphinium seedlings were grown in plant cells of different types. Transplanting was at the cotyledon stage or at the 2-true-leaf stage. Seedlings were exposed to low temperature treatment during the night. Cut flower length and weight improved by some low temperature treatments. Early transplanting decreased root weight (Miyake and Ubakata, 2003).

Breeding

These flower stems grow from 12 cm to roughly 2.5 m in height, the blossoms of which sprout from a raceme and consist of five sepals which grow together and form a hollow bowl shape with a spur at the tip, while four delicate petals grow at the center. These blossoms traditionally come in bright, creamy white, dainty pink, and a marvelous purple blue combination; however, some of the highbred flowers can be seen in shocking reds, oranges and yellows. The plants flower from late spring to late summer, and are pollinated by butterflies and bumble bees. Most species are toxic. Despite the toxicity, *Delphinium* species are used as food plants by the larvae of some *Lepidoptera* species including Dot Moth and Small Angle Shades. Baker's larkspur (*Delphinium bakeri*) and Yellow larkspur (*D. luteum*), both native to very restricted areas of California, are highly endangered species. Positive correlation was observed in *Delphinium grandiflorum* between the days to bolting and the days to flowering in northern areas of Japan. (Sasaki *et al.*, 2008).

Random amplified polymorphic DNA (RAPD) markers were used to assess genetic diversity in three species of toxic larkspurs (*Delphinium spp*.). Based on the presence or absence of the 144 polymorphic RAPD bands in individuals, a dendrogram was generated to assess the genetic similarity among the samples. The cophenetic values were 0.64 between D. *occidentale* and *D. barbeyi*, and 0.55 between the cluster of these two species and *D. glaucum*. These relationships are congruent with those based on morphological characters and support the contention that these are separate species. Allozyme diversity was evaluated in four closely related texa of the Delphinium series Fissa distributed throughout the western Mediterranean region. *Delphinium bolosii* and Delphinium mansaentium are narrowly endemic to the eastern Iberian Peninsula. *Delphinium fissum* sp *sorbidum* is found in a few population across Peninsula (Li, *et al.*, 2007). The effects of plant density on pollination and seed set were explored in the larkspur (*Delphinium nuttallianum*) and monkshood (*Aconitum columbianum*). At a Rocky Mountain Biological Laboratory site in the Colorado Rocky Mountains, flowers of *D. nuttallianum* are pollinated primarily by queen bumble bees, solitary bees, and hummingbirds. The results showed that the quantity of pollination service (pollinator visitation rate and pollen deposition) was at best weakly related to density. In contrast, seed set declined by approximately one-third in sparse populations relative to nearby dense populations. This decline may stem from the receipt of low-quality pollen, for example, inbred pollen. Alternatively, scarcity may indicate poor environmental conditions which lower seed set for reasons unrelated to pollination (Bosch and Waser, 1999).

Delphinium luteum, an endangered larkspur, is restricted to two wild populations near Bodega Bay, California, USA. The total number of individuals in these two populations is estimated to be

<100. Alloenzyme and random amplified polymorphic DNA (RAPD) markers are used to (1) assess levels and patterns of genetic diversity in one wild population and two cultivated populations and (2) to test the hypothesis that *D. luteum* is of hybrid origin between *D. decorum and D. nudicaule*. These data will be used to aid in developing a management plan to conserve the species. The wild population maintains high levels of genetic diversity. Genetic data indicate that both cultivated populations, especially the north Sonoma population, have several alloenzymes and RAPD markers not found in the wild population and could be used to establish new populations of *D. luteum* or to enhance the diversity and size of the wild population. The alloenzyme data did not reveal any fixed differences between *D. decorum and D. nudicaule*, although allele frequencies of the putative parental populations differed. At these loci, *D. luteum* resembled *D. nudicaule* more than *D. decorum*. Many unique RAPD markers distinguish each of the three species. The diagnostic markers from populations of *D. nudicaule* and *D. decorum* were not additive in the putative hybrid, and these data indicate that *D. luteum* is not of recent hybrid origin (Koontz, *et al.*, 2001).

Allozyme electrophoresis was conducted to survey the levels and distribution of genetic diversity in the tetraploid perennial larkspur *Delphinium montanum* (Ranunculaceae), which is endemic to the eastern Pyrenees of Spain and France and considered endangered. Seven populations were sampled, resolving 14 putative enzymatic loci belonging to eight enzymes. Banding patterns stained in gels revealed several enzymatic duplications attributable to autotetraploidy, such as the presence of both balanced and unbalanced heterozygotes and the lack of fixed heterozygosity. Variability in *D. montanum* (P=23.8 per cent, A=1.48, and H< sub>e</sub>=0.082) was lower than that expected for an autotetraploid species. This, in addition to the scarcity of loci showing three or four different alleles, could indicate that this species suffered genetic erosion through population bottlenecks, or, alternatively, that it is undergoing diploidization (Lopez-Pujol, *et al.*, 2007).

The relationship between individual heretozygosity and fitness was explored in the perennial larkspur *Delphinium bolosii,* an endangered species endemic to Catalonia (North-Eastern Spain). Allozyme electrophoresis was used to quantify the levels of heterozygosity at nearly to one hundred surveyed individuals in two extant populations. At the same time, eight parameters of vegetative and reproductive fitness were measured. A principal component analysis reduced the original fitness variables to three uncorrelated principal components, two associated with the maternal plant and the other one with the offspring. None of the components were significantly correlated with heterozygosity. The low number of variable allozyme markers and the likely influence of ecological factors could be responsible for lack of correlation between individual heterozygosity and fitness (Orellana, *et al.*, 2007).

The study was taken up to determine the differences in toxicity between Tall larkspur (*Delphinium barbeyi*) and Duncecap larkspur (*D. occidentale*) and to find whether these were genetically inherent within the species, or due to environmental influences. There was no difference in the concentration of the toxic alkaloids between the two species when grown in common gardens. Both species had higher concentration of toxic alkaloids when grown in the southern Rocky Mountain region characterized by summer monsoonal thunderstorms, compared to summer drought in the north. In a unique occurrence where Duncecap larkspur grew beside Tall larkspur, toxic alkaloids were very low in Duncecap larkspur, but total alkaloid concentration was often higher than in Tall larkspur (Ralphs and Gradner, 2001)

The diterpene alkaloid content was used to assess the chemical taxonomic diversity in three larkspur species. Samples (n=163) were collected from 18 different locations in five western states (Utah, Colorado, Wyoming, Idaho and California), extracted and analysed for diterpene alkaloids using electrospray mass spectrometry. *Delphinium glaucum* samples were easily grouped and were significantly different from all other groups ($P<0.005$) for two compounds. *Delphinium barbeyi* and *Delphinium occidentale* were found to be distinct groups, but more closely related. Samples representing a putative hybrid between *D. barbeyi* and *D. occidentale* were found to be closely related to *D. occidentale*, but were significantly different from all other groups. These data support the classification of *D. glaucum*, *D. barbeyi*, and *D. occidentale* as distinct species and suggest that a possible hybrid is more similar to *D. occidentale* than *D. barbeyi* (Gardner *et al.*, 2002).

Based on previous research, the plants of tall larkspur were separated into their respective parts, freeze-dried, extracted, and analyzed for toxic and total alkaloid concentration, and alkaloid pools were calculated. Concentration of toxic and total alkaloids in leaves and stems declined as the plants matured, while concentration in flowers and pods increased ($P<0.004$). Concentration of alkaloids in the root declined in the early growth, then increased at the end of the season ($P=0.002$). Alkaloid pools in the root decreased during early growth, with a corresponding increase of pools in foliar parts. In the late flower and pod stage, alkaloid pools in the leaves and stems declined rapidly, while the pool in the crown and roots tended to increase (Ralphs and Gradner, 2003).

Postharvest Handling and Seed Storage

The cut flowers of delphinium were treated with 0, 2, 4, 6 and 8 per cent sucrose in combination with 0.2 per cent Mm STS. Treatment with sucrose significantly improved the pigmentation of flower colour, increased the floret size and extended the vase life. These positive effects by STS and 4 per cent sucrose were remarkable when the cut flowers were held at 10° C for 48 hrs to stimulate long distance transport as reported by Kuroshima (2008).

Larkspur seed does not germinate when soil temperatures are held above 24° C. Germination will be improved with chilling seed (1.1-1.2 °C) for at least 14 days prior to planting. Seed treatment with potassium nitrate, thiourea, gibberellin, kinetin and boric acid inactivate the inhibitors in the seed or seed coat and thus increase the seed germination in marigold. (Selvaraju and Selvaraj, 1994, Singh and Karki 2003). The seeds of *Delphinium ajacis* treated with gibberellic acid (10 ppm) and kinetin (50 ppm) resulted in the highest germination in June (96.67 per cent) and Sept. (90.33 per cent). Whereas, kinetin (50 ppm), gibberellic acid (25 ppm) and boric acid (75 ppm) resulted in maximum and equivalent germination (59.67 per cent) in December. *Delphinium ajacus* kept under cold storage resulted in the highest germination (34.42 per cent) than in incubator (26.27 per cent) and ambient storage (25.92 per cent). The seed maintained above 75.00 per cent germination during the initial four months but significant decline was noticed in the proceeding months irrespective of packaging and storage temperature. Among the packaging, aluminum foil showed the maximum (30.78 per cent) and poly set bin the lowest (26.70 per cent) germination after 18 months of storage. The maximum germination was recorded in aluminium foil (33.00 per cent) in succeeding harvesting season in April. The maximum viability (49.99 per cent) was recorded under cold storage, followed by incubator (47.13 per cent) and ambient storage (41.23 per cent) after 18 months of storage. Among the packaging, the desiccator and aluminium foil were equally effective with 48 per cent viability after 18 months of storage. The polyethylene bag kept in cold store maintained the highest viability (76.67 per cent) in succeeding harvesting season (Dhatt, 2008).

The decrease in seed moisture content during storage improved the germination of ageratum and delphinium (Carpenter *et al.*, 1995). Suetlakova (1989) recorded the highest germination of *Delphinium inconspicuum* (43 per cent) and *D. oreophilum* (66 per cent) at 0-4 °C after 12 months of storage. *D. cuneatum* and *D. dictyocarpum* germinated at 0-6 °C and 23-27 °C, respectively, but both the species responded to stratification. The sealed containers maintained seed viability of sweet peas for 33 months (Barton 1939). Storage of pansy in open at 5 °C in saturated atmosphere showed considerable fall in viability after nineteen months. The air-dried seeds of annual delphinium in sealed containers retained their original germination power for nearly 12 years at 8 °C and of perennial delphinium at –15 °C. The air-dry seeds of regal lily at 9.9 per cent moisture content remained fully viable in the laboratory for one year and showed decline after two years. The storage in open at 5 °C resulted in 16 per cent germination after three years, whereas, storage in airtight containers at 5 °C showed viability for eight years (Barton 1953). The seeds of *Callistephus chinenesis* were stored at three temperatures (ambient, 5 °C and approximately –4 °C), 3 moisture levels (from 4 to 8 per cent), ambient conditions and sealed containers. The results showed that sealed storage at 5 °C and –4 °C retained viability for nine and sixteen years, respectively. The seeds lost viability in open containers at 5 °C at saturated atmosphere after 18 months (Barton 1953). Weibull (1955) conducted experiments on the storage of vegetable and flower seeds and found that –20 °C was optimum for storage of *Kochia scoparia*, *Myostis sylvatica* and *Callistephus chinenesis* and however, no difference between –20 °C and ordinary storage was found for *Delphinium hybridum* and *Delphinium ajacus*, The sealed storage was more effective means of maintaining seed viability than open storage (Harrington 1973).

Major Diseases and Insects

Damping Off

The most important diseases of larkspur cut flower production are damping off of seedlings by *Pythium* and *Rhizoctonia* fungi. *Pythium* attacks larkspur seedlings. *Pythium* may rot germinated seed or cause root rot that stunts or kills seedlings. *Pythium* produces a tan, water soaked lesions in roots and stem. The root cortex may slough off leaving the white root vascular cylinder exposed, like fine white threads. The fungus *Rhizoctonia solanii* can also cause damping off and seedling blight. Usually *Rhizoctonia* is favoured by warm conditions. When compared to *Pythium*, *Rhizoctonia* lesions are more defined, with tan, sunken areas.

Botrytis Blight

Botrytis can attack seedlings, but it is especially damaging when it spots and blights flowers. High humidity and a film of water on plant surface are necessary for botrytis to infect plant tissues. It produces a small, defined spots that may rapidly enlarge to blight the flower, stem and leaf. Weak plant tissues such as yellowing senescent foliage, flower petals and wounds are easily invaded by botrytis. Under humid conditions a grey mould is produced on blighted tissues.

Powdery Mildew

Some larkspur cultivars are susceptible to powdery mildew caused by the fungus *Erysiphe polygoni*. Symptoms include twisted and distorted foliage and white spots and patches on foliage, stems and buds. Flower buds may fail to open and flowers may be distorted which are not fit for sale. Sunny, warm and dry days and cool nights are favourable for powdery mildew. Powdery mildew caused by *Erysiphe aquilegiae* var. *ranunculi* is one of the major diseases affecting field grown delphinium. The cultivars Blue birds, Kings Arthur, Cameliard and Galahad were consistently more resistant (0- 1.1 disease severity) than Casa Blanca, Blue Shadow, Belladona and Bellamosum (2.9-5.0 ds). Spearmans' coefficients of rank correlation were significant for all pairs of experiments. (Wegulo and Vilchez 2008). The powdery mildew fungus on *Delphinium volkerfrieden* (belladonna type) was identified as *Erysiphe aquilegiae* variety *ranunculi* on the basis of morphological characters. The conidia of *E. aquilegiae* germinated at 5-30 °C and the optimum temperature for *in vitro* germination was 20-25 °C (Sato *et al.,* 2005).

Aphids

Aphids can rapidly become major pests in larkspur plantings, if the population is left unchecked. Large population of aphids may result in noticeable honeydew accumulation on foliage and flowers. Sooty mould fungus will grow on the honeydews. Since aphids feed on plant phloem, they compete with the plant for nutrients, thus heavy population may measurably reduce plant growth and vigour.

References

Barton L V. 1939. Storage of some flower seeds. Contr. Boyce Thompson Inst 10: 399-427.

Barton L V. 1953. Seed storage and viability. Contrib Boyce Thompson Inst Prof Paper 17: 87-103. (Cited by Hartman *et al.,* 1997) Plant propagation principles and practices. Prentice Hall of India, New Delhi.

Bhandary K K, Ramasubbu N, Joshi B S, Desai H K and Pelletier S W. 1990. Structure of delvestine: a norditerpenoid alkaloid from *Delphinium vestitum* Wall. Acta. Crystallographica C46: 1704–1707.

Bosch M and Waser N M. 1999. Effects of local density on pollination and reproduction in *Delphinium nuttallianum* and *Aconitum columbianum* (Ranunculaceae). American Journal of Botany. 86(6): 871-879.

Carpenter W J, Ostmark E R and Cornell J A. 1995. Evaluation of temperature and moisture content during storage on the germination of flowering annual seeds. Hort Sci 30 (5): 1003-1006.

Chowdhery HJ and Wadhwa BM. 1984. Flora of Himachal Pradesh. III Volumes. Howrah, Botanical Survey of India, India. pp. 16–18.

Christopher Brickell. 1996. The Royal Horticultural Society A- Z Encyclopedia of Garden Plants. pp. 350-352. Droling Kindersley, London.

CSIR [Council of Scientific and Industrial Research] 1986. The Useful Plants of India. Publications and Information Directorate, CSIR, New Delhi, India. pp. 165–166.

Dhatt K K. 2008. Studies on seed yield, quality and storage in flower seed crops under Punjab conditions. Ph D thesis Punjab, India.

Gardner D R and Pfister J A. 2000. Late season toxic alkaloid concentrations in tall larkspur (*Delphinium spp*) Journal of Range Management.53 (3): 329-334.

Gardner D R, Ralphs M H, Turner D L and Welsh S L. 2002. Taxonomic implications of diterpene alkaloids in three toxic tall larkspur species (*Delphinium spp*.). 30(2): 77-90.

Gonzalez-Coloma A, Guadano A, Gutierrez C, Cabrera R, Pena E, Fuente G and Reina M. 1998. Antifeedant *Delphinium* diterpenoid alkaloids. Structure-activity relationships. Journal of Agricultural Food Chemistry 46: 286–290.

Harrington J F. 1972. Seed storage and longeivity.(In) T T Kozlowski (Ed) Seed Biology. pp 145-240. Academic Press, New York.

Herbert Aswith. 1961. Complete guide to garden flowers: An encyclopedia of garden planning. p 256.

Hashimoto F, Tanaka M, Maida H, Shimuzo K and Sakata Y. 2000. Characterization of cyanic flower colour of delphinium cultivars. Journal of the Japanese Society for Horticultural Science. 69(4): 428-434.

Hosokawa K, Koiwa H and Yamamaru S. 2001. *In vitro* adventitious shoot formation on petioles of commercial cultivars of delphinium. Scientia Horticulturae 90(1-2): 143 – 150.

Ippolito P and Henry C. 1975. Flowers: A guide for your garden. New York.

Kuroshima M, Ichimura K and Ubukata M 2008. Effects of pulse treatment with sucrose in combination with STS on the quality and vase life of cut Delphinium flowers. Horticultural Research Japan. 7(2): 305-308.

Linda F. 1984. Encyclopedia of garden plants.

Miyake N and Ubukata M. 2002. A new method of raising seedlings of larkspur for autumn flowering. Bulletin-of-Hokkaido-Prefectural-Agricultural-Experiment Stations.(83): 31-34.

Neelam Jain and Bohra A. 2006. Antibacterial activity of plant part extracts of *Delphinium ajacis* Linn. against a human and a plant pathogenic bacterium. Advances in Plant Sciences. 19(1): 171-174.

Orellana M R, lOpez, P Blanche J and Bosch M. 2007. Genetic diversity in endangered dysploid larkspur *Delphinium bolosii* and its close diploid relatives in the seroes Fissa of the western Mediterranean area. Biological Journal of the Linnean Society. 92(4): 773-784.

Pfister JA and Gardner DR. 1999. Consumption of low larkspur (*Delphinium nuttallianum*) by cattle. Journal of Range Management 52: 378–383.

Pfister JA, Gardner DR, Panter KE, Ralphs MH, Manners GD, Stegelmeier BL and Schoch TK. 1999. Larkspur (*Delphinium* spp.) poisoning in livestock. Journal of Natural Toxins 8: 81–94.

Polunin O and Stainton A. 1984. Flowers of the Himalaya. Oxford University Press, New Delhi, India. pp. 6–8.

Prajapati ND and Kumar U. 2003. Agro's Dictionary of Medicinal Plants. Agrobios (India) Jodhpur, India. pp. 107.

Rahman A, Nasreen A, Akhtar F, Shekhani M S, Clardy J, Parvez M, and Choudhary M I. 1997. Antifungal diterpenoid alkaloids from *Delphinium denudatum*. Journal of Natural Food Products 60: 472–474.

Ralphs M H and Gardner D R. 2001. Alkaloid levels in Duncecap (*Delphinium occidentale*) and Tall larkspur (*D. barbeyi*) grown in reciprocal gardens: separating genetic from environmental influences. Biochemical Systematics and Ecology.29(2): 124

Ralphs M H and Gardner D R. 2003. Distribution of norditerpene alkaloids in tall larkspur plant parts through the growing season. Journal of Chemical Ecology.29(9): 2013-2021.

Raza M, Shaheen F, Choudhary MI, Sombati S, Rafiq A, Suria A, Rahman A and DeLorenzo RJ. 2001. Anticonvulsant activities of ethanolic extract and aqueous fraction isolated from *Delphinium denudatum*. Journal of Ethnopharmacology 78: 73–78.

Raza M, Shaheen F, Choudhary MI, Sombati S, Rafiq A, Rahman A and DeLorenzo RJ. 2004. Inhibition of sustained repetitive firing in cultured hippocampal neurons by an aqueous fraction isolated from *Delphinium denudatum*. Journal of Ethnopharmacology 90: 367–374.

Said H M. 1970. Hamdard Pharmacopoeia of Eastern Medicine. Hamdard National Foundation. Times Press, Karachi, Pakistan. pp. 49.

Sasaki K, Honda K, Saga K and Sameshima M. 2008. Flowering habit of *Delphinium grandiflorum* L. in the cool northern areas of Japan. Horticultural Research Japan. 7 (2): 249-253.

Sato S, Sugiyama S and Sato Y. 2005. Identification and change in disease incidence of powdery mildew of delphinium. Annual Report of The Society of Plant Protection of North Japan. 56: 51-54.

Satyanarayanan. 2006. Indian floriculture on fast track. Floriculture Today.9: 18-24.

Selvaraju P and Selvaraj J A. 1994. Effect of pre sowing treatments on germination and vigour of seed in marigold (*Tagetes erecta* L). Madras Agriculture Journal 81 (9): 496-497.

Sharma K D Singh B M and Jyoti Saroop. 2007. Population structure and genetic relationships among *Delphinium* species – a fast dwindling genus from north western Himalayas. Current Science 93(3): 386-391.

Sharma K D Singh B M and Jyoti Saroop. 2007. Current status of *Delphinium* species (Ranunculaceae) from north western Himalayas. Plant Genetic Resources Newsletter 149: 5-8.

Singh A K and Karki K. 2003. Effect of grading and GA_3 on germination and seedling growth attributes in balsam. Progressive Horticullture 35 (2): 158-160.

Stegelmeier B L, Panter,-K E, Pfister A, James L F Manners G-D, Gardner, D R, Ralphs, M H and Olsen, J D 1998. Experimental modification of larkspur (*Delphinium* spp.) toxicity. Toxic-plants and other natural toxicants. 205-210.

Suetlakova A A. 1989. Characteristics of seed germination in some *Delphinium* species. Byulleten' Glavnogo Botaniches kogo Sada No. 152: 54-57 Orn Hort Abstracts, 18 (1) No. 98.

Subramonium A and Pushpangdan P. 1999. Development of phytomedicines for liver diseases. Indian Journal of Pharmacology 31: 166–175.

Swarup V. 1997. Garden flowers. National Book Trust, India. p.60-62

Wegulo S N and Vilchez M. 2008. Evaluation of delphinium cultivars for resistance to powdery mildew. Hort Technology. 18(3): 407-409.

Weibull G. 1955. The cold storage of vegetable seed, further studies. Rep.14th Int Hort Congr. pp 647-667.

Zafar S, Ahmad MA and Siddiqui TA. 2001. Protective role of *Delphinium denudatum* (Jadwar) against morphine-induced tolerance and dependence in mice. Journal of Ethnopharmacology 78: 95–98.

2014, Future Crops Vol. 2 Pages 225–236
Editor: Professor K.V. Peter
Published by: DAYA PUBLISHING HOUSE, NEW DELHI

Chapter 11
Embelia ribes: An Endangered Medicinal Plant

Vasantha Kumar, T. Sane Anuradha, V.K. Rao and P.E. Rajasekharan
Indian Institute of Horticultural Research, Bangalore – 560 089
E-mail: thuppil@iihr.ernet.in

The genus *Embelia* consists of *Embelia ribes* Burm. f., *Embelia tsjeriam-cottam* (Roem. and Schult.) A. DC and *Embelia basal* (Roem. and Schult.) A. DC. Small to medium bushy shrub indigenous to India, *Embelia* berries are very important medication in Ayurvedic medicine and are used for thousands of years. *Embelia* fruit is used in India and in the Eastern Colonies for its anthelmintic properties. It is useful for elimination of tape worm, one or two teaspoonfuls of the powdered fruit being administered in milk on an empty stomach and followed by a purgative. It is used as a contraceptive, toothache, cathartic, anthelmintic, carminative and diuretic. Better conservation measures need to be taken for these species as they are red listed as "Endemic".

Botanical Identification

Genus Embelia is a dioecious, woody climbing shrub or *liana*, branchlets drooping consisting of white lenticels. Leaves are alternate, distichous or pseudoverticillate. Inflorescences are racemose, paniculate, umbellate or corymbose, bracteate at base which is rounded with glands pitted on either side of the midrib beneath the base. Pedicel is subtended by one persistent floral bract. Flowers are 4- or 5-merous, bisexual or functionally unisexual and plants dioecious, polygamous, or monoecious. Sepals are united at base. Petals are free or united at base, glandular granulose inside and often along margin. Stamens form adnate to base of petals, rarely free, normally exerted in staminate flowers, reduced in pistillate ones; filaments free; anthers dehiscing longitudinally, connectives usually punctate abaxially. Pistillodes are reduced and style included in staminate flowers; ovary is globose or ovoid; ovules ca. 4, uniseriate. Style is exerted; stigma discoid or capitate, sometimes slightly lobed. Fruit drup, prominently punctate, 1-seeded; endocarp crusty or rarely bony. Drup globose with vertical lines, glabrous, greenish yellow turns black upon ripening. Seeds are subglobose, covered by membranous remnants of placenta; endosperm ruminate; embryo terete, transverse.

Binomial Nomenclature

Embelia ribes Burm. f.

Synonyms

Embelia glandulifera Wight, *Embelia indica, Embelia robusta.*

Vernacular Name

English: Embelia, Bidang; Sanskrit: Vidanga, Vidang, Vella; Hindi: Vayavidang, Vayvidang, Babarang; Kannada: Vrishanasana, Vidanga; Malayalam: Vilal, Pu-valli; Tamil: Vaivilangam, Kattukkodi; Telugu: Vidangamu, Vellal; Marathi: Vyavadinga

Field Recognition Features

The plant is an extensive woody climber with slender hanging branches; the branchlets have distinct white lenticels. Its leaves resemble an elongated egg with minute sunken glands scattered all over. They are papery to touch. Leaves show numerous minute gland dots when seen against sunlight. Flowers are greenish white in colour, bisexual and have a mild fragrance. They measure the size of a tiny pinhead and are borne on a stalk about 18cm long. Drupes are smooth, round or egg shaped and measure up to 4 mm across; they turn bluish black on ripening.

Habitat

It is commonly seen in places that are up to the height of 1660 m amsl. It is generally seen in areas of eastern India, Central and lower Himalayas, Sri Lanka, Malaysia, South China and Singapore.

Threat Status

The Red list status of this species was assessed as "Vulnerable" for both Karnataka and Tamil Nadu and "Lower Risk- near threatened" for Kerala. Mhaskar *et al.* (2011) studied the distribution map of *E. ribes* for the north Western Ghats of Maharashtra. Population decline is critically discussed with respect to issues regarding misidentification, adulteration and status of rawdrug trade.

Origin and Distribution

It is an Indo–Malaysian species, commonly seen in places up to the height of 1660 m amsl. It is seen in Sri Lanka, Malaysia, South China and Singapore and throughout India mostly in the hill tracts of Central Himalayas, Arunachal Pradesh, Assam, Jammu and Kashmir, Himachal Pradesh, Uttar Pradesh, Maharashtra, Andhra Pradesh, Karnataka, Kerala and Tamil Nadu. The species are commonly seen in the wet forests of Karnataka, Kerala and Tamil Nadu. In Karnataka, it is distributed in Shimoga's Hulical and in Kerala in lower Ghats of Palghat, and reported in Peerumedu, Kadalar, Bodi to Devikulam, Lockhert gap, Devikulam to Munnar and in Meenmutty Kulamavu of Idukki district; in Mudianpara—Sholayar of Thrissur district; in Kaikatty, Mannarkhattu Mukkali, Kurukkankundu and Pathanthode of Palakkad district; Nilambur of Malappuram district; Chandanathode of Kannur district, Ponmudi of Thiruvananthapuram district, and in Pamba of Pathanamthitta District.

Macroscopic Identification

Fruit brownish-black, globular 2-4 mm in diameter, warty surface with a beak like projection at apex, often short, thin pedicel and persistent calyx with usually 3 or 5 sepals present, pericarp brittle enclosing a single seed covered by a thin membrane, entire seed, reddish and covered with yellowish spots (chitra tandula), odour slightly aromatic, taste, astringent.

Fruit

The dry fruits vary in colour from red to nearly black. The fruits of *E. ribes* are warty and inflorescence branches racemose and leaf margin punctate. The drug has a slightly astringent, aromatic taste.

Stem

T. S. of mature stem is somewhat circular in outline. The epidermis is single layered and is provided with unicellular glandular trichomes. Periderm is 5 to 6 layered followed by distinguishable cortical region of parenchymatous and schlerenchymatous tissues. Tannin cells are very prominent in parenchymatous region. Schizogenous cavities are also characteristics of stem. Schlerenchymatous strandarches into the secondary phloem tissue. The secondary xylem and secondary phloem are produced by the cambium, as usual. Tannin cells are very common in the phloem tissue. Vessels are characteristically parenchymatous of large lumen. Medullary rays are thin walled and biseriate. Pith cells are parenchymatous.

Root

T.S. of mature root is somewhat circular in outline with single layered epidermis. Epidermis possesses multicellular and unicellular glandular trichomes. Mature root cork is about 15- 17 layered having schizogenous cavities. Cork is followed by a wavy schlerenchymatous ring. Beneath this, secondary vascular tissues are seen. Biseriate medullary rays with tannin cells are very conspicuous.

Petiole

T. S. of petiole is somewhat shield shaped in outline; vascular bundle is crescent shaped; epidermis is single layered with thin cuticle. Cortical region is parenchymatous. Schizogenous cavities are a few when compared to that of stem. Tannin cells are very common.

Leaf

T. S. of leaf shows common dicotyledonous characters. Epidermis is single layered without any trichomes. Mesophyll consists of a single layered palisade, and spongy tissue with abundant intercellular spaces. Most of these cells are rich in tannin. Characteristic oil glands are very common in mesophyll especially near the midrib. Vascular bundle is broad 'C' shaped. Phloem fibres are very prominent. Stomata are of Rariunculaceous type.

Agrotechnology

It grows in all types of soils especially the well-drained light cotton soils (Black soils), sandy loam soils and sandy soils. It is propagated both by seeds and vegetative means.

Seed Propagation

Fresh seeds may be collected during May-August and sown directly in raised beds. Time for initial germination is about 20 days and germination per cent is 50. After 60 days in the nursery, the seedlings are ready for transplanting.

Vegetative Propagation

Through cuttings: Stem cuttings with 3 nodes of pencil size are treated with "Keradix" and planted 3 cm deep in polybags. Rooting commences only after 35-40 days. After a healthy root system is formed, the plants are ready for transplanting to the main field.

Management

Seedlings must be transplanted when it attains a height of 5cm. Irrigation is provided daily 3 times.

Conservation

The seeds of *Embelia ribes* are difficult to germinate *ex situ* and seed viability very low. Hence *in vitro* germination was tried in a petridish, using different concentrations of Gibberellic acid and Indole Acetic Acid to break seed dormancy and enriching with half strength Murashige and Skoog basal liquid medium. Feasibility of storage of seeds in liquid nitrogen was assessed, as a long-term conservation strategy. The results show with fresh seed moisture of 12.4 per cent the seed viability was 86.42 per cent with addition of Gibberellic acid 50 ppm, IAA 10 ppm and half strength MS basal liquid medium. Seed cryo-feasibility was 82.92 per cent and optimized seed moisture for cryostorage was 4.96 per cent (Shashikumar *et al.*, 2010).

Pharmacognosy of Fruit

Physico-chemical and Organoleptic Parameters

Physico-chemical Constants			*Organoleptic Characters*	
Parameters	*Values*	*Limit*	*Parameters*	*Properties*
TA	6.2 per cent	NA	**Taste**	Bitter
AIA	0.55 per cent	NA	**Colour**	Brown
ASE	1.0 per cent	NA	**Odour**	Mild
WSE	10.0 per cent	NA	**Texture**	Fibrous

TA: Total Ash; AIA: Acid Insoluble Ash; ASE: Alcohol Soluble Extractive; WSE: Water Soluble Extractive.

Limit as per Ayurvedic Pharmacopoeia of India.

TLC Finger Printing Profile

Prominent bands were observed with anisaldehyde spray and sulphuric acid spray at the Rf values 0.27, 0.91 and 0.32, 0.51, 0.56 respectively. One band is visible under long UV and short UV.

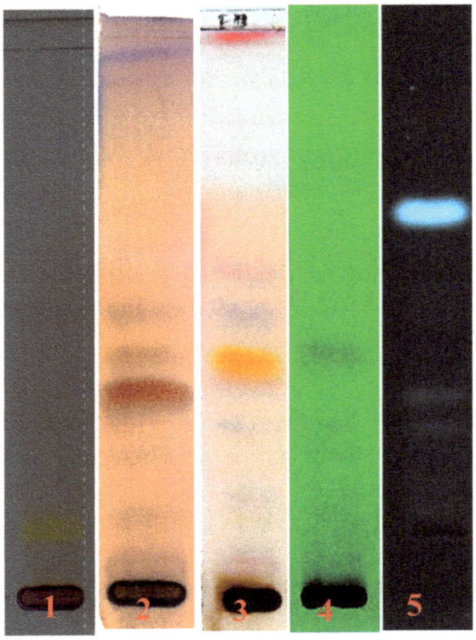

Methanolic extraction of the raw drug powder
Solvent system: Toluene: Ethyl acetate (8:2)
1. Observed under visible light
2. Sprayed with 10 % Sulphuric acid
3. Sprayed with Anisaldehyde reagent
4. Observed under short UV (254 nm)
5. Observed under long UV (366)

Anatomical Studies (Fruit)

Salient features T. S. of *Embelia ribes* (Fruit): Transverse section of fruit shows epicarp consisting of single row of tabular cells of epidermis, usually obliterated. In surface view cells are rounded with wrinkled cuticle. Mesocarp consists of a number of layers of reddish-brown coloured cells and numerous fibrovascular bundles and rarely a few prismatic crystals of calcium oxalate. Inner part of mesocarp

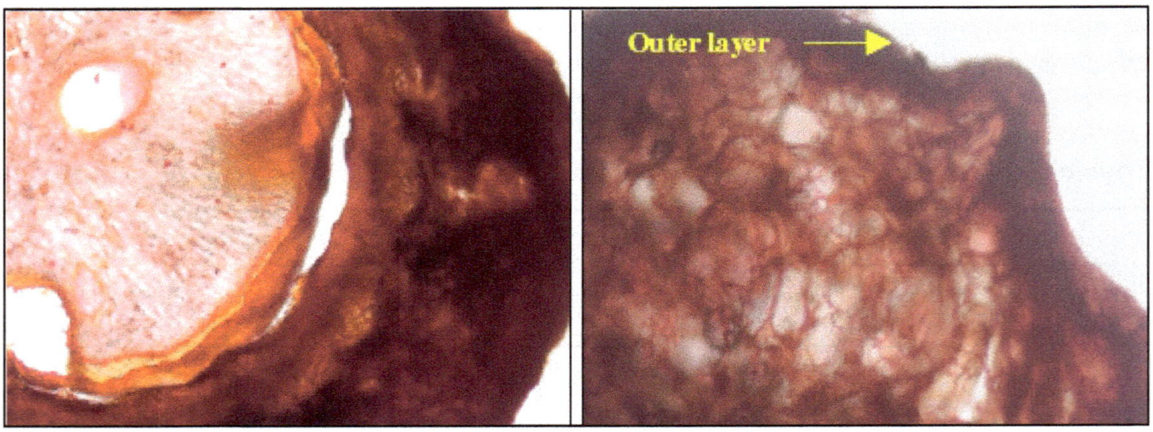

T.S. of *Embelia ribes* fruit.

and endodermis are composed of stone cells, endodermis consisting of single layered thick-walled, large, palisade-like stone cells, seed coat composed of 2-3 layered reddish-brown coloured cells. Endosperm cells irregular in shape, thick-walled, containing fixed oil and proteinous masses.

Microbial Limit Test

Results of microbial assay of *Embelia ribes -fruit* showed 75000 colonies for aerobic bacteria and 1500 colonies for yeast and moulds which are below the limits (TABC d" $1x10^7$; TYMC d" $1x10^5$) of international guidelines.

Parts Used

Fruits, roots and leaves

Phytochemistry

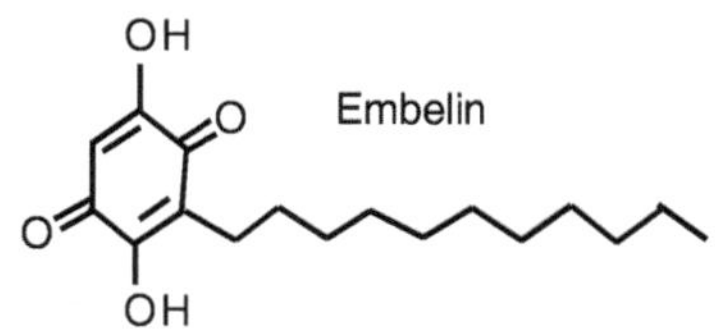

Embelin

The chief constituent of the drug is embelic acid, $C_{18}H_{28}O_4$; this substance, present to the extent of about 2.5 per cent., is obtainable in golden-yellow lamellar crystals, melting at 142°c. Embelic acid is insoluble in water, but soluble in alcohol, ether, chloroform, and benzene. A quinone was also detected in the drug (McErlean *et al.*, 2007). Sitosterol, daucosterol, gomphilactone derivative and quinone derivatives were isolated by Lin *et al.* (2006). Phytochemical investigation of seeds of Embelia ribes Burm. by Hao *et al.* (2005) resulted in isolation of new compounds embelinol, embeliaribyl ester, embeliol and embelin.

Medicinal Uses

Roots, fruits and leaves are used. Root infusion is given for coughs and diarrhoea. Decoction of dried fruits is used for fevers and diseases of chest and skin. Dried fruit is anthelmintic, astringent. Tonic is used in scorpion-sting and snake bite. Leaves are useful in pruritis, skin diseases and leprosy.

Ayurvedic and Siddha Uses

The fruit is hot, dry, with a sharp bitter taste; good appetizer; carminative, anthelmintic, alexiteric, laxative, alterative; cures tumours, ascites, bronchitis, mental diseases, dyspnoea, diseases of the heart, urinary discharges, used in snake bite, jaundice, hemicranias, worms in wounds.

It is used for treatment of Katurasam, Ushnaveeryam, Lagui, Diksham, In Sollam, Admanam, Udaram and Krimi vatavibandam.

Therapeutic Uses

Sula, Krmiroga, Udararoga, Adhmina

Properties and Action

Rasa: Katu, Tikta, Guna, Laghu, Ruksa, Tiksna, Virya : Usna, Vipaka: Katu, Karma: Anulomana, Dipana, Krminasana, Vatakaphapaha.

Ayurvedic Formulations

The plant is used in about 75 formulations. Fruits are used in Ayurvedic preparations like Vidangarista, Vidanga Lauha, Vidangadi Lauha, Abhayarishtam, Ayaskrithi, Pippallyasavam, Anuthailam and Kachuradithailam. The fruit in combination with other drugs is prescribed in snake bite (Sushruta) and scorpion-sting (Charaka, Sushruta); but it is not an antidote to either snake venom or scorpion venom.

Unani Uses

The fruit is bitter; good for plethoric constitutions, analgesic, purgative, vulnerary, anthelmintic; cures bronchitis by thickening phlegm and drying it; dries discharges from wounds; reddens the urine; removes bad humours from the body.

Preparations

Decoction: dose half to 1 ounce; Liquid extract: dose 1 to 4 drachms; Powder: dose 1 to 4 Drachms; Paste; confection; ammonium embelate (a salt): dose 3 to 6 grains.

Dosage

5-10 g of the drug in powder form. The dose is a tea-spoonful of the powder, twice a day for a child, and a dessert-spoonful for an adult; it can hardly be called a purgative; taste is rather pleasant, slightly astringent, and faintly aromatic. The worm is expelled dead. A purgative should be given to prepare the patient for the drug. It is a common practice to put a few berries of the plant in milk given to young children; they are supposed to prevent flatulence (Dymock). 180 grains (a tola) of the powdered seeds administered at bed time in curdled milk, followed by a dose of castor oil on the following morning, was found an efficacious remedy in Tape worm (Sakharam arjun). Decoctions of the root twice or thrice daily proved to be a very effective medicine for influenza epidemic.

Pharmacological Activity

Cardiovascular Health

In a study, cardioprotective effect of aqueous extract of fruits of *Embelia ribes* Burm (ER) was evaluated in a rat model having acute myocardial infarction, induced by isoproterenol. The results provide evidence for the first time, that aqueous ER extract pretreatment ameliorated myocardial injury and enhanced the antioxidant defense against ISO-induced myocardial infarction in rats and exhibited cardioprotective property (Bhandari *et al.*, 2008). Abana, a herbomineral preparation, showed that Abana treatment could contribute to restoring myocardial integrity and cardiac function disturbed by isoproterenol-induced ischemia.

Dyslipidemia

In the study, the lipid-lowering and antioxidant potential of ethanolic extract of *E. ribes* Burm was investigated in streptozotocin induced diabetes in rats. This is the first pilot study to provide biochemical evidence of potential of *E. ribes* in diabetic dyslipidemia (Bhandari *et al.*, 2002).

Hypolipidaemic Activity

Chronic administration of Abana, an Indian herbomineral preparation, showed a significant Hypolipidaemic activity in rats.

Antidiabetic Activity

Forty days of orally feeding the aqueous *Embelia ribes* extract (100 and 200 mg/kg), suggest that aqueous *E. ribes* extract exhibits a significant blood glucose and blood pressure lowering potential (Bhandari and Ansari, 2008). The wide range of medicinal properties of *Embelia ribes* may be, in part, due to scavenging of oxidizing free radical property. The antioxidant defense by ethanolic extract of

Embelia ribes on streptozotocin induced diabetes in Wister rats was studied (Bhandari *et al.*, 2007). The study concluded that *Embelia ribes* enhances the antioxidant defense against reactive oxygen species produced under hyperglycaemic condition and this protects beta cells against loss, and exhibits antidiabetic property.

Antioxidant Activity

The study was aimed to find out the protective effect of ethanolic extract of *E. ribes* fruits on homocysteine, lactate dehydrogenase (LDH) and lipid profile in serum, lipid peroxidation (LPO) and non-enzymatic antioxidant glutathione (GSH) levels in brain homogenates and histopathological examination of brain tissue in methionine induced hyperhomocysteinemic rats. The results provided clear evidence for the first time, that ethanolic *E. ribes* extract treatment enhances the antioxidant defense against methionine-induced hyperhomocysteinemia and oxidative stress in brain (Ansari and Bhandari,2008). In a study (Joshi *et al.*, 2007), free radical scavenging reactions and antioxidant activity of embelin were studied. Studies suggest that embelin acts as a competitive antioxidant in physiological conditions.

The nitric oxide (NO) scavenging activities of traditional polyherbal drugs like abana, chyavanaprasha, geriforte, septilin, mentat and triphala were examined using sodium nitroprusside as a NO donor *in vitro* (Ganesh *et al.*, 2004). All the drugs tested demonstrated direct scavenging of NO and were superior to *Gingko biloba*, which was used as a positive control. The extracts of various polyherbal drugs exhibited dose-dependent NO scavenging activities and the potency was in the following order: abana > chyavanaprasha > triphala > geriforte > septilin > mentat > Gingko biloba.

Wound Healing Activity

Ethanol extract of the leaves of *Embelia ribes* Burm. (Myrsinaceae) and its isolated quinone compound embelin were screened for wound healing activity by excision, incision and dead space wound models on Swiss Albino Rats (Kumara Swamy *et al.*, 2007). The histological examination of the granulation tissue of embelin treated group showed increased cross-linking of collagen fibers and absence of monocytes. The wound healing effect was comparatively evaluated with the standard skin ointment Framycetin.

Anticancer Activity

Embelin, identified primarily from the *Embelia ribes*, is one such compound shown to exhibit chemopreventive, anti-inflammatory, and apoptotic activities through an unknown mechanism. Ahn *et al.*(2007) reported that the embelin might mediate its activity through modulation of NF-kappa B activation. The down-regulation was associated with enhanced apoptosis by cytokine and chemotherapeutic agents. Together, the results indicate that embelin is a novel NF-kappa B blocker and potential suppressor of tumour genesis.

The other activities by *E. ribes* include analgesic activity, antifertility property (Prakash *et al.*, 1979), antispermatogenic effect (Gupta *et al.*, 1989), antiandrogenic activity (Agrawal *et al.*, 1986), contraceptive activity (Gnanasam *et al.*, 2002), effect on hormones (Nelly *et al.*, 1997), neuroprotection (Ansari *et al.*, 2008), improvement of cognition (Trivedi, 1999; Bhardwaj and Srivastava, 1995), memory enhancing property (Parle and Vasudevan, 2007) and effects of Abana on intestine (Pasnani *et al.*, 1988).

Toxicity Studies

Balasinor *et al.*(2007), evaluated the postnatal developmental toxicity and reproductive performance of the progeny exposed in uterus to pippaliyadi. Pippaliyadi yoga or pippaliyadi vati is an ayurvedic contraceptive used in India since ancient times. It is a combination of powdered fruit berries of *Embelia ribes* Burm. f. (Myrsinaceae), *Piper longum* L. (Piperaceae) and borax in equal proportion. The study suggests that *in utero* exposure to pippaliyadi does not have any adverse effect on the postnatal development and reproductive performance of the F1 progeny. The toxicity of embelin was assessed in female cyclic rats. Its administration at a dose of 120 mg/kg body weight did not cause any changes in the weight of liver, kidney and spleen, however, the wet weight of the adrenals showed a remarkable increase. Biochemical constituents like protein and glycogen did not show any change in these organs except in the adrenal where a significant increase was observed. The activity of acid and alkaline phosphatase was increased in the kidney and adrenal. Administration of embelin for 6 weeks caused severe pathological changes in the liver and kidney which mainly included disintegration, necrotic changes and perinuclear vacuolation. Marked tubular damage was observed in kidneys (Prakash, 1994). Toxicity study by Johri *et al.*(1990) revealed the effects of potassium embelate, a new analgesic compound. The results did not indicate adverse effects suggesting that potassium embelate is a safe compound. Low *et al.*(1985) studied the toxicity effects by evaluating the visual deficits and retinotoxicity caused by the naturally occurring anthelmintic, *Embelia ribes* and *Hagenia abyssinica*. Embelin was administered at a dose of 0.001 g (0.02 g/kg)/day for 9 days and suggested that embelin may be responsible for the visual defects.

References

Agrawal S, Chauhan S and Mathur R. 1986. Antifertility effects of embelin in male rats. *Andrologia,*; 18 (2): 125-31.

Ahn KS, Sethi G and Aggarwal BB. 2007. Embelin, an inhibitor of X chromosome-linked inhibitor-of-apoptosis protein, blocks nuclear factor-kappaB (NF-kappaB) signaling pathway leading to suppression of NF-kappaB-regulated antiapoptotic and metastatic gene products. *Mol. Pharmacol.,*; 71(1): p-209-19.

Anonymous, The Wealth of Asia – a Compact Disc-alphabet E. Council of Scientific and Industrial Research CSIR, New Delhi.

Anonymous., The Ayurvedic Pharmacopoeia of India. Part I, Volume 1., Government of India, Ministry of Health and Family Welfare, Department of Ayush, New Delhi, pp. 163-65.

Ansari MN, Bhandari U, Islam F and Tripathi CD., 2008. Evaluation of antioxidant and neuroprotective effect of ethanolic extract of *Embelia ribes* Burm in focal cerebral ischemia/reperfusion-induced oxidative stress in rats. *Fundam Clin Pharmacol.*, 22(3): 305-14.

Ansari MN and Bhandari U., 2008. Protective effect of *Embelia ribes* Burm on methionine-induced hyperhomocysteinemia and oxidative stress in rat brain. *Indian J Exp Biol.*, 46(7): 521-527.

Balasinor N, Bhan A, Paradkar NS, Shaikh A, Nandedkar TD, Bhutani KK and Roy-Chaudhury M. 2007. Postnatal development and reproductive performance of F1 progeny exposed in utero to an ayurvedic contraceptive: Pippaliyadi yoga. *J. Ethnopharmacol.*, 109 (3): 406-11.

Bhandari U, Ansari MN and Islam F., 2008. Cardioprotective effect of aqueous extract of *Embelia ribes* Burm fruits against isoproterenol-induced myocardial infarction in albino rats. *Indian J Exp Biol.*, 46 (1): 35-40.

Bhandari U and Ansari MN., 2008. Antihyperglycaemic activity of aqueous extract of *Embelia ribes* Burm in streptozotocin-induced diabetic rats. *Indian J Exp Biol.*, 46 (8): 607-13.

Bhandari U, Jain N, Ansari MN and Pillai KK., 2008. Beneficial effect of *Embelia ribes* ethanolic extract on blood pressure and glycosylated hemoglobin in streptozotocin-induced diabetes in rats. *Fitoterapia*;79(5): 351-5.

Bhandari U, Jain N and Pillai KK. 2007. Further studies on antioxidant potential and protection of pancreatic beta cells by *Embelia ribes* in experimental diabetes. *Exp Diabetes Res.*, 2007; 15 803.

Bhandari U, Kanojia R and Pillai KK. 2002. Effect of ethanolic extract of *Embelia ribes* on dyslipidemia in diabetic rats. *Int J Exp Diabetes Res.*, 3(3): 159-62.

Bhardwaj, S. K. and Srivastava, K. K. 1995. Effect of a Composite Indian Herbal Preparation, CIHP (III) on Avoidance Learning during Endurance Performance of Rats. *Indian Journal of Experimental Biology*, (33), 580-584.

Ganesh Chandra Jagetia, Shaival Kamalaksh Rao, Manjeshwar Shrinath Baliga and Kiran S. Babu. 2004. The evaluation of nitric oxide scavenging activity of certain herbal formulations in vitro: a preliminary study. *Phytotherapy Research*, 18(7): 561 – 565.

Gnanasam S. K., Annamalai T., Narendranath K. A., Ramachandran S., Saravanan M. and Sridhar S. K., 2002. Evaluation of antifertility activity of polyherbal formulation. Indian drugs, vol. 39(6): 330-332.

Gupta S, Sanyal SN and Kanwar U. 1989. Antispermatogenic effect of embelin, a plant benzoquinone, on male albino rats *in vivo* and *in vitro*. *Contraceptio*, 39(3): 307-20.

Hao K, Ali M, Siddiqui AW. 2005. New compounds from the seeds of *Embelia ribes* Burm. *Pharmazie*, 60 (1): 69-71.

Johri RK, Dhar SK, Pahwa GS, Sharma SC, Kaul JL and Zutshi U. 1990. Toxicity studies with potassium embelate, a new analgesic compound. *Indian J Exp Biol*. 28 (3) : 213-7.

Joshi R, Kamat JP, Mukherjee T. 2007. Free radical scavenging reactions and antioxidant activity of embelin: biochemical and pulse radiolytic studies. *Chem Biol Interact.*, 167(2): 125-34.

Lin P, Li S, Wang S, Yang Y and Shi J. 2006. A nitrogen-containing 3-alkyl-1,4-benzoquinone and a gomphilactone derivative from *Embelia ribes*. *J Nat Prod.*, 69 (11): 1629-32.

Low G, Rogers LJ, Brumley SP and Ehrlich D. 1985. Visual deficits and retinotoxicity caused by the naturally occurring anthelmintics, *Embelia ribes* and *Hagenia abyssinica*. *Toxicol Appl Pharmacol.*, 81(2): 220-30.

McErlean CS and Moody CJ. 2007. First synthesis of N-(3-Carboxylpropyl)-5-amino-2-hydroxy-3-tridecyl-1,4-benzoquinone, an unusual quinone isolated from *Embelia ribes*. *J Org Chem.*, 72 (26): 10298-301.

Mhaskar, M., Joshi, S., Chavan, B., Joglekar, A., Barve, N. and Patwardhan, A 2011. Status of *Embelia ribes* Burm f. (Vidanga), an important medicinal plant of commerce from north Western Ghats of India, *Cur. Sci.*, 100, (4): 547- 552.

Nelly N. Mungai, Dominic W. Makawiti and Victor N. Konji. 1997. Effect of different doses and routes of administration of embelin on plasma testosterone levels. *Phytotherapy Research*,11(7): 532 - 534.

Parle Milind and Vasudevan M. 2007. Memory Enhancing Activity of Abana®: An Indian Ayurvedic Poly-Herbal Formulation. *Journal of Health Science,* 53(1): 43-52.

Pasnani, J. S., Hemavathi, K. G. and Rajani, A. P. 1988. Effects of Abana, an Ayurvedic Preparation on Rabbit Atrium and Intestine. *Journal of Ethnopharmacology,* (24): 287-302.

Prakash A. O., 1994. Short term toxicity of embelin in female rats. *Phytother. Res.,* 8(5): 257-264.

Prakash A. O. and Mathur R. 1979. Biochemical changes in the rat uterine tissue following *Embelia ribes* Burm. extracts. *Indian Journal of Pharmacology,* 11 (2): 127-134.

Prakash A. O. 1985. Biological evaluation of some medicinal plant extracts for contraceptive efficacy in females. *Contracept Fertil Sex (Paris),* 13(4): 649-55.

Prakash, A. O. and Mathur, R. 1979. Studies on oestrous cycle of albino rats: response to *Embelia ribes* extracts. *Planta Medica,* 36 (2): 134-141.

Prakash, A. O., 1979. Effect of *Embelia ribes* on uterine weight of normal and ovariectomized rats. *Planta Medica,* 35 (4): 370-372.

Shashikumar S, Renuka Jain, Venugopal R B, Vijay Danapur and Villoo Morawala Patell.2010. Cryopreservation of *Embelia ribes* N. Burm. and allied species a Rare, Endangered and Threatened medicinal plant., National Conference on Production of Quality Seeds and Planting Material-Health Management in Horticultural Crops, March 11-14, 2010 New Delhi, India, TS11: 0-6

Swamy Kumara HM, Krishna V, Shankarmurthy K, Abdul Rahiman B, Mankani KL, Mahadevan KM, Harish BG and Raja Naika H. 2007. Wound healing activity of embelin isolated from the ethanol extract of leaves of *Embelia ribes* Burm. *J Ethnopharmacol,* 109 (3): 529-34.

Trivedi, B. T. 1999. A Clinical Trial on Mentat. *Probe,* 38(4): 226.

2014, Future Crops Vol. 2 Pages 237–256
Editor: Professor K.V. Peter
Published by: DAYA PUBLISHING HOUSE, NEW DELHI

Chapter 12
Propagation and Cultivation Techniques of *Embelia ribes* (Vidanga)

A. Patwardhan[1,2]*, *M. Mhaskar*[1,2], *A. Joglekar*[1,2], *M. Tadwalkar*[1,2], *R. Wagh*[1,2] *and R. Vasudeva*[3]
[1]*Department of Biodiversity, Abasaheb Garware College, Deccan Gymkhana, Karve Road, Pune – 411 004, Maharashtra*
[2]*Research and Action in Natural Wealth Administration (RANWA), 16, Swastishree Society, Ganesh Nagar, Kothrud, Pune – 411 052, Maharashtra*
[3]*Department of Forest Biology and Tree Improvement, College of Forestry, University of Agricultural Sciences, Dharwad, Sirsi Campus – 581 401, Karnataka*
**E-mail: ankurpatwardhan@gmail.com*

Biological resources are now viewed as 'resource capital' of a nation. Medicinal plants are among the most valued resources of a country. The current volume of markets dealing with medicinal plants at international level is reported US $ 100 billion. India ranks 2nd in terms of the volume and value of medicinal plants exported (32000 tones – US$ 265 million), after China. India with its rich repository of medicinal plant species (1/4th of the world) meets health care needs of more than 80 per cent of the population of this country (Ravikumar and Ved, 2000). In the past a few decades, many species of plants all over the world are gaining international importance because of the medicinally or economically important biomolecules they possess. Out of these plants, 25 per cent are found in Western Ghats of India, a mega biodiversity hot-spot (Myers *et al.*, 2000) being an important site of wild collection of medicinal plants. Recently Ved and Goraya (2008) reported that about 960 plant species from India are traded for their medicinal properties and of them about 178 species are consumed in volumes exceeding 100 metric tons. Their consolidated consumption accounts for about 80 per cent of the total industrial demand of all botanicals in the country. *E. ribes* (Vavding or Vidanga) is one such species high in demand for its various uses in Ayurveda and modern medicine alike.

Unfortunately despite its huge importance, the entire demand for the species is met by wild collections alone. This has lead to its indiscriminate harvest from the wild and hence the very existence of this species is under severe threat. Virtually there are no efforts to develop and standardize the

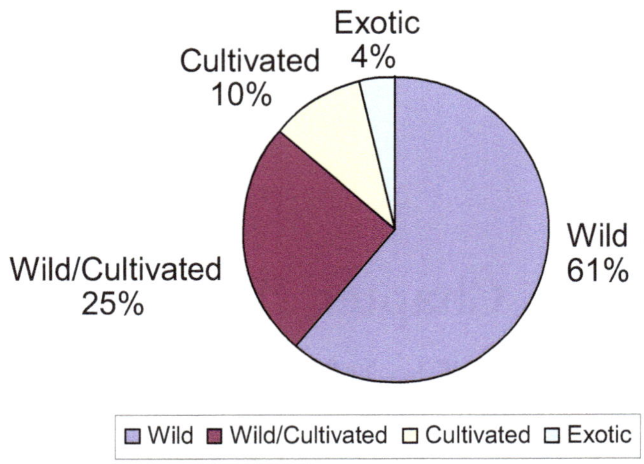

Figure 12.1: Share of wild collected and cultivated origin of MAPs in trade.

nursery and planting techniques attempted in this country. The present chapter takes an overview of status of *E. ribes* in general, and focuses on various aspects of standardization of propagation techniques and development of mass multiplication protocols. Comparative account of *E. ribes* with *E. basaal* (syn. *E. tsjerium-cottam, E. robusta*) is also presented as the later is also used and traded as *Vidanga*. Cultivation and domestication will help tackle problems inherent to medicinal plant industry/herbal medicines like unavailability of Quality Planting Material (QPM: seeds and seedling), misidentification, instability in the supply of material, and contaminants.

Distribution and Taxonomy

The genus *Embelia* belongs to family Myrsinaceae (order Ericales) which is one of the 'primuloid' families (Ma and Saunders, 2003). It is mainly distributed in warmer regions of northern and southern hemisphere. Most of the genera and species are tropical. Genus *Embelia*, represented by more than 100 species, is distributed in Tropical Asia, Africa and Australia and Pacific islands. Species of *Embelia* like *E. ribes* and *E. basaal* are known for their medicinal use since thousands of years in parts of Southeast Asia. *E. ribes* is an Indo Malayan species, distributed in India, Sri Lanka, Singapore, Malaysia and S. China. It is sparsely distributed in the evergreen to moist deciduous forests of Arunachal Pradesh, Assam, Maharashtra, Andhra Pradesh, Karnataka, Kerala and Tamil Nadu and is now confined to the informally protected '*sacred groves' and restricted* to a few remnant forest pockets and MPCAs (Medicinal Plant Conservation Areas) that Forest Department has established (Figure 12.2). Within India, it is found throughout up to an altitude of 1750 m amsl in these regions. It is 'Vulnerable-VU' in Karnataka and Tamil Nadu, 'Near Threatened – NT' in Arunachal Pradesh and Kerala and 'Data Deficient -DD' in Assam, Meghalaya, Sikkim and Maharashtra (CAMP, 2001). In Karnataka, *E. ribes* is distributed in evergreen forests. In northern parts of Western Ghats, *E. ribes* is distributed on edges of disturbed evergreen and semi evergreen forests (Mhaskar *et al.*, 2011). Recently Hareesh *et al.* (2009) reported that in humid tropics of the central Western Ghats, only 8 individuals out of 24 identified in a 1 km transect bear fruits. Although size class distribution of lianas suggested a near 'reserve J' pattern, the regeneration was restricted to mildly disturbed spots. Further there was lack of regeneration of Class-IV seedlings indicating poor regeneration in its type habitat.

Figure 12.2: Distribution map of *E. ribes*.
Source: FRLHT-IAIM Database 1991-2011.

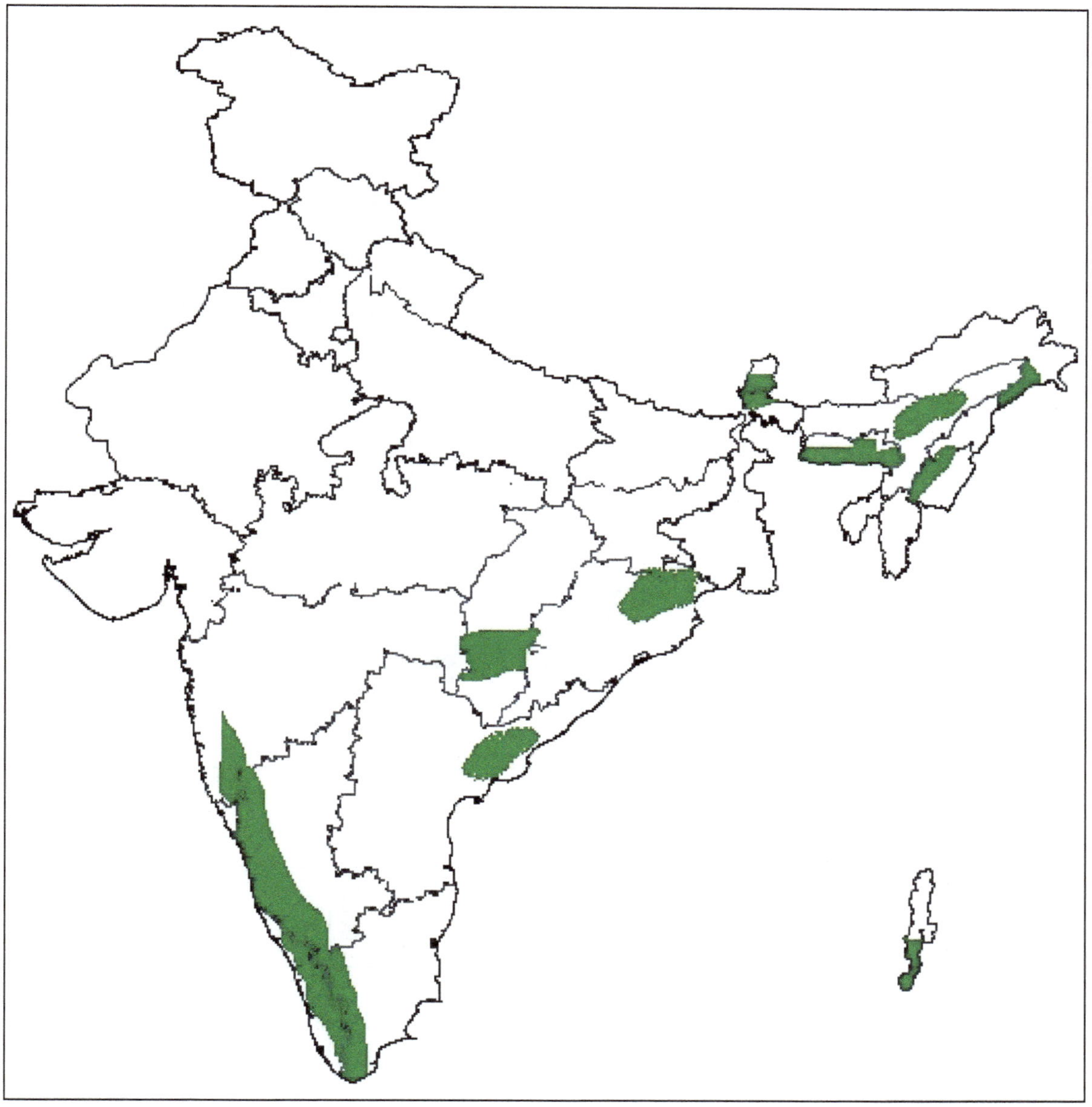

Contd...

Figure 12.2–*Contd...*

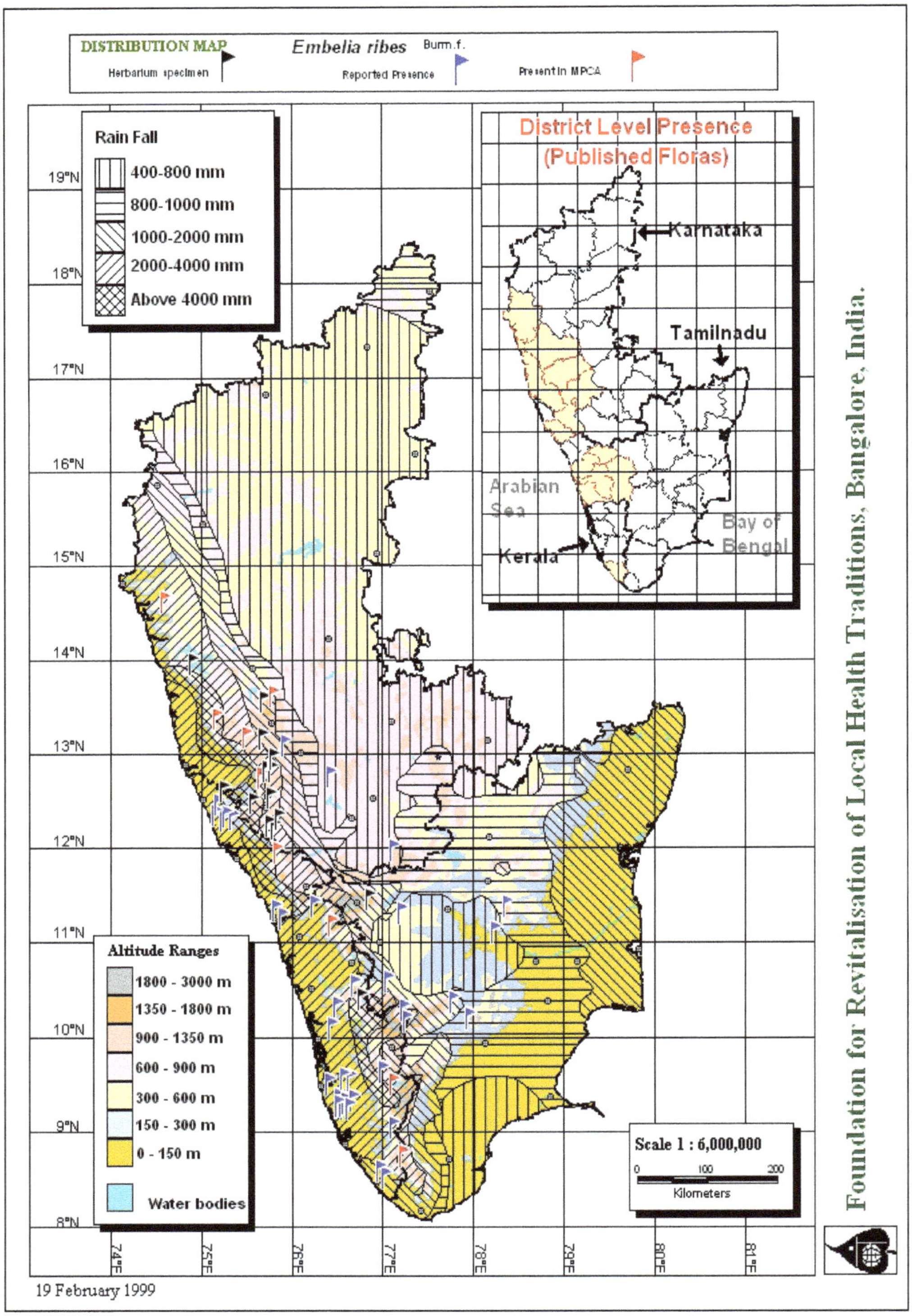

Embelia ribes Burm. f.

Common Name

False pepper

Vernacular Name

Vidanga, Vavding, Karkanni, Murudvel, Vaivarang, Baibidanga

It is a dioecious woody climber with lenticellular bark. Leaves are entire, elliptic, lanceolate, thinly coriaceous and gland dotted beneath. Base is rounded and apex acute. The whole surface is covered with scattered minute sunken glands. Flowers are in terminal, pubescent, paniculate racemes and greenish yellow in colour. Pentamerous flowers have hairy sepals; calyx is connate at the base and petals are papillose. Stamens 5 slightly exerted. Ovary is ribbed. Fruits are globose, smooth succulent, tipped with persistent style. The drupes are green when young and black when ripe with usually one seed; seeds globose (Figure 12.3).

Flowering and Fruiting period

February – July

Drooping habit | Lenticellular bark | Branch | Sunken glands near midrib

Male | Flower | Bisexual Flower Fruits

Figure 12.3: *Embelia ribes.*

Comparative Account of *E. ribes* with *E. basaal*

There are at least 3-4 botanical species, currently traded as Vidanga, one of them is *E. basaal*. Comparison of *E. ribes* with its 'substitute' and 'adulterant' species was critically discussed by Mhaskar *et al.* (2011). If an NTFP (Non Timber Forest Produce) is coming from two closely related plant species, correct identification forms the first and foremost criteria for sustainable utilization of the resource. *E. basaal* is one such species where it is very often mistaken with *E. ribes*. This becomes more serious when one species is threatened and possesses high conservation significance. Figure 12.4 gives various features of *E. basaal* and will help clearing confusion in identification of these two *Embelia* species that may arise because of misidentification.

Figure 12.4: *Embelia basaal.*

Use of *E. ribes* in Traditional and Modern Medicine

Embelia, especially *E. ribes* and *E. basaal* possess deworming activity in humans and are utilized since ages in traditional medicine. It is employed in India, since ancient times, as anthelmintic and administered as powder, usually with milk, followed by a purgative. It is effective in the treatment of ascariasis (Anonymus, 1952). *E. ribes* is also used in Unani medicine as an antihelminthic and has antifertility action. The root of *E. ribes* is boiled in water and the decoction is effective against chest pains. Besides, paste of powdered leaves, mixed with honey is consumed to treat ulcers of mouth (Sharma 1997). Roots are ground with lime juice and taken orally with honey against cough (Rajakumar and Shivanna, 2010).

It is used in about 75 formulations. Fruits are used in Ayurvedic preparations like Abhayarishtam, Ayaskrithi, Pippallyasavam, Anuthailam and Kachuradithailam (lyer, 1983). Ayurvedic products like Vidangadi Churna, Vidanga lauha, Vidanga Taila and Vidangarishta are also available. The fruits are acrid, astringent, anthelmintic, depurative, digestive, diuretic, carminative and contraceptive (Shah and Khanna, 1961; Warrier *et al.*, 1994). The paste is locally applied against ringworm and other skin infections. Seed powder is used to cure headache. The roots and leaves are astringent, thermogenic and stomachic (Warrier *et al.*, 1994). Powder from dried bark of the root is a reputed remedy for toothache. The plant is used as an anti-inflammatory drug to relieve rheumatism and fever (Khan *et al.*, 2010b). The fruit cures tumors, ascites, bronchitis, jaundice and mental disorders.

Major Phyto-constituents

The phytochemical characterization of *E. ribes* has predominantly focused on the composition of fruits and/or seeds. It yields embelin and other high valued secondary metabolites which have a wide range of clinical applications. Reported constituents include fatty acids glycerides (linoleic acid and palmitic acid), gallic acid and simple carbohydrates (glucose, fructose), quercitol and glycerol (Kuichi *et al.*, 1998; Lakshmanan *et al.*, 1990; Mitra, 1995). Kuichi *et al.* (1998) reported the Embelin content of *E. ribes* fruit as 2.4-4 per cent w/w (isolated yield) using cold ether extraction. From water extracts of *E. ribes* fruit, gallic acid (~1 per cent w/w/fruit), glucose, fructose and glycerol were isolated (Kuichi *et al.*, 1998).

Table 12.1 gives details of various phyto-constituents extracted from different parts of *E. ribes*. Details on pharmacological activity are presented in Table 12.2.

Table 12.1: Chemical constituents and activities reported from *E. ribes*.

Plant Part Used	*Phyto-constituents*	*Activity*
Fruits	Embelin, embolic acid, rapanone (Fieser and Chamberlin, 1948)	Anti-helminthic, against tumour, bronchitis, mental disorders, jaundice (Kirtikar and Basu, 1987), Anti-fertility (Seth *et al.*, 1982), analgesic (Atal *et al.*, 1984), antibacterial(Narang *et al.*, 1961), anti-inflammatory(Ahn *et al.*, 2007), antioxidant (Bhandari *et al.*, 2007)
	Embelin	Against diabetes (Bhandari *et al.*, 2002), Cardio protective activity (Bhandari *et al.*, 2002), (Bhandari *et al.*, 2008a) Contraceptive (Chaudhury *et al.*, 2001)
	Sitosterol and daucosterol (Haq *et al.*, 2005), embelinol, embeliaribyl ester, embeliol (Haq *et al.*, 2005), (Latha, 2007), vilangin (Rao *et al.*, 1965), christenbine (Chopra *et al.*, 1956)	–
Whole plant	Embelin	Anti-inflammatory (Kapoor *et al.*, 1983)
Roots	Embelin	Influenza epidemic (Menon, 1999)
Seeds	Embelin	Antibiotic, Anti-tuberculosis (Guhabakshi *et al.*, 2001)
Leaves	Embelin	Skin diseases and Leprosy (Sharma *et al.*, 2002)

Table 12.2: Pharmacological activities reported from *E. ribes*.

Sl.No.	*Constituents*	*Pharmacological Activity*	*Activity Studied against*	*References*
1.	Ethanolic extract of *E. ribes*	Hepatoprotective activity	Mice	Tabassum *et al.*, 2003
		α-amylase inhibitor activity	–	Prashanth *et al.*, 2001
		Anti-diabetic activity	Rats	Bhandari *et al.*, 2008b Bhandari *et al.*, 2007
		Methionine-induced hyperhomocysteinemia and oxidative stress	Rats	Ansari *et al.*, 2008
		Antihyperglycemic activity, antioxidant property	Rats	Bhandari *et al.*, 2002
2.	Ethanol extract of leaves of *Embelia ribes* Burm.	Wound healing property	Swiss Albino Rats	Kumara Swamy *et al.*, 2007
3.	Aqueous extract of *Embelia ribes*	Cardio-protective effect	Wistar albino rats	Bhandari *et al.*, 2008a
		Antihyperhomocysteinemic property	Rats	Bhandari *et al.*, 2008c
		Methionine-induced hyperhomocysteinemia and oxidative stress	Rats	Bhandari *et al.*, 2008c
4.	Aqueous, 80 per cent ethanolic and hexane extracts of leaves of *E. ribes*	No antimicrobial activity	–	Jiapiyasakul *et al.*, 2007
5.	Aqueous and ethanol extracts of *Embelia ribes*	Antibacterial activity	Gram positive strains	Khan *et al.*, 2010a
6.	Embelin	Antioxidant activity	Rats	Singh *et al.*, 2009
		Anticonvulsant activity	–	Mahendran *et al.*, 2010
		Decrease in testicular weight and variable degrees of spermatogenic arrest	Oral treatment in male dogs	Dixit *et al.*, 1983
7.	Embelin from *E. ribes*	Antioxidant property	–	Joshi *et al.*, 2007
		Effect on reproductive system	Male albino rats	Satyavati *et al.*, 1984; Anderson 1986; Arora *et al.*, 1971; Krishnaswamy *et al.*, 1980a, b; Purandare *et al.*, 1979; Shah 1971; Chaudhary *et al.*, 2001; Sharma *et al.*, 2001; Agarwal *et al.*, 1986; Mathur 1986; Gupta *et al.*, 1989

Contd...

Table 12.2–*Contd...*

Sl.No.	*Constituents*	*Pharmacological Activity*	*Activity Studied against*	*References*
		Anti-cancer property	Wistar rats	Sreepriya and Bali, 2005; Retuer *et al.*, 2010; Ahn *et al.*, 2007; Chitra *et al.*, 1995
		Anti., oxidant activity	–	Jain *et al.*, 2007
8.	Fruits of *E. ribes*	Anti., nematodal activity	Goats	Javed *et al.*, 1990
		Mollusicidal activity	*Lymnaea acuminata*	Rao *et al.*, 2001; Singh *et al.*, 2001
9.	Methanolic extract of *E. ribes* and Embelin	Anti-bacterial activity	–	Gajjar *et al.*, 2009; Tambekar *et al.*, 2009; Chitra *et al.*, 2003
10.	Dichloromethane extract of dried fruits of *Ardisia colorata* Roxb	Ant-itoxic, high inhibitory effect	Cancer cells	Prasertsang *et al.*, 2003
11.	Diethyl ether, petroleum ether, methanol extracts of *E. ribes* and embelin	Anti-fungal activity	–	Suthar *et al.*, 2009
12.	Benzene extract of *P. longum* and methanol extract of *E. ribes* berries	Pregnancy inhibition	Animals	Kholkute *et al.*, 1979
13.	Petroleum ether, solvent ether, methanol and water extract of *E. ribes*	Anti-fungal activity	–	Rathi *et al.*, 2010
14.	Seed oils of *Embelia ribes*	Anti-helminthic and antibacterial activity	*Pheritima posthuma*	Jalalpure *et al.*, 2007; Tambekar *et al.*, 2009
15.	Hexane extract of fruits of *E. ribes*	Anti-proloferative activity	Human cells	Joy and Lakshmi, 2010
16.	Ayurvedic formulation of Embelin and Embelin-borax complex	Anti-fertility activity	–	Singh *et al.*, 2007
17.	Formulation of Diarex containing Embelin	Anti-diarrhoeal	Humans	Srivastava *et al.*, 1988
18.	Zycox, containing *H. antidysentrica, B. aristata, E. ribes* and *A. calamus*	Anti-cocidal activity	–	Tipu *et al.*, 2006
19.	Powdered seeds of *E. ribes*, fruit of *P. longum* L. and borax powder	Ayurvedic contraceptive	–	Chaudhary *et al.*, 2001
20.	Potassium embelate	Analgesic activity	Rats and mice	Atal *et al.*, 1984; Zutshi *et al.*, 1989

Contd...

Table 12.2–*Contd...*

Sl.No.	*Constituents*	*Pharmacological Activity*	*Activity Studied against*	*References*
21.	–	Trypsin inhibition	–	Med. Arom Plant Abstr, 1995
22.	–	Adaptogenic property	–	Handa *et al.*, 1986; Paranjape and Kulkarni, 1995; Med Arom Plant Abrstr, 1995
23.	–	Antihyperlipidemic activity	–	Bhandari *et al.*, 2002

Trade

Vidanga is one of the most traded plants in volume with demand of > 100 metric tones/annum. The commercial importance of this plant drug is also confirmed by the study conducted by CERPA, New Delhi, under the aegis of Dept. of ISM and H (Indian System of Medicine and Homoeopathy) in 2000-2001. This study estimated the domestic consumption of *Vidanga* to be about 400 tons, valued approximately at Rs. 40 million, for 1999-2000 and is one of the top 50 traded plant drugs in India. Demand for vidanga seeds saw a steep increase in this period, as its export market, especially to Middle Eastern countries, increased tremendously and rates went up to Rs. 200/kg. Table 12.3 gives details of trade of *Embelia* species in leading markets in India in 2000-01. A study by FRLHT estimated the annual growth rate of about 23 per cent and projected the demand to touch 1000 tones by 2004-2005. But, for last a few years, a decline in demand was seen in the market. Studies indicated that Embelia ribes to be the authentic species. Trade data pointed out that > 95 per cent of the traded species are *E. tsjeriam-cottam*. This is because being a large shrub, maximum fruit bearing capacity and throughout distribution, coupled with non-availability of genuine *E. ribes*, market of *E. basaal* has increased. *E. basal* and *Myrsine africana* L. are used as substitutes as they possess embelin, while Maesa indica Roxb. is used as an 'adulterant' of *E. ribes* owing to its similarity with fruits. The local community in Maharastra identify M. indica also as Vavding (Mhaskar *et al.*, 2011). Curiously, Talbot (1911) mentioned that dried fruits of this species were earlier employed by the local communities of Uttara Kannada district of Karnataka as an adulterant with black pepper because of its morphological similarity.

Table 12.3: Trade of *Embelia ribes* and *E. basaal* in leading markets of India in 2000-01.

Sl.No.	*Plant Name*	*Trade Name*	*Plant Part*	*Estimated Consumption (Tones)*	*Price Rs/kg*	*Market*
1.	*E. ribes*	Baibidang, black, Vai-vidang	(Fr.-B), (Fruit)	4-5	150-160	Hyderabad
2.	*E. ribes*	Vai-Vidang	Fruit	>10	135-160	Delhi
3.	*E. basaal*, *E. ribes*	Vaividang (Red), Vavding (Kala), Vavding (Lal-Kala)	Fruit	>100	60-120 60-250, 110-160	Mumbai
4.	*E. basaal*, *E. ribes*	Bai-Bidang,	Fruit	10-80	110-160	Kolkata
5.	*E. ribes*	Vai - Vidang	Fruit	100 - 500	85	Dhamtari, Chattisgarh
6.	*E. ribes*	Vai - Vidang (black)	Fruit	100 - 500	70	Jagdalpur
7.	*E. ribes*	Vai - Vidang (black)	Fruit	100 - 500	70	Keshkal, Odisa
8.	*E. ribes*	Vai - Vidang (black)	Fruit	100 – 500	80	Raipur, (Chattisgarh)
	E. basaal	Vai - Vidang (red)	Fruit	100 – 500	95	Raipur
9.	*E. ribes*	Vai - Vidang (black)	Fruit	<100	65	Bagdihi (Odisa)

Propagation Techniques and Cultivation

The ideal time for seed collection is from June to August. Mature seeds of *E. ribes* are purple to black unlike red colored seeds of *E. basaal*. Propagation can be achieved through seed germination and through stem cuttings.

By Seeds (Figure 12.5)

Matured and ripened fruits are packed in polythene covers and placed under the artificial light source for 8 h per day. The cycle is continued for a period of 4-5 days. Pre sowing treatment of (10 per cent) H_2SO_4 for 10 min + GA_3 (4000 ppm) is the best treatment for *Embelia ribes*. By this method,40 per cent seed germination can be achieved successfully. Sand media are sufficient to achieve the seed germination. Sand bed with at least 9 cm thickness prepared in large tray is sufficient for raising seedlings under nursery conditions. Seeds are sown in a row inside the handmade furrows. Uniform and thin layer of sand should be spread over the seeds. Immediate watering should be done. The trays thus prepared should be placed inside the poly

Figure 12.5: Various steps in germination of *E. ribes* through seeds.

tunnel. Watering should be done regularly or as and when required. Weeding should be done if required. Compost/Vermi compost and organic manure are preferred. Seed germination starts after 50 days of sowing with 40 per cent germination.

Through Stem Cuttings (Figure 12.6)

Pencil sized stem cuttings of 1.0-1.5 cm thickness bearing 3 nodes are planted 3 cm deep in polybags and kept in shade. Cuttings are treated with two types of root regulators such as Indole Butyric Acid (IBA) (200ppm) and commercial quick root solution before planting. Post-Monsoon season is the best time for collection of stem cuttings. Pre treated stem cuttings are placed in root trainers and transferred to poly chambers. Rooting media is prepared from sand, coir pith and perlite. Microclimate with ambient condition is created inside the micro-poly chambers. By this method nearly 10 per cent of the stem cuttings sprout immediately. The rooting commences only after 35-40 days and within 3 weeks one or two axillary buds above the leaf scars develop and grow into branches. After the formation of healthy root system, it is ready for transplanting.

Figure 12.6: Various steps in vegetative propagation of *E. ribes* through stem cuttings.

Seedling Growth and Nutrient Management

Once the plants attain considerable growth, it should be transferred to polythene bags of bigger size. Periodical watering and weeding should be done. Additional boosting nutrients are necessary. The plants are also resistant to diseases and pests and hence protective sprays are not necessary. Seedling should be nurtured under the nursery conditions until they are transferred to fields. About 50 kg seeds are required for preparing 10,000 plants.

Development in the Nursery

After thirty days seedlings should be transferred to polythene covers as soon as the radical growth begins. Polythene bags of 12 x18cm size are ideal for transplanting. A potting mixture made out of sand, soil and Farm yard manure in a ratio of 1: 1: 0.5 acts as good rooting media. SeedlingS should be protected from scorching sunlight by using shade nets. Watering, weeding and shifting should be done as and when required. Capton (0.05 per cent) is sprayed over the seedlings to protect plants from fungal infection.

Embelia can be cultivated in all types of soils, the well drained light cotton soils or sandy loam soils are the best suited for cultivation.

Transplanting to the Fields

The seedlings raised in polythene bags under nursery conditions can be transferred to natural conditions immediately. Area with partial shade is preferred for this species. June and July are ideal for planting. Since it is a woody climber supporting tree is required. Saplings should be planted next to tree species as it requires shade in the initial stages. Saplings of one year and above can be easily transferred to the fields. Pits of 15 to 30 cm size should be made and filled with farm yard manure, sand and forest soil (1: 1: 0.5). Soil should be tightened properly. The pit size should be always 2-3 times wider than the root balls. Saplings can be planted at a spacing of 2-5 m distance. Saucers should be made around the planting place so that water supplied or obtained naturally will trickle down into the hole and helps in growth of saplings. The plantation can be irrigated usually at an interval of 15 days. Harvesting is done after two years. The fruits are collected, dried in shade and stored in clean porous Jute-bags. The fruits are collected manually and the plant is allowed to grow further. Weeding and thinning of plants may be done as and when required usually after 15- 30 days for better growth of plants.

Tissue Culture of *E. ribes*

Direct shoot organogenesis from leaf explants of *E. ribes* was reported by Raghu *et al.* (2006). The *in vitro* procedure involved three steps that included induction of shoot initials from leaf tissue, regeneration and elongation of shoots from shoot initials, and rooting of shoots. The induction of shoot initials was achieved on Murashige and Skoog (MS) solid medium supplemented with different concentrations of thidiazuron (TDZ). The best medium for shoot induction was MS with 0.272μM TDZ. Numerous shoot primordia were observed within 2–3 weeks on leaf margin as well as on the midrib region, without any callus phase. In second step, the shoot clumps are seperated from the leaf explant on transfer to MS basal medium, resulting in differentiation of 90 per cent of the shoot initials into well-developed shoots. The 2 to 3cm long shoots rooted on half-strength MS basal medium supplemented with 4.90μM indole-3-butyric acid (IBA) and 3 per cent (w/v) sucrose in the third stage. The rooted plants could be established in soil with 70 per cent success. This protocol could be utilized for *in vitro* propagation and conservation of this important threatened medicinal plant.

Discussion

E. ribes is listed in the 'Priority Species List' for cultivation by the National Medicinal Plant Board and the Maharashtra State Horticultural and Medicinal Plant Board. The population of this species has declined > 50 per cent in last decade because of immature and over harvest of fruits, narrow genetic base, dioecious breeding system and extrinsic factors like habitat loss, fragmentation, and degradation. Reducing harvest pressure on wild population and promoting viable commercial cultivation with the community and government participation are huge challenges. Efforts are on to develop suitable agronomy practices so that the pressure on wild populations will decrease. Once the 'package of practices' (POP) is developed, the species can be further inducted to suitable agro-forestry systems. Participatory cultivation through 'Contractual Farming Schemes' of National Horticulture and Medicinal Plant Board can be explored. Cultivation and domestication will also help tackle problems, inherent to herbal medicines like misidentification, instability in supply of material and adulterants.

Acknowledgements

We thank Rufford Small Grant Programme, UK for their support. We also thank the Principle Chief Conservator of Forests (Territorial and Wild Life wing of Maharashtra), CF (WL) Kolhapur and other forest department staff. We are also thankful to Dr. Ved and Mr. Vijay Barve of FRLHT who kindly shared the distribution map of *E. ribes*. Thanks are also due to Dr. B. S. Janagoudar, Dean, College of Forestry, Sirsi for support and encouragement.

References

Agarwal S, Chauhan S and Mathur R. 1986. Antifertility Effects of Embelin in Male Rats. Andrologia 18(2): 125-32.

Ahn KS, Sethi G and Aggarwal BB. 2007. Embelin, an inhibitor of X chromosome-linked inhibitor-of-apoptosis protein, blocks nuclear factor-kappaB (NF-kappaB) signaling pathway leading to suppression of NF-kappaB-regulated antiapoptotic and metastatic gene products. Mol. Pharmacol 71(1): 209-219.

Anderson EF. 1986. Ethnobotany of Hill tribes of Northern Thailand. II, Lahu medicinal plants. Econ Bot 40(4): 442-450.

Anonymous. 1952. The Wealth of India: a Dictionary of Indian Raw Materials and Industrial Products, Vol. III, CSIR, New Delhi, 167.

Ansari M N and Bhandari U. 2008. Protective effect of *Embelia ribes* Burm on methionine-induced hyperhomocysteinemia and oxidative stress in rat brain. Indian Journal of Experimental Biology 46: 521-527.

Arora RB, Ghatak N and Gupta SP. 1971. Antifertility effects of *Embelia ribes*. J. Res. Indian Med 6: 107.

Atal CK, Sidique MA, Zutshi U, Amla V, Johri RK, Rao PG and Kour AS. 1984. Non acrotic, orally effective, centrally acting analgesic from an Ayurvedic drug. J. Ethnopharmacol 11(3): 309–318.

Bhandari U, Ansari MN and Islam F. 2008a. Cardioprotective effect of aqueous extract of *Embelia ribes* Burm fruits against isoproterenol- induced myocardial infarction in albino rats. Indian J Exp Biol 46: 35-40.

Bhandari U, Ansari MN, Islam F and Tripathi CD. 2008c. The effect of aqueous extract of *Embelia ribes* Burm on serum homocysteine, lipids and oxidative enzymes in methionine induced hyperhomocysteinemia. *Indian J Pharmacol* 40(4): 152-157.

Bhandari U, Jain N and Pillai KK. 2007. Further studies on antioxidant potential and protection of pancreatic β-cells by *Embelia ribes* in Experimental diabetes. Exp Diabetes Res Article ID 15803, 1-6.

Bhandari U, Jain N, Ansari MN and Pillai KK. 2008b. Beneficial effect of *Embelia ribes* ethanolic extract on blood pressure and glycosylated hemoglobin in streptozotocin-induced diabetes in rats. Fitoterapia 79: 351-355.

Bhandari U, Kanojia R and Pillai KK. 2002. Effect of Ethanolic Extract of *Embelia ribes* on Dyslipidemia in Diabetic Rats. Intl. J. Experimental Diab. Res 3: 159–162.

Chaudhury MR, Chandrasekaran R and Mishra S. 2001. Embrotoxicity and teratogenecity studies of an Ayurvedic contraceptive: Pippaliyadi vati. J. Ethnopharmacol 74(2): 189–193.

Chitra M, Shyamala Devi CS and Sukumar E. 2003. Antibacterial activity of embelin Fitoterapia 74: 401-403.

Chitra M, Sukumar E and Devi CS. 1995. [^{3}H]-Thymidine Uptake and Lipid Peroxidation by Tumor Cells on Embelin Treatment: An *in vitro* Study. Oncology 52(1): 66-68.

Chopra RN, Nayar SL and Chopra IC. 1956. Glossary of Indian Medicinal Plants, CSIR, New Delhi Vol 10.

Dixit VP and Bhargava SK. 1983. Reversible contraception like activity of embelin in male dogs (*Canis indicus* Linn). Andrologia 15(5): 486-94.

Fieser LF and Chamberlin EM. 1948. Synthesis of embelin, rapanone and related quinones by peroxide alkylation. J. Am. Chem. Soc 70: 71–75.

Gajjar UH, Khambolja KM and Patel RK. 2009. International Journal of Pharm Tech Research 1(4): 1594-1597.

Guhabakshi DN, Sensarma P and Pal DC. 2001. A Lexicon Medicinal Plants of India, Naya Prakashan, 135–136.

Gupta S, Sanyal SN and Kanwar U. 1989. Antispermatogenic effect of embelin, a plant benzoquinone, on male albino rats *in vivo* and *in vitro.* Contraception 39(3): 307-320.

Handa SS, Sharma A and Chakraborti KK. 1986. Natural products and plants as liver protecting drugs. Fitoterapia 57: 307-45.

Haq K, Ali M and Siddiqui AW. 2005. New compounds from the seeds of *Embelia ribes* Burm. Pharmazie 60: 69–71.

Hareesh TS, Hegde VD, Gunaga S, Gunaga RP and Vasudeva R. 2009. Occurrence and regeneration status of *Embelia ribes* Burm. F.: an endangered and medicinally important liana of central western ghats. Journal of Non- Timber Forest Products 16(1): 5-10.

Iyer SR. 1983. Ayurveda Yogasamgraham, Vaidyaratnam P. S. *Varier's Arya Vaidya Sala*, Kottakkal.

Jain SC, Menghani E and Jain R. 2007. Biomarkers as a Tool for Validation of Herbs and Spices. Journal of Food Safety 9: 1-6.

Jalalpure SS, Alagawadi KR, Mahajanshetti CS, Shah BN, Salahuddin, Singh V and Patil JK. 2007. *In vitro* Anthelmintic Property of Various Seed Oils Against *Pheritima posthuma*. Indian Journal of Pharmaceutical Sciences 69(1): 158-160.

Javed I and Akhtar MS. 1990. Screening of *Veronia anthelmintica* seed and *Embelia ribes* fruit mixed in equal parts against gastrointestinal nematodes. Pakistan Journal of Pharmaceutical sciences 3(2): 69-74.

Jiapiyasakul P and Areekul V. 2007. Antimicrobial Activity in Some Indigenous Plant Extracts. Food Innovation Asia 2007. The 9th Agro-industrial Conference. Bitec Conventional Center, Bangkok. 14-15.

Joshi R, Kamat JP and Mukherjee T. 2007. Free radical scavenging reactions and antioxidant activity of embelin: Biochemical and pulse radiolytic studies. Chemical Biology Interact 167(2): 125-134.

Joy B and Lakshmi S. 2010. Antiproliferative Properties of *Embelia Ribes*. The Open Process Chemistry Journal 3: 17-22.

Kapoor VK, Chawla AS, Kumar M and Kumar P. 1983. Search for anti-inflammatory agent in Indian laboratories. Indian Drugs 30: 481–488.

Khan MA, Naidu MA and Akbar Z. 2010a. *In-vitro* Antimicrobial Activity of Fruits Extract of *Embelia ribes* Burm. International Journal of Pharmaceutical and Biological Archives 1(3): 267-270.

Khan MI, Ahmed A, Akram M, Mohiuddin E, Usmanghani K and Shah SM. 2010b. Monograph of *Embelia ribes* Burm. f. Plant Science 4: 503-505.

Kholkute SD, Kekare MB and Munshi SR. 1979. Antifertility effects of the fruits of *Piper longum* in female rats. Indian J. Exp. Biol 17: 289.

Kirtikar KR and Basu BD. 1987. Indian Medicinal Plants, International Book Distributors, Dehradun, India 2: 1511.

Krishnaswamy M and Purushothaman KK. 1980a. Antifertility properties of *Embelia ribes*. Indian J Exper Biol 18(6): 638-639.

Krishnaswamy M and Purushothaman KK. 1980b. Antifertility properties of *Embelia ribes* (embelin). Indian J Exper Biol 18(11): 1359-1360.

Kuichi F, Suzuki N, Fukumoto Y, Goto Y, Mitsui M and Tsuda Y. 1998. Chemical transformation of embelin through dimerization during preparation of a decoction. Chemical and Pharmaceutical Bulletin 46(8): 1225-1228.

Kumara Swamy HM, Krishna V, Shankarmurthy K, Abdul Rahiman B, Mankani KL, Mahadeven KM, Harish BG and Raja Naika H. 2007. Wound healing activity of embelin isolated from the ethanol extract of leaves of Embelia ribes Burm. J. Ethnopharmacol 109 (3): 529-534.

Lakshmanan KK and Naryanan ASS. 1990. Antifertility herbals used by the tribals in Anaikkatty Hills, Coimbatore district, Tamil Nadu India. Journal of Economic and Taxonomic Botany 14(1): 171-173.

Latha C. 2007. Microwave-assisted extraction of embelin from *Embelia ribes*. Biotechnol. Lett 29: 319–322.

Ma OSW and Saunders RMK. 2003. Comparative floral ontogeny of Maesa (Maesaceae), Aegiceras (Myrsinaceae) and Embelia (Myrsinaceae): taxonomic and phylogenetic implications. Plant Syst. Evol 243: 39–58.

Mahendran S, Thippeswamy BS, Veerapur VP and Badami S. 2010. Anticonvulsant activity of embelin isolated from *Embelia ribes*. Phytomedicine 18(2-3): 186-188.

Mathur R. 1986. Histochemical localization of alkaline phosphatase in the uterus of rats: response to a few indigenous plant extracts. Acta Eur. Fertil 17(4): 297-300.

Menon K. 1999. *Embelia ribes* – a medicine for influenza. Indian Forester 45: 210.

Mhaskar M, Joshi S, Chavan B, Joglekar A, Barve N and Patwardhan A. 2011. Status of *Embelia ribes* Burm f. (Vidanga), an important medicinal species of commerce from northern Western Ghats of India. Current Science 100(4): 547-552.

Mitra M. 1995. 'Vidanga' (*Embelia ribes*) - an Ayurvedic drug can help family planning. Applied Botany Abstracts 15(4): 267-82.

Myers N, Mittermeier R, Mittermeier C, Da Fonseca G and Kent J. 2000. Biodiversity hot-spots for conservation priorities. Nature 403: 853-858.

Narang GD, Garg LC and Mehta RC. 1961. Preliminary studies on antibacterial activity of *E. ribes* (myrisinaceae). J. Vet. Anim. Husbandry 5(1): 73–79.

Paranjpe P and Kulkarni PH. 1995. Comparative efficacy of four Ayurvedic formulations in the treatment of acne vulgaris - A double-blind randomised placebo-controlled clinical trial. Current Research on Medicinal and Aromatic Plants. 343-347.

Prasertsang P and Chavasiri W. 2003. Bioactive phenolic lipids from the fruits of *Ardisia colorata* Roxb. In 29th Congress on Science and Technology of Thailand.

Prashanth D, Padmaja R and Samiulla DS. 2001. Effect of certain plant extracts on alpha- amylase activity. Fitoterapia 72(2): 179-181.

Purandare TV, Kholkute SD, Gurjar A, Joshi UM, Dattatreyamurthy B, Sheth AR, Sheth XR, Swamy XR, Yayaraman S and Munshi SR. 1979. Semen analysis and hormonal levels in bonnet macaques administered *Embelia ribes* berries, an indigenous plant having contraceptive activity. Ind. J Exp. Biol 17: 935-936.

Raghu AV, Geetha SP, Martin G, Balachandran I and Ravindran PN. 2006. Direct shoot organogenesis from leaf explants of *Embelia ribes* Burm. f.: a vulnerable medicinal plant. Journal of Forest Research 11(1): 57-60.

Rajakumar N and Shivanna MB. 2010. Traditional herbal medicinal knowledge in Sagar Taluk of Shimoga district, Karnataka, India. Indian Journal of Natural Products and Resources 1(1): 102-108.

Rao IG and Singh DK. 2001. Combinations of *Azadirachta indica* and *Cedrus deodara* Oil with Piperonyl Butoxide, MGK-264 and *Embelia ribes* against *Lymnaea acuminata*. Chemosphere 44: 1691-1695.

Rao TVP and Venkateswarlu VV. 1965. Some natural and synthetic methyl bisbenzoquinones. Bull. Natl. Inst. Sci. India 28: 14.

Rathi SG, Bhaskar VH and Patel PG. 2010. *In-Vitro* Anti-fungal Screening of *Embelia ribes* Plant Extract through EUCAST Method. International Journal of Pharma Sciences and Research 1(2): 134-138.

Ravikumar K and Ved DK. 2000. Illustrated field guide of 100 red listed medicinal plants of conservation concern in southern India. Foundation for Revitalisation of Local Health Traditions, Bangalore.

Retuer S, Prasad S, Phromnoi K, Kannappan R, Yadav VR and Agarwal BB. 2010. Embelin suppresses osteoclastogenesis induced by receptor activator of NF-êB ligand and tumor cells *in vitro* through inhibition of the NF-êB Cell signaling pathway. Mol. Cancer. Res. 8(10): 1425-36.

Satyavati GV. 1984. Indian plants and products with antifertility effects (A review of literature between 1975-1982. Ancient Sci Life 3 (Suppl 4): 193-202.

Seth SD, Johri N and Sundaram KR. 1982. Antispermatogenic effect of Embelin from *Embelia ribes*. Indian J. Pharmacol 14: 207–211.

Shah CS and Khanna PN. 1961. Pharmacognostic comparison of *Embelia ribes* Burm. and E. robusta C. B. Clarke fruits. The Indian Journal of Pharmacy 23(10): 275-276.

Shah NK. 1971. A Study of an Indigenous Drug, Maswin as an Oral Contraceptive. Curr. Med. Pract 15(2): 614-616.

Sharma GK. 1997. Ethnomedicinal flora: Ayurvedic system of medicine in a remote part of the Indo-Tibetan Himalayas. Journal of the Tennessee Academy of Science 72(3-4): 53-55.

Sharma PC, Yelne MB and Dennis TJ. 2002. Database on medicinal plants used in Ayurveda. CCRAS 478–489.

Sharma RS, Rajalakshmi M and Jeyaraj DA. 2001. Current status of fertility control methods in India. J. Biosci 26(4): 391-405.

Singh A and Singh DK. 2001. Molluscicidal activity of *Lawsonia inermis* and its binary and tertiary combinations with other plant derived molluscicides Ind. J. Exp. Biol 39: 263-268.

Singh D, Singh R, Singh P and Gupta RS. 2009. Effects of Embelin on Lipid Peroxidation and Free Radical Scavenging Activity against Liver Damage in Rats. Basic Clin Pharmacol Toxicol

Singh IP, Bharate SB, Singh A and Bhutani KK. 2007. Fate of Embelin in Pippalyadi Yoga, an Ayurvedic oral contraceptive Structure of Embelin borax complex and evaluation of anti-fertility activity. Indian Journal of Chemistry 46(B): 320-325.

Sreepriya M and Bali G. 2005. Chemopreventive effects of embelin and curcumin against N, nitrosodiethylamine/phenobarbital, induced hepatocarcinogenesis in Wistar rats. Fitoterpia 76: 549-555.

Srivastava RK, Singh RC, Bharadwaj SK, Kapoor KN, Sharma MK and Munna Lal MD. 1988. A Comparative Study of Antidiarrhoeal Drugs most commonly used - Allopathic, Ayurvedic and Homeopathic Drugs 3: 181-184.

Suthar M, Patel R, Hapani K and Patel A. 2009. Screening of *Embelia ribes* for Antifungal Activity. International Journal of Pharmaceutical Sciences and Drug Research 1(3): 203-206.

Tabassum N and Agarwal SS. 2003. Hepatoprotective activity of *Embelia ribes* against paracetamol induced hepatocellular damage in mice. JK-Practitioner 10(1): 43-44.

Talbot WA. 1911. Forest flora of the Bombay Presidency and Sind. Bishensingh Mahendra Pal Singh, Dehradun Vol 2. 146-147.

Tambekar DH, Khante BS, Chandak BR, Tltare AS, Boralkar SS and Aghadte SN. 2009. Screening of antibacterial potentials of some medicinal plants from Melghat forest in India. Afr. J. Trad. Complementary and Alternative Medicines 6(3): 228-232.

Tipu MA, Akhtar MS, Anjum MI and Raja ML. 2006. New Dimension of Medicinal Plants as animal feed. Pakistan Vet. J 26(3): 144-148.

Ved DK and Goraya GS. 2007. Demand and Supply of Medicinal Plants in India. NMPB, New Delhi and FRLHT, Bangalore, India.

Warrier PK, Nambiar VPK and Rarnankutty C. 1994. Indian Medicinal Plants, Orient Longman, Hyderabad, India 1(4).

Zutshi U, Johri RK and Atal CK. 1989. Possible interaction of potassium embelate: A putative analgesic agent, with opiate receptors. Ind. Journal of Experimental Biology 27: 656-657.

2014, Future Crops Vol. 2 Pages 257–277
Editor: Professor K.V. Peter
Published by: DAYA PUBLISHING HOUSE, NEW DELHI

Chapter 13
Jujube

Sunil Kumar Sharma*
IARI Regional Station, Agricultural College Estate,
Shivajinagar, Pune – 411 005
**E-mail: sunilksharma@yahoo.co.in*

Jujubes are fruit plants under the genus *Ziziphus* Tourn. which belong to the Rhamnaceae. There are two types of jujubes, *Z. mauritiana* Lam. (Indian jujube or ber), and *Z. jujuba* Mill. (Chinese or common jujube). These two species of genus *Ziziphus* are cultivated over vast areas of Asia and Africa. Multipurpose value of jujubes including food, honey production, forage and environmental protection was reported from various parts of the world (Sharma, 2009, von Maydell, 1986). The International Centre for Underutilised Crops highlighted *Z. mauritiana* as a priority species for enhanced research attention in the 1990s. India included *Z. mauritiana* in its national programme on underutilised crops (Pareek, 2001). Haryana State Council for Science and Technology sanctioned a project on conservation of biodiversity of 'ber' germplasm to Haryana Agricultural University. The project included an extensive survey to gather the endangered germplasm developed at various research stations and create a repository at the University. Other national programmes also recognized the importance of jujube species in the fruits tree based farming systems (Sharma, 2003, 2004; Tagiev, 1992). In general, jujubes are popular as fresh fruits and other processed products. Genetic diversity of *Ziziphus* is high in India and about 20 species are found naturally. Understanding biodiversity in jujube is likely to help in developing cultivars to suit the diverse agroclimatic conditions, and tolerant to pests and diseases.

Origin and Distribution

Indian Jujube

The Indian jujube is cultivated for almost 4000 years in India. Its precise natural distribution is uncertain due to extensive cultivation. It is thought to be cultivated in southern Asia, Africa and Europe. It is likely that the original wild species was spread from India through Myanmar. It is thought to have been in Africa for only a few centuries and it is likely that introduction to China and Indonesia is also recent. It was introduced into Guam about 1850 but is not often planted there except as an ornamental, and to Australia late in the nineteenth century (Grice, 1998). It reached America during the nineteenth century. *Z. mauritiana* was introduced to the Negev desert of Israel (Nerd *et al.*, 1990). It has been known in southern Europe for more than 2000 years. The tree is naturalized in uncultivated areas in Barbados, Jamaica and Puerto Rico. Main Asian regions where jujube is cultivated are Lebanon, Iran, Pakistan, northern India, Bangladesh, Nepal, Korea, Philippines, Malayan region, and southern and central China. It is also cultivated in Persia, Arabia, North Africa, Mediterranean France and Spain.

Chinese Jujube

Z. jujuba was likely to be domesticated in the Yellow River area which is the main area of cultivation for this species today. Vavilov (1951) considered the primary centre of origin in China and the cultivation must have spread through Central Asia. The wild species has a very wide area of distribution from China to Pakistan. It was taken from the Levantine Coast to Europe and presumed to spread along the North African coast in the seventh century AD. It has, however, developed a secondary centre of diversity in West Asia and is naturalised in many areas along the Black Sea Coast (Tutin, 1968). It was first introduced from Europe to USA in 1837 by Robert Chisholm and planted in Beaufort, South Carolina and introduced to California and neighbouring states from Southern France (Rixford, 1917). Grafted Chinese varieties were introduced in 1906. In the twentieth century, superior Chinese materials were introduced to Japan and North Africa.

Taxonomy

Species of the genus *Ziziphus* were classified based on characteristics of the inflorescences (Hooker, 1875 and Brandis, 1906). Liu and Cheng (1995) suggested that the species are grouped into two Sections, one further divided into two series:

1. Ziziphus - Occurring in temperate zones. Plants are glabrous and with deciduous fruiting branchlets.
2. Perdurans - Occurring in subtropical and tropical zones. Plants are pilose and without deciduous fruiting branchlets.
   - 2.1 Series Cymosiflorae - Occurring widely in subtropical and tropical zones. Flowers in axillary cymes. Ovary and fruit glabrous with thick and hard endocarp.
   - 2.2 Series Thyrsiflorae - Occurring in South and Southeast Asia. Flowers in terminal or axillary thyrses. Ovary and young fruit pilose with very thin endocarp.

The Indian jujube belongs to Section Perdurans, Series Cymosiflorae and the Chinese jujube belongs to Section *Ziziphus*.

The Major Cultivated Species

Indian Jujube

The species has a wide range of morphological variations ranging from shrubs to small or medium sized trees which might be erect, semi-erect or spreading. Plant height varies from 3-4 to 10-16 m. Trees are semideciduous and branched. The bark has deep longitudinal furrows and is greyish brown or reddish in colour. Plant is usually spinous. Branchlets are densely white pubescent, and tend to be zig-zag. Fruiting branches are not deciduous. Leaf laminae are elliptic to ovate. The apex is rounded, obtuse or subacute to emarginated, the base is rounded and mostly symmetrical. Margins are minutely serrulate. There are three marked nerves almost to the apex. Lower surface is whitish due to persistent dense hairs. Leaves are petiolate and stipules are mostly spines. Flowers have sepals which are dorsally tomentose and have a 2-celled ovary. Two styles are one mm long and connate for half their length. Flowers are borne in cymes or small axillary clusters. Fruit - greenish, yellow or sometimes reddish - is a glabrous globose or oval edible drupe varying greatly in size. The pulp is acidic and sweet. The species is distributed throughout the warm subtropics and tropics of South Asia. *Z. nummularia* and *Z. mauritiana* were distinct genotypes (Diwakar *et al.,* 1992). It has spread South-Eastwards through Malaysia and eastwards through IndoChina and southern China. It is widespread in Africa and southern Arabia while in Africa, it has naturalised. It adapts well to hot tropical climates with varying rainfall, and tolerating poor soils.

Chinese Jujube

Plants are shrubs or small trees up to 8-10 m high with stiff branches. Tree forms have a small canopy extending 3.5-4.5 m. Trunks may be short or long depending on genotype. Branches are armed with paired spikes. Older parts of trees can lose their spines. Branchlets are flexuous, green and glabrous when young. Fruiting branchlets are deciduous. Leaves are oblong, obtuse, glandular and have 3-nerves. Stipules form the spines. Flowers are a few in a small axillary cluster or cyme. Styles are two, connate for half their length. Fruit is an ovoid-oblong edible drupe 1.5-2.3 cm long, dark reddish brown to black. Pulp is sour to sweet. Chinese jujube spread westwards to the Mediterranean, throughout the Near East and SW Asia and spread eastwards to Korea and Japan. Like *Z. mauritiana* this species also got naturalised in many Asian countries. It is mostly cultivated in China, India, Central Asia and southwest Asia.

Z. jujuba var. *spinosa* Hu ex H. F. Chow is typified by possession of small sour fruits and is usually a spiny shrub or small tree (Wang and Sun, 1986). *Z. jujuba* var. *inermis* has unarmed branches and styles not connate (Brandis, 1874). Chinese jujube is adapted to subtropical and warm temperate areas. It prefers a relatively dry climate during growing season but cool during dormancy. It tolerates lower temperatures than Indian jujube and can survive up to -10°C.

The Minor Cultivated Species

The following two species of *Ziziphus* are also cultivated on a small scale (Table 13.1):

Z. spina-christi

Plant is a shrub, often with intertwined branches, or small tree up to 10-15 m tall. Bark deeply furrowed and scaly, white-brown to pale grey. Branchlet are densely pubescent. Fruiting branchlets are not deciduous. Tends to produce a very deep tap root. Mostly spinous with paired spines. Leaves are ovate-lanceolate, margin crenulated, rounded at base and nearly symmetrical, obtuse. They are minutely pubescent beneath and become glabrescent at maturity. Flowers in short axillary cymes. Peduncles 1-3 mm at flowering and 3-6 mm at fruiting. Sepals woolly dorsally. Fruits subglobose to globose fleshy glabrous drupes, yellow, reddish or redbrown, usually about 2 cm long and 1 cm wide. Flesh astringent to taste. The species is evergreen where water is adequate, but deciduous when conditions are drier. It is the best to consider a typical variety, var. *spina-christi.* It is a species of the Middle East through Arabia and West Africa to N. E. Africa, Ethiopia and Eastern Africa. It is wild in the Middle East, especially Iran, Saudi Arabia and in Turkey. It was introduced to Africa by Arab traders along the Mediterranean coast and also via the Horn of Africa (von Maydell, 1986; von Sengbusch and Dippolo, 1980). It is a minor cultivated plant in India, Pakistan, Egypt, Syria, the Mahgreb, Saharan oases and Zanzibar. It can survive with half the annual rainfall needed by *Z. jujuba* and could be a source of drought resistance.

Table 13.1: Various species of the genus *Ziziphus.*

	Type	Species	Synonyms
Major cultivated species	Indian jujube	*Z. mauritiana* Lam.	*Z. jujuba* (L.) Lam., *Z. jujuba* (L.) Gaertn. (including var. *stenocarpa* Kuntze and forma, *aequilatrifolia* Engl.), *Z. tomentosa* Poir., *Z. rotundata* D. C., *Z. aucheri* Boiss., *Z. insularis* Smith, *Z. sororia* Roem. and Schult., and *Z. orthocantha* D. C.)
	Chinese jujube	*Z. jujuba* Mill.	*Z. sativa* Gaertn., *Z. vulgaris* Lam., *Z. flexuosa* Wall., *Z. nitida* Roxb., *Z. sinensis* Lam., *Z. zizyphus* (L.) Karst., *Z. mairei* Dode, *Z. officinarum* Med., *Z. chinensis* D. C., *Z. chinensis* Watt.)
Minor cultivated species		*Z. spina-chisti* (L.) Desf.	*Z. africana* Mill., *Z. amphibia* A. Chev., *Z. nabeca* (Forsk.) Lam., *Z. inermis* A. Chev., *Z. sphaerocarpa* Tul., *Z. spina-christi* Willd.)
		Z. lotus (L.) Lam	*Z. nummularia* Aubrev., *Z. saharae* Blatt. and Trab., *Z. lotus* (L.) Desf. subsp. s*aharae* Maire, *Z. sylvestris* Mill., *Z. parviflora* Del.
Wild Species	Asia	*Z. jujube* (*Z. rugosa*) *Z. nummularia* (Burm. F.) Wight and Arn. *Z. oenoplia* Mill.	*Z. rotundifolia* Lam.
	Africa	*Z. abyssinica* Hochst. ex A. Rich.	*Z. mucronata* Willd

Z. lotus

This species is a spiny shrub growing up to 1.5 m tall and resembling *Z. jujuba*. However, fruiting branchlets are not deciduous and twigs are grey. Internodes on branchlets are less than 1 cm long. Leaves are suborbicular or broadly elliptic to ovate, shallowly glandular-crenate, pubescent beneath and less so above. Flowers are solitary or 2/3 together. Fruits are subglobose fleshy drupes about 1 cm diameter and deep yellow. It is cultivated in Asia Minor, Arabia, Egypt, North African coast, Cyprus, Greece, Portugal, Spain, Italy and France. Indian jujube, *Z. lotus* could be used for earliness of fruit maturity as well as drought tolerance. *Z. lotus* and wild *Z. nummularia* of India are two distinct species.

Wild Species

Asia

There are numerous wild species of *Ziziphus* in Asia and they tend to cluster in two regions: China and the Indian subcontinent. There are 14 species in China. *Z. jujube* which belongs to Section *Ziziphus* of the genus. They are found mostly in the central and lower parts of the Huanghe River valley, although the primary centre of diversity in China for *Ziziphus* species is south Yunnan and the southeast of Guangxi Province. Pareek (2001) noted the distribution of wild and naturalised *Z. mauritiana* from the lowlands to 1500 m in the Himalayas and in Sri Lanka. It is associated with dry areas where tree types can be found. It was recorded as wild species in the Siwalik forests, east of the Ganges and in forests of Central India (Brandis, 1906). *Z. jujube* (*Z. rugosa*) can be found naturalised in the central and eastern sub-Himalayan region, the Central Provinces and western side of the Peninsula. Other species are found in the north-west desert region of India and only a few in the Himalayas. The species is a thorny shrub producing red, edible fruits. Some wild species with wide distributions, such as *Z. oenoplia* have become weedy in places such as India, Sri Lanka, Myanmar or Malaysia. A few additional wild species cluster are found in Malaysia (Ridley, 1922) and in Indonesia (Martin *et al.*, 1987). Wild species are sources of diversity for jujube improvement. Pareek (2001) noted the attributes of several wild species of jujube (Table 13.2).

Table 13.2: Exploitable attributes of wild species in ber.

Wild Relatives	*Exploitable Attributes*
Z. nummularia and *Z. lotus*	i) Drought tolerance ii) Dwarf tree stature and extensive root system iii) Early fruit maturity
Z. jujube	i) Resistance to low temperature damage ii) Excellent dehydration quality of fruits iii) High vitamin C and P contents in fruits
Z. mistol	i) Resistance to low temperature damage
Z. mauritiana var. *Rotundifolia*	i) Vigorous tree frame ii) Wood of marginal timber value

Source: Pareek, 2001.

Africa

Some wild species are also found in Africa. *Z. abyssinica* is found as scattered trees in grassland 400-2200 m from Senegal to Ethiopia and south to Zimbabwe and Mozambique. *Z. mucronata* grows in open woodland from 0-200 m above sea level from Senegal to Arabia and south to S. Africa and Madagascar. While *Z. spina-christi* is distributed from 0-1300 m and is indigenous in semi-desert wadis from 600-1000 m above sea level in the Horn of Africa and North Africa. *Ziziphus* was probably introduced through seed propagation in East Africa where it has reverted to wild types.

The Ziziphus Genepool

The two major cultivated species are widely distributed in congenial climates in Asia and they cover vast regions. Most of these distributions comprise wild or naturalised materials superimposed on the patterns of distribution in the areas where the species are cultivated. The wild populations are heterozygous and extremely variable from which farmers have selected the best trees in terms of production and propagated them vegetatively, but are now doing so through grafting. A study of wild populations of Chinese jujube has identified promising types to use as germplasm (Liu and Wang, 1991). As a result of long time of local selection, many cultivars arose, some of which have become widely recognised. Very little is known about the patterns of genetic variation in the wild populations of any of the other cultivated species of jujube.

Chromosome Numbers

Indian jujube is usually polyploidy with counts of $n = 12, 20, 24, 30, 36$, or 48. Khoshoo and Singh (1963) looked at a range of cultivars and found $n = 24$ in most of them, but in two it was $n = 48$ and in one it was $n = 30$. In some wild material, Nehra *et al.* (1983) found $n = 48$; and also in naturalised 'wild' material it was the same count. *Z. lotus* appears to be diploid, *Z. oenoplia* tetraploid and *Z. nummularia* shows a polyploid series. The possibility exists that the genus is tribasic with $x = 10, 12$, or 13 (Darlington and Wylie, 1955). It is now generally thought that Indian jujube shows a range of polyploids: diploid, triploid, tetraploid, pentaploid and octoploid (Mehetre and Dahat, 2000). Chinese jujube also represents a polyploid series, chromosome counts tending to represent 2n = 45, 60, 90. Chinese jujube exhibited high diversity in chromosome karyotypes, shape, size and surface sculpture of pollen, leaf length and flower diameter, shape, colour, weight of fruit, growth period and soluble solids and ascorbic acid of fruits. The level of ploidy appears to be important for some cultivars of Indian jujube. Powdery mildew resistant genotypes were diploids and a seedless form was octoploid. However, other diploid cultivars were susceptible to powdery mildew.

Biodiversity in *Ziziphus* spp.

Economically important species are *Z. nummularia, Z. oenoplia, Z. rugosa, Z. sativa, Z. vulgaris* and *Z. xylopyrus*. The ability of *Ziziphus* species and different varieties/types within *mauritiana* to cross freely has allowed the build up of rich gene pool which depicts heterozygosity in their adaptability to soil and climate; morphological, physiological and phenological traits; chromosome number; tolerance/resistance to biotic and abiotic stresses and genomic DNA. Promising germplasms with distinctive traits such as desired adaptation (Gola, Umran, Thar Bhubhraj, Thar Sevika), diverse quality traits (Banarasi Karaka, Illaichi), high and stable yield (Seb, Ponda), tolerance to biotic and abiotic stress (Tikadi, Katha, Bawal-Sel-1, Sanaur-2) were identified which have comparative advantages to improve the productivity of ber (Awasthi and More, 2009).

Variability in Jujube Cultivars

A wide variation due to cross pollination is exhibited by the jujubes in vegetative, leaf, floral, fruit and quality traits.

Variations in Vegetative Characters

The more appropriate vegetative characters for classification are leaf area and branching habit, while the most dependable fruit characters are apex type, stalk and stylar flesh cavities and shape (Bal, 1992). *Z. mauritiana* genotypes Umran, Illaichi, Desi-1 and Desi-3 plants have a spreading habit, whereas, Kathapal and Desi-2 plants exhibit a semi-spreading habit. The cultivars of Gola group (Gola Gurgaon No. 3, Bhadurgarhia Gola, Dankan Gola, and Kakrola Gola) produce erect plants. Leaf margins of some cultivated and wild forms of ber are serrated except in Desi-3 and Jharber (Gupta *et al.*, 2003). The minimum stomatal density, internodal length and plant height were recorded in Gola budded onto *Z. nummularia* and was the most dwarfing combination. Whereas, scion combination of Ponda budded onto rootstock *Z. mauritiana* ecotype- Assam-Gauhati was the most vigorous. The rootstock *Z. mauritiana* ecotype-291 was the most compatible and moderate in plant growth, stomatal density and internodal length thereby giving moderate vigour to trees (Verma *et al.*, 2001). Paired spines were observed in many cultivars of *Z. mauritiana* like Umran, Kathaphal, Gola Gurgaon No. 3, Bhadurgarhia Gola, Dankan Gola and Kakrola Gola (Gupta *et al,*. 2003). The maximum tree height was recorded in Desi Alwar while tree spread was in Sanori No.5 (Saran, 2005). Maximum height of a 15 year old tree of *Z. mauritiana* was recorded in LR-13 (6.59 m) while the girth was maximum (3.32 m) in LR 11 (Kundi *et al.*, 1989).

Reproductive Variability

Flowering: The peak period of flowering and fruit set in *Z. mauritiana* cultivars Banarsi Karaka, Ponda, Illaichi, Gola and Tikdi was September-October. Tikadi had the shortest duration of flowering (47 days) and fruit set (36 days) but the highest number of fruits/branch (239), fruit set (28 per cent), number of fruits reaching maturity/branch (48) and fruit retention (20 per cent) (Sharma *et al.*, 1990).

Figure 13.1: Difference in fruit size caused by the influence of rootstock on scion.

Figure 13.2: Selection of rootstock can influence the plant vigour.

Figure 13.3: Natural variation in fruit size in cv Gola.

Whereas, Saran (2005) found the longest period of bloom in Katha Rajasthan (78 days) and the minimum period of bloom in the Safeda Rohtak (23 days). Flowering occurred from 57 to 75 days, depending on cultivar (Dhaliwal and Bal, 1998). Only one flowering season was observed in Hyderabad from May to July and its total duration varied from 68-94 days. Cultivars Gola, Mundia and Akola flowered early; Umran and Seb in midseason; Banarsi and Kaki were late flowering (Babu and Kumar, 1988). The number of flowers per branch is quite high. Umran had the highest number of hermaphrodite flowers (22.2 per cent) followed by Gola Gurgaon with 20.1 per cent (Darbara and Jindal, 1982). Cultivars of *Z. mauritiana* differed in the time of flowering in Maharashtra (Desai *et al.*, 1986). Pollen germination was more than 50 per cent in most of the ber cultivars in 25 per cent sucrose solution whereas Illaichi cultivar was found sterile. Pollen grains differed in size, shape and exine characteristics. Size and shape of pollen within cultivars were quite uniform. Differences in exine pattern, size and P/E ratios could be used for identification of *Ziziphus* genotypes (Diwakar *et al.*, 1996). Size of pollen grains of *Z. mauritiana* Illaichi, Umran and four wild forms was different (Nehra *et al.*, 1984). Pollen diameter in seven cultivars of ber ranged from 20.05 mμ in Darakhi 1 to 32.04 mμ in Seedless. Similarly, considerable differences were also observed for pollen diameter among the different ploidy levels (Pradeep and Jambhale, 2000). Some cultivars have cross incompatibility in *Z. mauritiana* (Teaotia and Chauhan, 1964).

Fruit set: Umran was compatible as female parent with Sanuar 2, but Sanuar 2 did not set fruit with Umran (Mehrotra and Gupta, 1985). The requirement for cross pollination, incompatibility and pollen sterility means that fruit set depends on physiological and environmental conditions. The mode and time of anthesis was also cultivar specific. Anther dehiscence started about two hours after anthesis and continued for two to four hours. Peak receptivity of the stigma appeared to be just as the flower opened (Dhaliwal and Bal, 1998). Fruit set in *Z. mauritiana* cultivars took place almost at the same time in Tikadi, Gola, Seb, Umran and the hybrid Umran x Seb cultivars, but Gola fruits matured earlier than other cultivars in the region. Fruit drop studied in seven *Z. mauritiana* cultivars indicated the lowest drop (24.1 per cent) in Ponda and the highest in Illaichi (68.6 per cent) (Vashishtha and Pareek, 1979). Kakrola Gola had considerably higher drop (up to 80.6 per cent) than Kaithali and Umran which did not differ significantly, showing 7.2 and 12.1 per cent drop respectively (Panwar, 1980). Neeraja *et al.* (1995) recorded the highest fruit drop in Seb (87.94 per cent) followed by Umran (83.43 per cent) and Gola (82.14 per cent). Gola was the earliest (108 days) and Umran was the latest (147 days) to ripen under Gurgaon conditions (Singh *et al.*, 1983). Based on harvesting time, Saran *et al.* (2007) classified ber cultivars into three categories early, medium and late. Umran, Katha Bombay, Chhuhara, Illaichi, 2g-3, Kathaphal, Jogia, Ponda, BS-2 and Desi Alwar are late bearing varieties while Gola, Gola Gurgaon No. 2, Gola Gurgaon No.3, Safeda Rohtak, Seo, Katha Rajasthan, Laddu and Akhrota were early bearing varieties and Kaithali, Dandan and Mirchia varieties bear during the mid-season. Kathaphal is late ripening while Gola is early and Kaithali comes in midseason. Seo, Sanaur No.2 and Umran are recommended as early, mid and late cultivars respectively for commercial cultivation (Chadha *et al.*, 1972) in northern India.

Fruit and seed: Considerable variation in fruit length and breadth in different cultivars of *Z. mauritiana* like Banarsi Karaka (5.4 and 3.4 cm), Dandan (5.1 and 3.0 cm), Jogia (4.9 and 3.8 cm) and Umran (4.8 and 3.8 cm) were reported by Ghosh and Mathew (2002). Cultivars Seo Bahadurgarhia, Nari Keli, Desi Alwar and Banarsi Karaka produced fruits weighing > 20 g each (Godara, 1980). Umran produced the heaviest fruits (39.8 g), Gola Gurgaon showed the highest content of pulp (97.2 per cent) and total soluble solids in Gurgaon, Haryana (Singh and Jindal, 1980). Pewandi was the best variety for commercial cultivation, having the largest fruit, a high percentage of edible pulp and good skin colour. BS-2 had the highest pulp/stone ratio; and the maximum fruit and stone size was recorded in Ponda (Saran, 2005). Cultivars Dabailing, Daguazao and Linyi Lizao are large fruited, with fruit sizes of 23-26.1 g, Jinsi 3 and Jinsi 4 are small fruited varieties (Chen *et al.*, 2003). The best *Z. jujuba* varieties for uniformity of fruit size are Ta-Pai and Hsueh-pai (Sin'ko, 1974). A lot of variation exists in ber genotypes for pulp/stone ratio. Pulp/stone ratio was maximum in Sanur-6 and minimum in Punjab Chhuhara (Bharad *et al.*, 2002), The pulp/stone ratio was reported the highest in Umran (Dhingra *et al.*, 1973). Significant genotypic differences were found in the seed characters of *Z. mauritiana* varieties. Saran (2005) observed variation in stone size of 35 cultivars of ber and reported that maximum (3.72 cm) was in Ponda and the minimum (0.53 cm) in Illaichi. Seeds of *Z. mauritiana* trees taken from 5 districts in Yunnan Province, China and Narkum, Myanmar, were tested for their morphological characteristics, germination characteristics, and the growth patterns of the seedlings and young trees. A close relationship was found between these characters and their geographical distribution and climatic condition (Wang and Wang, 1994). In Chinese jujube, one group of cultivars gave 85 per cent seed germination and the other gave 98 per cent germination (Kim and Kim, 1984). The thinnest seed coat wall was observed in Godhan (Singh, *et al.*, 1973). However, the quickness and extent of germination depend upon viability and after-ripening status of the seed, presence of endogenous inhibitors, weathering of the stony endocarp and environmental conditions like the temperature, moisture, salinity and alkalinity of the growing medium.

Yield

Variations in yields of *Z. mauritiana* cultivars (55.5 to 116.11 kg/tree) were observed by Kumar *et al.* (1986). In the semi-arid subtropical climate of northern India under irrigated conditions, the fruit yield per tree ranges from 80 to 200 kg depending on the varieties and management practices (Bakhshi and Singh, 1974). Fruit yield was the highest in Gola (38.4 kg/tree) and the lowest in the local cultivar Sukavani (3.54 kg/tree) at Sardar Krushinagar in Gujarat (Chovatia *et al.*, 1992). Highest yields were obtained from Umran (210 kg/tree) followed by Sanaur No. 2, Dandan and ZG 2 (Gupta, 1977). Thornless gave the highest yield/tree (74.4 kg) followed by Sanaur 5 (71.3 kg) and Sanaur 4 (Singh and Tomar, 1988) at Bhatinda. Chinese jujube cultivars yield 50 to 300 kg fruits per year depending on cultivar, location and age of tree (Ciminata, 1996). On the basis of yield, ripening date, frost resistance and storage quality in 7 ber cultivars Ta-Yantszao, Da-bai tszao, Ya-tszao, 93, 58, 107, 52 and 48 are recommended for cultivation (Sin'ko, 1977). In China - Taiwan, fruit yields of 158.6 kg per tree in cultivar Kaolang-1 and 140.8 kg per tree in cultivar Telong were obtained (Chiv, 1997).

Chemical Variability

A wide range of varietal variability for quality traits was detected in 30 cultivars of ber at Hisar (Bisla and Daulta, 1986). Ber germplasm collection and evaluation at Amer, Bharatpur, Deeg, Tijara and Jhunjhunu areas of Rajasthan, India indicated significant variation in physicochemical characteristics of 23 genotypes, and ten produced fruits of excellent quality (Lal *et al.*, 2003). Variation in 10 local ber genotypes grown in West Bengal, were noted for soluble solids, total sugars, acidity and ascorbic acid content; ranges of these parameters were 9.53-19.13 per cent, 4.94-12.30 per cent, 0.38-2.60 per cent and 17.25-51.98 mg/100g respectively (Ghosh and Mitra, 2004). The cultivar Umran has 19 per cent TSS and 1.2 per cent acidity. In Kathaphal, the TSS is 23 per cent and acidity was 0.77 per cent while in Gola, the TSS is 17-19 per cent and 0.46-0.5 per cent acidity. In Kaithali, TSS is 18 per cent and acidity 0.5 per cent (Daulta and Chauhan, 1982). Godara (1980) evaluated 16 cultivars for quality characters. The fruits of Mundia Murhara and Chhuhara had the highest TSS content (22.8 and 22.4 per cent respectively). HB-1 and LB had excellent flavour and HB-2, KB1, SB and SB-2 were rated good at Bangalore. Nazuk had the highest total soluble solids (28.9 per cent) followed by Illaichi under rainfed conditions of Bawal, Haryana (Yamdagni *et al.*, 1985). The vitamin-C content was the highest in Narikelee, followed by Kaithali 165 and 125 mg/100g fruit pulp respectively (Gupta, 1977). Godhan was the most nutritive with regard to total soluble solids and ascorbic acid content, whereas Kharki was the sweetest at Hoshangabad. The highest acidity (0.49 per cent) was exhibited by Soni while the lowest acidity (0.25 per cent) was exhibited by Kabra and Amrabati. The highest total sugar content (14.48 per cent) was exhibited by Bekanta and the lowest (8.5 per cent) by Karka (Gupta *et al.*, 2004). Singh and Jindal (1980) found the highest content of pulp (97.2 per cent) TSS (21.4 per cent) and total sugar (10.7 per cent) in Gola Gurgaon, and Kaithli had the highest (113.5 mg/100 g) ascorbic acid content. The TSS content was highest in Chonchal and Illaichi and the latter cultivar had the highest ascorbic acid content. Cultivars grown in China, the United States and Pakistan contained 500-600, 300-500 and 45.2-160.8 mg Vitamin C/100 g respectively (Munier, 1973). Late ripening varieties generally contain more ascorbic acid than early ripening ones (Ahmad and Malik, 1971). Lomakina (1976) evaluated jujube varieties in South-West Turkmenistan. The better cultivars were Ya-tszao and Ta-Yan-tszao. Umran was noted for high sugar and protein contents, followed closely by Kathaphal (Khera and Singh, 1976). Of the 17 free amino acids represented in wild samples, glutamic acid, amino-butyric acid, threonine, proline and alanine were well represented. While in cultivars, the number of amino acids per sample was in the range 3-10. Of the 22 amino acids represented in the cultivars, arginine, cysteine, cystine, alanine and proline were common. The more widely represented amino acids in cultivars (arginine, cysteine and cystine) were absent in wild taxa. The commonly represented amino acids in wild samples (glutamic acid, amino-butyric acid and threonine) were not common in cultivars.

Genetic Variability

As per the study conducted on genetic variability, heritability and correlation coefficients in four year old ber plants of 12 popular ber cultivars in Raichur, Genotypic Coefficient of Variation (GCV) and Phenotypic Coefficient of Variation (PCV) were higher for stomatal frequency. GCV and PCV values were the lowest for relative water content (RWC). High values of heritability and genetic gain were observed for stomatal index. High estimates of GCV, PCV, heritability and genetic advance were recorded for stone size, pulp stone ratio, fruit weight and yield indicating effectiveness of improvement through simple selection (Saran, 2005).

Disease Resistance

Forty ber accessions were screened for resistance to black leaf spot disease in Faizabad. Tikri during 1997-98 and 1999-2000; Seedless during 1998-99; and ZG-3 during 1999-2000 were found immune. Guli, Seedless and Ber selection-5 during 1997-98; Guli, Darackhi-2, Ber selection-2, 3, 4 and 5 during 1998-99; and Jalandher, Kali, Bagwadi, Banarasi Peondi, Illaichi, Villaiti, Sanour-3, Chhohara, Katha, Seedless, Darakhi-2 and Ber selection-5 during 1999-2000 were found resistant. Other accessions showed susceptible reaction against the disease (Kumar *et al.,* 2003).

Hybridisation

The taxonomic literature on *Ziziphus* refers to hybridization between certain wild species. It is highly likely that a number of the taxa generally accepted will turn out to be stabilised hybrid segregates. Further insight into this could be useful in relation to genetic resources for use in breeding. Improvement programmes have not yet successfully used hybridisation with wild species, although such interspecific hybridisation would be of potential value to expand adaptation of cultivars to wider ecological areas and to introduce resistance to pests and diseases.

Important Jujube Cultivars

The majority of cultivars of both Indian and Chinese jujubes are selections from heterogeneous populations. Superior genotypes were protected and used by local people. Modern cultivars propagated clonally and some have become very widespread, *e.g.*, Umran in Indian jujube; or Sui Men or Li in Chinese jujube. The cultivated ber has more than 300 varieties but only a few are commercially important (Pareek and Nath, 1996). Over 180 named cultivars have been mentioned in the literature (Pareek, 2001). Important traits of Indian cultivars are given in Table 13.3. A Chinese work published over 300 years ago listed 43 cultivars (Locke, 1948). In China, there are at least 400 cultivars of Chinese jujube (Hayes, 1945) but Qu and Wang, (1993) reported more than 700 cultivars. These can be divided into two groups: the sour type mainly used as rootstocks, medicines or animal fodder, and the cultivated type (Ciminata, 1996).

Table 13.3: Important traits in Indian jujube cultivars.

Cultivars	*Characters*
Umran	Yield, stem girth, fruit size, fruit weight, flesh thickness, flowers/cyme and pulp/stone ratio.
Ponda	Yield, spread, number of flowers per cyme, fruit weight, fruit size and flesh thickness.
Sanori 5	Yield, number of flowers per cyme, fruit weight, size of fruit and flesh thickness.
Laddu	Yield, spread and number of flowers per cyme
Gola	Spread, fruit weight, flesh thickness and stone weight
Chhuhara	Spread, stem girth

Source: Saran, 2005

There is very little information available on the genetic diversity of jujube. The reduction of genetic variability makes the crop vulnerable to diseases and other adverse factors. Ten morphological traits, *viz.*, powdery mildew resistance, leaf length, leaf breath, leaf area, fruit weight, fruit length, fruit breath, stone weight, stone length, stone breath were analyzed for thirty-five important ber Indian genotypes at Hisar. They were classified into VII distinct clusters (Table 13.4). Cluster I was the largest with nine genotypes followed by cluster IV (six genotypes). Clusters VI and VII were the most divergent with an inter cluster distance of 5.162, suggesting that the parents for hybridization could be selected especially for powdery mildew resistance (Sanori No. 5, Noki and Mirchia) and smaller stone size (Illaichi and Kishmish) from these diverse clusters to develop disease resistant and smaller stone size breeding material for Indian jujube (Saran *et al.*, 2007). Important traits have been identified by various workers which can be used by breeders are listed in Table 13.3.

Umran

This cultivar is commercially cultivated on a large scale in Punjab, and Haryana. It was developed from germplasm from Rajasthan at the Fruit Research Station at Bahadurgarh, Patiala, Punjab. The fruits are large sized (30-80 g), oval in shape and have a roundish apex. They have an attractive golden yellow colour which later turns into a chocolate brown at full maturity. The fruit matures in the mid-season (February to March) and ripens during mid-March to mid- April. The fruit is sweet with 19 per cent total soluble solids (TSS) and 0.12 per cent acidity. It has a pleasant flavour and excellent dessert quality. Umran fruits have good keeping quality and can withstand long transportation. The fruit is also known locally as Ketha, Ajmeri and Chamdi.

Table 13.4: Classification of germplasm of ber in different clusters.

Clusters	*No. of Cultivars*	*Name of Germplasm*
I	8	Kaithali, Govindgarh selection, Sanori No. 1, Noki, Chonchal, Mirchia, Sandhura Narnul and Jogia
II	6	Seo, Laddu, Safeda Rohtak, Dandan, Popular Gola and Seo Bhadurgarh
III	1	Illaichi
IV	5	Katha Rajasthan, Ponda, Gola, Sanori No. 5 and Sua
V	6	Katha Bombay, Bhadurgadhia, Thornless, Umran, Banarsi Karaka and Desi Alwar
VI	2	Mundia Murhara and Bawal Selection-2
VII	1	Chhuhara
VIII	4	ZG-3, Akhrota, Gola Gurgaon No. 3 and Gola Gurgaon No. 2
IX	2	Kathaphal and Kishmish

Source: Saran, 2005.

Kathapal

This is a late ripening cultivar of ber well known in Gujarat that has small to medium size fruits. The fruits remain green on one side while the other side develops a reddish yellow tinge at maturity. The average fruit weighs 10 g. They have a high TSS (23 per cent) and an acidity of about 0.77 per cent.

Gola

Gola is an early-maturing cultivar grown in Uttar Pradesh, Gujarat, Punjab, Rajasthan, Haryana and Delhi. It starts to bear fruit in the first week of January. The fruits are attractive, roundish and golden yellow with an average weight of 20 g. The white flesh is juicy, semi-soft and has delicious taste. The fruit pulp has a TSS of 17 to 19 per cent and 0.46 to 0.51 per cent acidity. The ratio of pulp to stone is 14.

Kaithali

This is a mid-season cultivar of Punjab that ripens during March. It was developed from the collection from the Kaithali area of Kurukshestra in Haryana. The fruit size is medium with an average weight of 18 g, oval in shape and has a tapering apex. The fruit pulp is soft, has a TSS of 18 per cent and 0.5 per cent acidity. Fruit does not withstand transportation and has a poor keeping quality.

Tikadi

This ber cultivar is a late maturing type that ripens during February and March in Rajasthan. The fruits are small, weighing on an average 10 g. The fruits become edible, with a creamy soft flesh, when the skin turns to red during a short (7 to 10 days) ripening period of the fruits. The ripe fruit has a high TSS (25 per cent) with a large stone (the pulp to stone ratio of ripe fruit is 6.9).

Jogia

The fruits of this cultivar are also well known in Rajasthan and have a light purple tinge when they are unripe, but are still edible at this time. Because of this, they have an extended harvest period up to early February. The skin surface has coarse ridges and is greenish yellow in colour. The flesh is white, soft, juicy and sweet with a TSS of 19 per cent and a pulp to stone ratio of 14.

Mundia

This is an early and high yielding cultivar that matures during mid-January. The fruits are large, juicy and bell shaped with an average weight of 40 g. They have a yellowish green skin with smooth depressions. The flesh is white and soft with a TSS of 20 per cent and a pulp to stone ratio of 23.

Genetic Erosion

Accelerated selection and wider adoption of clonally-propagated cultivars and the common practice of top worked wild trees with improved cultivars may lead to a degree of genetic erosion. However, wild and naturalised populations persist in such large geographic areas in the primary centres of diversity, and numerous areas in other parts have become secondary centres, (*e. g.*, tropical Africa for *Z. mauritiana*, Central Asia and South-west Asia and parts of Africa for *Z. jujuba*). Many rural communities use jujubes for fencing, wind-breaks and other purposes. These activities provide adequate protection and this is not a major cause for concern.

Existing Germplasm Collections

Collections of germplasm are geared to maintenance of cultivars and other selections to support the national improvement efforts. More information are required for characterisation and genetic affinities to develop strategic planning to rationalise existing collections so that synonyms of cultivars can be eliminated. The accessions should truly represent specific patterns of genetic variability and be related to improvement needs. Many of the traits considered in improvement are polygenic and there will always be temptation of maintaining a larger number of accessions than are actually required. A large number of Institutions in India are holding ber germplasm and are working on their evaluation and improvement (Table 13.5).

Table 13.5: Indian Institutions holding ber germplasm collections (number of accessions).

Institutions	*Ziziphus Species*	*Ber Cultivars*
Central Institute for Arid Horticulture, Bikaner-334006, Rajasthan, India	7	162
Central Arid Zone Research Institute, Jodhpur-342003, Rajasthan, India	—	68
Indian Institute of Horticultural Research, Bangalore-560089, Karnataka, India	3	32
Central Horticultural Research Station, Godhra-389 001, Gujarat, India.	—	22
Haryana Agricultural University, Hisar-125 004, Harayana, India	—	74
Narendra Dev University of Agriculture and Technology, Kumarganj, Faizabad-224229, U. P, India	—	32
Mahatma Phule Agricultural University, Rahuri-413 722, Maharasthra, India	—	87
Gujarat Agricultural University, Sardar Krushinagar-385506, Gujarat, India	—	75
Fruit Research Station, Punjab Agricultural University, Bahadurgarh, Patiala-147001, Punjab, India	—	42
Indian Agricultural Research Institute, New Delhi-110012, India	4	52
Dryland Agriculture Research Station, Haryana Agricultural University, Bawal-123501, Haryana, India	—	36

Source: Pareek, 1988 and Pareek and Sharma, 1993.

Chinese jujube is maintained at a number of sites in Hebei, Shanxi, Henan and Shantung provinces and the Jujube Institute at the Hebei Agricultural University, Baodong, Hebei and the Chinese Academy of Forestry. Other collections of jujubes are to be found at the Apsheron Experimental Station for Subtropical Crops, Azerbaijan, where in the 1990s the best 40 seedlings of ber were selected from 3762 jujube seedlings, and 25 were included among the elites. The 11 best forms were given varietal names. There are 65 forms of *Ziziphus jujuba* at the Turkmen Experiment Station which include several local, foreign and Soviet bred varieties of jujube. A number of accessions of *Z. jujuba* are held at the Research Institute of Plant Production, Prague, Czechoslovakia. The Fruit-Tree Research Station (FTRS) in Japan has a collection of *Z. jujuba* maintained at the Okitsu Branch (Moriguchi *et al.*, 1994). The National Germplasm System of the USA also holds significant accession numbers of the two major jujubes. Mediterranean Turkey maintains a number of accessions of *Z. jujuba* at Meneme, Izmir through its national genetic resources programmes. Bangladesh maintains 35 accessions but only two of these cultivars are under cultivation on farms (Saha, 1997).

Rootstock Resources

In addition to cultivars, more attention needs to be given to wild sources of current or potential use as rootstocks and their maintenance in the collections. A little research on using rootstocks, other than those of cultivars has been done because very limited genetic material of each wild species used are tested for rootstocks (Table 13.6, Bal *et al.*, 1997). *Z. nummularia* is a popular rootstock in the drier parts of India. There are three other species used as rootstocks. (i) *Z. xylopyra* is an erect, small tree producing a woody, inedible fruit in South India and Sri Lanka, (ii) *Z. rugosa* is a straggly bush found in the Central Hills and eastern parts of India, and (iii) *Z. oenoplia* is a scrambling shrub with small black fruits often used for tanning and found in south India, Sri Lanka and Myanmar. Chinese jujube rootstocks are the most frequently wild materials related to cultivars. A number of other wild species are tried in China with extreme climates (Ming and Sun, 1986).

Table 13.6: Species used for rootstocks for Indian jujube.

Compatibility	*Species*
Most successful	*Z. mauritiana* cultivars
Can be widely used	*Z. mauritiana* var. *rotundifolia* (wild/naturalised), *Z. abyssinia*
Less successful but mostly compatible; often cultivar specific	*Z. nummularia, Z. xylopyrus, Z. spina-christi, Z. mucronata, Z. oenoplia, Z. jujuba*

Source: Pareek, 2001.

Conservation Methodologies

A little attention is paid to storage of seeds of jujubes for long-term conservation because of the heterozygosity and the loss of particular trait of cultivar if propagated by seed. Jujube seeds show an orthodox behaviour when dried and stored at low temperatures. The Millennium Seed Bank of the Royal Botanic Gardens, Kew, UK stores wild species of *Ziziphus* from arid zones of Africa. The rationalisation of a number of existing germplasm collections could transfer them into field genebanks easily. A systematic evaluation of existing resources is essential to utilise the variable genepools. Insufficient information on performance of accessions is due to enormity of the task. Thus, proper characterisation and evaluation of germplasm and dissemination of the information to breeders and others are important.

In situ Conservation

In situ conservation has two approaches (i) biosphere reserve and (ii) habitat approach. The natural biosphere reserve is a useful solution for endangered species. The habitat approach refers to management of target species in its original habitat through protected areas, managed forests, natural reserves, preservation plots, wildlife sanctuaries, habitat or national parks and agro-ecosystems. At present, much of the conservation is through maintaining cultivars on-farm (Le, 1998). Many practicable opportunities exist for conservation of *Ziziphus in situ* in natural habitats or in the areas where it grows naturally. This can be achieved by protecting areas from human interference. A gene sanctuary conserves the existing genetic diversity present in the population, it also allows for new gene combinations which appear with time. There is a pressing need to establish gene sanctuaries in India for the conservation of *Z. mauritiana* and in China for *Z. jujuba*.

In vitro Conservation

In vitro genebanks can be useful for clonally propagated material using slow growth in tissue culture and long-term cryopreservation of tissues and / or embryos. No such facilities exist for jujubes yet. However, a technique for *in vitro* storage of plants including *Z. jujuba* was developed in the Nikitsky Botanical Gardens, Ukraine (Mitrofanova *et al.*, 2002).

Current Scenario and Research Needs

The following aspects related to preservation of biodiversity require immediate attention:

- ☆ A large number of selections and cultivars are available for both Indian and Chinese jujubes. There is a great deal of confusion in their naming. An accurate assessment of the taxonomy of the genus *Ziziphus* is needed for better characterisation using molecular fingerprinting of all existing germplasm. A genepool approach needs to be taken rather than one based solely on morphological types. One may come across the species which may have the required productivity, fruit quality or pest and disease resistance attributes which can be of great use in improvement programmes. Therefore, a long term, well-planned research is needed for understanding the broader taxonomy of *Ziziphus* and the patterns of genetic variation in the cultivated species and their wild forms.

- Very little is known about the diversity in the primary and secondary centres of the cultivated species. The objective should be to rationalize the size of germplasm collections to best serve improvement needs.
- The cultivated jujubes represent polyploid series and there appears to be a range of self and cross-incompatibilities in *Ziziphus* species. Systematic research should be undertaken to know about the distribution of ploidy levels.
- As jujubes are cultivated more in small or medium sized plantations, there is a need to develop cultivars to suit the diverse agroclimatic conditions, and tolerant to pests and diseases. In this respect, knowing more about the genetic relations between *Z. mauritiana* and *Z. jujuba*, and the possibilities of their crossing, would be helpful.
- Adequate networking is required across regions to develop cooperative goals and research. Network linkages can be developed with concerned organisations to maintain specific wild germplasm *in situ* when specific ecotypes or genotypes have been identified and characterised.

Acknowledgement

This manuscript is mainly based on the monograph 'Ber. International Centre for Underutilised Crops' (Azam-Ali *et al.*, 2006) whose copyright are with International Centre for Underutilised Crops, University of Southampton, Southampton, SO17 1BJ, UK [©2006 Southampton Centre for Underutilised Crops], and other sources including authors own experiences and research output.

References

Awasthi, O. P. and More, T. A. 2009. Genetic diversity and status of *ziziphus* in India. *Acta Hort.* 840: 33-40.

Azam-Ali, S., Bonkoungou, E., Bowe, C., deKock, C., Godara, A. and Williams, J. T. 2006. Ber. International Centre for Underutilised Crops, Southampton, UK.

Babu, R. H. and Kumar, P. S. 1988. Studies on the growth and flowering of certain cultivars of ber (*Ziziphus mauritiana* Lamk.) of Hyderabad. *Haryana Journal of Horticultural Science*, 17: 24-27.

Bakhshi, J. C. and Singh, P. 1974. The ber - a good choice for semi-arid and marginal soils. *Indian Horticulture* 19: 27-30.

Bal, J. S. 1992. Identification of ber (*Ziziphus mauritiana* L.) cultivars through vegetative and fruit characters. *Acta Hort.* 317: 245-253.

Bal, J. S., Singh, M. P. and Sandhu, A. S. 1997. Evaluation of rootstocks for ber (*Ziziphus mauritiana* Lamk.) cv. Umran. *Journal of Research, Punjab Agricultural University* 34: 60-63.

Barbeau, G. 1994. Inventory of tropical fruit trees in Central America and the West Indies. *Fruits* Paris 49: 383-389, 469-474.

Bisla, S. S. and Daulta, B. S. 1986. Correlation and path analysis studies of some quality attributes, disease and yield in ber (*Zizyphus mauritiana* Lamk.). *HAU Journal of Research* 16: 348-351.

Brandis, D. 1874. *The Forest Flora of Northwest and Central India*. Herbarium of the Royal Botanic Gardens, Kew.

Brandis, D. 1906. *Indian Trees*. Fifth impression 1971. Bishen Singh Mahendra Pal Singh, Dehra Dun, India. 169-172.

Chadha, K. L., Gupta, M. R. and Bajwa, M. S. 1972. Performance of some grafted varieties of ber (*Ziziphus mauritiana* Lamk.) in Punjab. *Indian Journal of Horticulture* 29: 137-50.

Chen-YouZhi, Lin-ZhenHai, Li-MingChun, Wang-ZhiMing and Xie-LinZhong 2003. The performance of 10 jujube varieties in Zhaoyuan area, Shandong province. *China Fruits* 6: 26-27.

Chiv, C. Y. 1997. Effects of various lighting sources on the flowering and yielding date of Indian jujube. *Taichung District Agricultural Improvement Station, Special Publication* No. 38: 251-256.

Chovatia, R. S., Patel, D. S., Patel, A. T. and Patel, G. V. 1992. Growth, yield and physico-chemical characters of certain varieties of ber (*Ziziphus mauritiana* Lamk.) under dryland conditions. *Gujarat Agriculture University Research Journal*, 17: 56-60.

Ciminata, P. 1996. The Chinese jujube, *Ziziphus jujuba*. *WANATCA Year Book* 20: 34-36.

Darbara Singh and Jindal, P. C. 1982. Studies on flowering and sex ratio in some ber (*Ziziphus mauritiana* Lamk.) cultivars. *HAU Journal of Research* 12: 292-294.

Darlington, C. D. and Wylie, A. P. 1955. *Chromosome Atlas of Flowering Plants*. Allen and Unwin, London.

Daulta, B. S. and Chauhan, K. S. 1982. Ber – a fruit with rich food value. *Indian Horticulture* 27: 7-9.

Desai, U. T., Ranawade, D. B. and Wavhal, K. N. 1986. Floral biology of ber. *Journal of the Maharashtra Agricultural University* 11: 76-78.

Dhaliwal, J. S. and Bal, J. S. 1998. Floral and pollen studies in ber (*Zizyphus mauritiana* Lamk.). *Journal of Research Punjab Agricultural University* 35: 36-40.

Dhingra, R. P., Singh, J. P. and Chitkara, S. D. 1973. Varietal variations in physico-chemical characters of ber (*Ziziphus mauritiana* Lamk.). *Haryana Journal of Horticultural Science* 2: 61-65.

Diwakar-Hegde, Sharma, V. P. and Hegde, D. 1992. A numerical taxonomic study of the ber *Zizyphus* spp. *New Botanist* 19: 21-26.

Diwakar-Hegde, Sharma, V. P. and Hedge, D. 1996. Palynological studies in ber (*Z.* spp.). *New Botanist* 23: 145-151.

Ghosh, S. N. and Mathew, B. 2002. Performance of nine ber (*Z. mauritiana* Lamk.) cultivars on top working in the semi-arid region of West Bengal. *Journal of Applied Horticulture* 4: 49-51.

Ghosh, D. K. and Mitra, S. 2004. Post harvest studies on some local genotypes of ber (*Z. mauritiana* Lamk.) grown in West Bengal. *Indian Journal of Horticulture* 61: 211-214.

Godara, N. R. 1980. *Studies on floral biology and compatibility behaviour in ber* (Ziziphus mauritiana *Lamk.*). Ph. D. thesis, Haryana Agricultural University, Hisar, India.

Godara, N. R., Chauhan, K. S. and Bisla, S. S. 1980. Evaluation of mid-season ripening ber (*Ziziphus mauritiana* Lamk.) germplasm. *Haryana Journal of Horticultural Science* 9: 101-105.

Grice, A. C. 1998. Ecology in the management of Indian jujube (*Ziziphus mauritiana*). *Weed Science*, 46: 467-474.

Gupta, M., Mazumder, U. K., Vamsi, M. L. M., Sivakumar, T., and Kandar, C. C. 2004. Anti-steroidogenic activity of the two Indian medicinal plants in mice. *Journal of Ethnopharmacology* 90: 29.

Gupta, M. R. 1977. Physico-chemical characters of some promising ber cultivars grown at Bahadurgarh (Patiala). *Punjab Horticultural Journal* 17: 131-134.

Gupta, R. B., Suneel Sharma, Sharma, J. R. and Sareen, P. K. 2003. Cytological studies in some varieties of genus *Ziziphus*. *National Journal of Plant Improvement* 5: 19-21.

Hayes, W. B. 1945. *Fruit Growing in India*. Kitabistan, Allahabad, India.

Hooker, J. D. 1875. *Flora of British India. Part I*. L. Reeve and Co., London.

Khera, A. P. and Singh, J. P. 1976. Chemical composition of some ber cultivars (*Ziziphus mauritiana* L.). *Haryana Journal of Horticultural Science* 5: 21-24.

Khoshoo, T. N. and Singh, N. 1963. Cytology of northwest Indian trees. *Ziziphus jujuba* and *Z. rotundifolia*. *Silvae Genetica*, 12 Heft. 141-180.

Kim, W. S. and Kim, Y. S. 1984. Changes in carbohydrates, proteins, RNA and hydrolytic enzyme activity of jujube seeds during germination. *Journal of the Korean Society for Horticultural Sciences*, 25: 109-115.

Kumar, P, Singh, H K and Saxena, R P. 2003. Ber germplasm screening and management of black leaf spot disease under Eastern U. P. conditions. *Journal of Applied Horticulture*. 2003: 5: 43-44.

Kumar, S. S., Babu, R. S. and Reddy, Y. N. 1986. Duration of fruit maturity seasons and yield of certain cultivars of ber (*Ziziphus mauritiana* Lamk.) at Hyderabad. *Journal of Research APAU*, 14: 88-89.

Kundi, A. H. K., Wazir, F. K., Abdul, G. and Wazir, Z. D. K. 1989. Physico-chemical characteristics and organoleptic evaluation of different ber (*Ziziphus jujuba* Mill.) cultivars. *Sarhad Journal of Agriculture* 5: 149-155.

Lal, G., Jat, R. G., Dhaka, R. S., Sen, N. L. and Agarwal, V. K. 2003. Evaluation of ber germplasm in semi arid regions of Rajasthan. *Journal of Ecophysiology* 6: 145-148.

Le T. T. H. 1998. Some aspects of fruit production and genetic conservation in Vietnam. In *Report of the Second Meeting of MESFIN on Plant Genetic Resources*, INIA, Portugal: 287-299.

Liu, M. J. and Cheng, C. Y. 1995. A taxonomic study of the genus *Ziziphus*. *Acta Hort*. 390: 161-165.

Liu, M. J. and Wang, Y. H. 1991. Promising varieties and forms of Chinese wild jujube. *China Fruits* 4: 23-35.

Locke, L. F. 1948. The Chinese jujube: A promising fruit tree for the southwest. *Bulletin of the Oklahoma Agricultural Experimental Station* 319: 78-81.

Lomakina, M. I. 1976. Chemical composition of the fruit of subtropical and pome fruits in south-west. *Turdy po Prikladnoi Botanike Genetike-i-Selektsii* 57: 150-154.

Malik SK, Chaudhury R, Dhariwal OP and Bhandari DC. 2010. Genetic Resources of Tropical Underutilized Fruits in India. NBPGR, New Delhi, p.168.

Martin, F. W. Campbell, C. W. and Ruberte, R. M. 1987. *Perennial edible fruits of the tropics: an inventory*. Agriculture Handbook, USDA. 642, 247.

Mehetre, S. S. and Dahat, D. V. 2000. Cytogenetics of some important fruit crops: *A review of Maharashtra Agricultural Universities* 25: 139-168.

Mehrotra, N. K. and Gupta, M. R. 1985. Pollination and fruit set studies in ber (*Ziziphus mauritiana* Lamk.). *Journal of Research, Punjab Agricultural University* 22: 671-674.

Ming, W. and Sun, Y. 1986. Fruit trees and vegetables for arid and semi arid areas in northwest China. *Journal of the Arid Environments*, 11: 3-16.

Mitrofanova, I. V., Movcha, O. P., Shishkin, V. A. and Mitrofanov, V. I. 2002. Gene-pool collection in vitro in Nikitsky Botanical Gardens – National Scientific Center. Nikitskogo Botanicheskogo Sada 85: 30-33.

Moriguchi, T., Teramoto, S. and Sanada, T. 1994. Conservation system of fruit tree genetic resources and recently released cultivars from Fruit Tree Research Station in Japan. *Fruit Varieties Journal* 48: 73-80.

Munier, P. 1973. Le jujubier et sa culture [French]. *Fruits*, 28: 377-388. 218.

Neeraja, G., Reddy, S. A. and Babu, R. S. H. 1995. Fruit set, fruit drop and fruiting behaviour in certain ber (*Ziziphus mauritiana* Lamk.) cultivars. *Journal of Research, AP Agricultural University* 23: 17-21.

Nehra, N. S., Chitkara, S. D. and Singh, K. 1984. Studies of morphological characters of some wild forms and cultivated varieties of ber. *Punjab Horticultural Journal*, 24: 49-59.

Nehra, N. S., Sareen, P. K. and Chitkara, S. D. 1983. Cytological studies in genus *Ziziphus*. *Cytologia*, 48: 103-107.

Nerd, A. Aronson, J. A. and Mizrahi, Y. 1990. Introduction and domestication of rare and wild fruit and nut trees for desert areas. Advances in new crops. Proceedings of the first national symposium 'New crops: research, development, economics', Indianapolis, Indiana, USA, 23-26 October 1988. Timber Press, Portland, Oregon, USA: 355-363.

Panwar, J. S. 1980. Pattern of fruit drop in ber (*Ziziphus mauritiana* Lamk.) in the arid conditions of Haryana. *HAU Journal of Research*, 10: 57-59.

Pareek, O. P. 2001. *Fruits for the Future 2: Ber*. International Centre for Underutilised Crops, University of Southampton, Southampton, UK.

Pareek, O. P. and Sharma, S. 1993. Genetic resources of under-exploited fruits. In. *Advances in Horticulture, Vol. I - Fruit Crops. E*ds. Chadha, K. L. and Pareek, O. P. Malhotra Publishing House, New Delhi. 189-225.

Pareek, O. P. and Nath, V. 1996. Ber. In *Coordinated Fruit Research in Indian Arid Zone - A two decades profile (1976-1995*). National Research Centre for Arid Horticulture, Bikaner. 9-30.

Pradeep, T. and Jambhale, N. D. 2000. Ploidy level variations for stomata, chloroplast number, pollen size and sterility in ber (*Zizyphus mauritiana* Lamk.). *Indian Journal of Genetics and Plant Breeding* 60: 519-525.

Qu, Z. Z. and Wang, Y. H. 1993. *Chinese fruit trees record Chinese jujube*. The Forestry Publishing House of China, Beijing.

Ridley, H. N. 1922. *The Flora of the Malay Peninsula. Vol. I. Polypetalae*. L. Reeve and Co., London, UK.

Rixford, G. P. 1917. *Ziziphus* spp. In *The Standard Cyclopaedia of Horticulture*. ed. Bailey, L. H., The MacMillan Co. NY., USA.

Saha, N. N. 1997. *Conservation and utilization of fruit plant genetic resources: Bangladesh perspective.* M. G. Hossain, R. K. Aurora and P. N. Mathur (eds.) Dhaka: BARC-IPGRI.

Saran, P. L. 2005. *Studies on genetic divergence in ber* (Ziziphus mauritiana *Lamk.*) *germplasm.* Ph. D. thesis, CCS Haryana Agricultural University, Hisar, India.

Saran P. L., A. K. Godara, R. and P. Dalal. 2007. Biodiversity among Indian jujue (*Ziziphus Mauritiana* Lamb) genotype for powdery mildew and other traits. *Not. Bot. Hort. Agrobot*. Cluj 35: 15-21.

Sharma, S. K. 2003. Optimizing the productivity of natural Sehima pasture by introducing fruit tree and legume components. *Range Mgmt. and Agroforestry*: 24: 38-41.

Sharma, S. K. 2004. Hortipastoral based land use systems for enhancing productivity of degraded lands under rainfed and partially irrigated conditions. *Ugandan Journal of Agricultural Sciences* 9: 320-325.

Sharma, S. K. 2009. Development of fruit tree based agroforestry systems for degraded lands – a review. Range Mgmt and Agroforestry 30: 98-103.

Sharma, V. P., Raja, P. V. and Kore, V. N. 1990. Flowering, fruit set and fruit drop in some ber (*Ziziphus mauritiana* Lamk.) varieties. *Annals of Agriculture Research Institute* 11: 14-20.

Singh, D. and Jindal, P. C. 1980. Physico-chemical characters of some promising ber cultivars grown at Gurgaon (Haryana). *Haryana Journal of Horticultural Science* 9: 114-117.

Singh, K. K., Chadha, K. L. and Gupta, M. R. 1973. *Ber cultivation in Punjab.* Punjab Agricultural University, Ludhiana.

Singh, R. and Tomar, N. S. 1988. Performance of some jujube (*Ziziphus mauritiana*) cultivars in semi-arid zone of Punjab. *Indian Journal of Agricultural Science* 58: 382-383.

Singh, R. R., Jain, R. K. and Chauhan, K. S. 1983. Flowering and fruiting behaviour of ber (*Ziziphus mauritiana* Lamk.) under Gurgaon conditions. *HAU Journal of Research* 13: 112-114.

Sin'ko, L. T. 1974. Economic and technological evaluation of *Zizyphus* fruits. *Byulleten'-Gosudarstvennogo-Nikitskogo-Botanicheskogo-Sada* 2: 28-31.

Sin'ko, L. T. 1977. Agrobiological characteristics of Chinese jujube in the Crimea [in Russian]. *Trudy Gosudarstvennogo Nikitskogo Botanicheskogo Sada* 73: 98-125.

Tagiev, T. M. 1992. Promising areas for cultivating jujube in Azerbaijan [in Russian]. *Sadovodstvo i Vinogradstvo* 1: 19-20.

Teaotia, S. S. and Chauhan, R. S. 1964. Flowering, pollination, fruit set and fruit drop studies in ber (*Ziziphus mauritiana* Lamk.). II. Pollination, fruit set, fruit development and fruit drop. *Indian Journal of Horticulture* 2: 40-45.

Tutin, F. G. 1968. *Ziziphus.* Flora Europea Vol. 2 eds. Tutin, T. G., Heywood, V. H., Borges, N. A., Moore, D. M., Valentino, D. H., Walter, S. M. and Webb, D. A. Cambridge. 243.

Vashishtha, B. B. and Pareek, O. P. 1979. Flower morphology, fruit set and fruit drop in some ber (*Ziziphus mauritiana* Lamk.) cultivars. *Annals of the Arid Zone*, 18: 165-169.

Vavilov, N. I. 1951. *The Origin, Variation, Immunity and Breeding of Cultivated Plants*. Chronica Botanica, NY, USA.

Verma, M. K., Sharma, V. P., Saxena, S. K. and Jindal, P. C. 2001. Stomatal density as influenced by stionic combinations of ber (*Ziziphus mauritiana* Lamb). *Indian Journal of Horticulture* 58: 350-353.

von Maydell, H-J. 1986. *Trees and Shrubs of the Sahel - their characteristics and uses*. GT2, Eschborn, Germany.

von Sengbusch, V. and Dippolo, M. F. 1980. *Das Entwicklungs potentia afrikanischer Heilpflanzen – Kamerun, Tschod, Gagun*. Möckmuhl: IFB, Germany.

Wang, M. and Sun, Y. W. 1986. Fruit trees and vegetables for arid and semiarid areas in northern China. *Journal of Arid Environments*, 11: 3-16.

Wang-Yun and Wang, Y. 1994. A preliminary study on the geographical provenance of *Zizyphus mauritiana*. *Forest Research* 7: 334-335.

Yamdagni, R., Gupta, A. K. and Ahlawat, V. P. 1985. Performance of different cultivars of ber (*Ziziphus mauritiana* L.) under rainfed conditions. *Annals of the Arid Zone*, 24: 175-177.

2014, Future Crops Vol. 2 Pages 279–291
Editor: Professor K.V. Peter
Published by: DAYA PUBLISHING HOUSE, NEW DELHI

Chapter 14

Jamun (*Syzygium cumini*)

Bratati Chaudhary* and Kunal Mukhopadhyay
Department of Biotechnology, Birla Institute of Technology,
Mesra, Ranchi – 835 215
**E-mail: bratatichaudhary@yahoo.co.in*

Syzygium cumini L. (syn. *Eugenia jambolana, Eugenia cumini* and *Syzygium jambolana*) a polyembryonic species (family Myrtaceae), is a tropical fruit tree of great economic importance. The genus *Syzygium* comprises of about 1100 species and has a native range that extends from Africa and Madagascar through Southern Asia east to the Pacific. Many species are poorly known and many more have not been described taxonomically (Ayyanar and Babu, 2012). The fruits are commonly known as Jamun (Hindi), java plum, black plum, jambul and Indian blackberry (Figure 14.1). It is a large, evergreen tree cultivated throughout India for its edible fruits.

Classification

Kingdom:	Plantae
Division:	Magnoliophyta
Class:	Magnoliopsida
Order:	Myrtales
Family:	Myrtaceae
Genus:	*Syzygium*
Species:	*cumini*

Fruits of *S. cumini* are rich source of a water soluble plant pigment - anthocyanin. Anthocyanin gives some fruits and flowers their pink, red, magenta, purple and dark blue colours. These pigments are synthesized in the cytosol and are localized in the vacuoles of the plant cells (Grotewold, 2006). Anthocyanins have antioxidant ability, potentially serving as an *in vivo* defense against the plant hypersensitive pathogen response. These are important considerations as these secondary metabolites can replace a range of synthetic food colourants in confectionery and artificial beverages.

Figure 14.1: *Syzygium cumini* fruits.

A growing number of scientific and epidemiological reports suggest that anthocyanins or the anthocyanin extracts exhibit a wide range of protective effects with potential benefits for human and animal health (Chalabi *et al.*, 2008; Espin *et al.*, 2007). Besides, anthocyanins have shown antagonistic activity to a few bacteria, virus and fungi and thus also protecting food from microbial spoilage (Chattopadhyay *et al.*, 2008). Most of the bioactivity and health properties of anthocyanins are proved experimentally with through *in vitro* studies. The health benefits are mainly due to antioxidant activities towards free radicals and reactive oxygen species. The free radical scavenging activity of anthocyanins which has a protective role on the animal cell death gives an anti aging property to fruits or vegetables. Reason for an increasing demand of these flavonoids is their role as attractive natural colourants which also add nutritive value to the food consumed by consumers.

Religious and Cultural Importance

According to the Hindu mythology, Ramayana, Lord Rama survived on the fruits in the forest during his exile from Ayodhya. This is the reason Hindus regard Jamun as the 'fruit of gods' especially in Gujarat. Lord Krishna is described as having the colour of Jamun since he has dark complexion. In Southern Asia, the tree is venerated by Buddhists, and is commonly planted near Hindu temples because it is considered sacred to Lord Krishna. The leaves and fruits are employed in worshipping the elephant-headed God, Lord Ganesha. In Maharashtra, the leaves of Jamun (locally known as jambhul in Marathi) tree are used for decoration purpose in marriage ceremonies.

Occurrence and Natural Habitat

S. cumini is one of the most widely distributed forest trees of India. It is found in all types of forests of South East Asia, Malaysia and Australia. The jambolan is native to India, Burma, Sri Lanka and the Andaman Islands. It was long ago introduced into and became naturalized in Malaya. The tree was introduced from India and tropical Asia to southern Africa for its edible fruits. The Bureau of Plant

Industry of the United States Department of Agriculture received jambolan seeds from the Philippines in 1911, from Java in 1912, from Zanzibar and again from the Philippines in 1920. The tree flourishes in California, especially in Santa Barbara, though the climate is not congenial for production or ripening of the fruit. It is now grown throughout the tropics and subtropics, the best forms are frequently cultivated in Java and Florida.

It is present in both moist and dry places, occurring in the tropical wet evergreen forests, tropical semi evergreen forests, tropical moist deciduous forests, swamps, tropical dry deciduous, tropical dry evergreen, subtropical broad leaved hills, and subtropical pine forests. The tree prefers moist, damp or marshy land. *S. cumini* is an evergreen tree especially in moist localities, the new leaves start even before old leaves fall. In dry localities, it becomes leafless for a short time in summer season. It tolerates prolonged flooding and drought, once established.

Botanical Description of the Tree

Syzygium is derived from the Greek term, Syzgios (paired), on account of the leaves and twigs which in several species grow at the same point. *S. cumini* is a medium-sized tree (Figure 14.2) 10-30 m high, with a straight to crooked, short, stout trunk, usually 40-100 cm in diameter. Crown is irregular or globular with many branches. The Bark is brown or dark grey, fairly smooth, inner bark with thin green outer layer, mottled light brown, astringent and bitter to taste. Leaves are entire with narrow transparent margin, 7-18 cm long, 3-9 cm broad, opposite, thick, coriaceous, glabrous, broadly obovate,

Figure 14.2: *S. cumini* tree in its natural habitat.

Figure 14.3: Immature green fruits of *S. cumini.*

elliptic or elliptic-oblong, upper surface dark green, lower surface yellowish to dull green. The flowers are white or pink, many in clusters, small, slightly fragrant and about 5mm in diameter. The flower appears during March to April and fruits ripen during May to July.

The tree starts bearing fruits within 4-7 years. The fruits are ovoid-oblong or elliptical berries, numerous, crowded in clusters, almost stalkless along twigs at the back of leaves. The fruits are green at first (Figure 14.3), turning pink and then finally purple-black and are usually 1-2.5 cm long but can also be 5 cm. The seed in each berry is strongly astringent and slightly bitter which tends to colour the tongue purple. The irregularly shaped seeds are compressed together into a mass resembling a single seed.

S. cumini is pollinated by honey bees, house flies and wind. Natural regeneration is profuse around mother trees, as seeds fall in large quantities. Germination takes place on moist ground; each fruit usually produces 1-4 seedlings clustered together in dense masses. The flowering starts in April and the fruit formation takes place about 32 days after flowering. The fruits are devoured by birds, squirrels and humans and are therefore widely dispersed.

Varieties of Fruit

The most common variety of *S. cumini* fruit is often oblong with a purple to blackish colour. The pulp is grey to pink in colour with a single seed in the center. There are some seedless varieties also available in Varanasi, India, which can be either white or purple in colour. Fruits with purple flesh are more astringent than the white-fleshed types. The most common type grown in North India is known as Raja Jamun. It is a large tree having oblong fruits of deep purple colour. A type having large sized fruits is known as Paras in Gujarat.

Economic Importance

Almost all parts of the tree are used for various purposes. The various important phytochemicals found in *S. cumini* are listed in the Table 14.1. Its dense foliage provides shade (Figure 14.4) and hence it is grown just for its ornamental value. Ripe fruits are usually consumed fresh as it is very juicy, almost odorless, with a pleasant, slightly bitter, astringent taste. Fruit pulps are also used to make jams, jellies, juice, vinegar and puddings. Fruits are also used to make wine in vast quantities in Philippines.

The leaves are used as fodder and as food for tassar silkworms in India. The leaf distillates yield an essential oil used as fragrance in soaps and is blended with other chemicals to make inexpensive perfumes. *S. cumini* flowers are rich in nectar and are useful in apiculture for the yield of a high quality honey.

Figure 14.4: Dense foliage of *S. cumini* tree.

Table 14.1: Phytochemicals present in *S. cumini.*

Sl.No.	*Plant Parts*	*Phytochemicals*
1.	Edible pulp	Vitamin C, Vitamin A, Riboflavin, Nicotinic acid, Choline, Folic acid, Malaic acid, Sugar, Amino acids, Na, K, Ca, P, Fe, Zn, Mn
2.	Seed	Glycoside jamboline
3.	Seed and bark	Gallic acid
4.	Stem and bark	Tannins, Resin, Phytosterols
5.	Seed and leaves	Essential oils
6.	Flower	Terpenoids

S. cumini wood has a specific gravity of 0.77 and burns well, giving off about 4800 kcal/kg. Another interesting feature of the wood is that it is very strong and also water as well as termite resistant. It is used to make boats, railway sleepers, agricultural implements, tool handles, cart wheels, furnitures and as props for shafts and galleries in mines. The wood is used to install motors in the wells. It is also used for building bridges and for making musical instruments, especially guitars. It is a fast-growing tree, which provides excellent firewood and charcoal.

The seed is used as an alternative natural healing system in the Ayurveda, Unani and Chinese medicines. The bark contains good amount of lignin (Mir *et al.*, 2009). Bark of *S. cumini* yields a brown dye due to its high tannin content, used in tanning leather and preserving fishing nets. Seeds and bark are well known in the Far East for treatment of dysentery and in control of hyperglycemia and glycosuria in diabetic patients. The astringent bark may be used as a gargle. Fruits are used as a relief for colic, while the wood yields a sulphate pulp that has medicinal uses.

Health Benefits

S. cumini is a healthy fruit with absolutely no trace of sucrose. It is therefore, the only fruit with minimum calories. The family Myrtaceae has some of the taxonomically informative molecules like mallic acid, oxalic acid, gallic acid, ellagic acid, betulic acid, tannins, flavonoids and essential oils. These compounds are present at different parts of the tree and can either act in combination or individually to cure some of the diseases and health problems. All parts of the plant are used for traditional medicinal purposes and resulted in considerable chemical analysis of the plant. Most of the active compounds which attribute to the medicinal properties were isolated and identified.

One of the very common traditional uses of *S. cumini* is the powdered seed for controlling blood sugar level in diabetic patients. The seeds are rich in proteins and calcium. The glycoside jamboline is the main compound in the seed which helps in controlling blood sugar level by switching off the mechanism of starch converting to sugar when there is optimum amount of sugar already present in blood.

Effect of fruit, seed and bark extracts were studied *in vitro* and *in vivo* in many animal cell lines and animal models. Table 14.2 lists some of the major health benefits of different parts of *S. cumini* plant.

Table 14.2: Health benefits from *S. cumini* studied *in vitro.*

Sl.No.	*Health Benefits*	*Plant Parts*	*References*
1.	Diabetes	Seed, Leaf, Bark, Pulp	(Kumar *et al.*, 2008a) (Leelavinothan and Saravanan, 2006) (Farswana *et al.*, 2009) (Bopp *et al.*, 2009)
2.	Anti-allergic	Leaf	(Brito *et al.*, 2007)
3.	Anti-oxidant	Leaf, Fruit, Seed	(Ruan *et al.*, 2008) (Zhang and Lin, 2009) (Loganayaki and Manian, 2010) (Faria *et al.*, 2011)
4.	Anti-viral	Leaf	(Bhanuprakash *et al.*, 2008)
5.	Anti-bacterial	Leaf, Seed	(Gowri and Vasantha, 2010) (Meshram *et al.*, 2011) (Banerjee and Narendhirakannan, 2011)
6.	Anti-cancer	Fruit	(Li *et al.*, 2009) (Barh and Viswanathan, 2008) (Afify *et al.*, 2011)
7.	Anti-inflammation	Leaf, Seed	(Kumar *et al.*, 2008b) (Roy *et al.*, 2011) (Modi *et al.*, 2010)
8.	Gastric ulcer	Seed	(Chaturvedi *et al.*, 2007)
9.	Anti-hyperlipidemic	Fruit	(Rekha *et al.*, 2010)
10.	DNA damage	Leaf	(Jagetia and Baliga, 2002)

Traditional Uses as Medicine

Herbal drugs are used since dawn of civilization due to its easy availability, non toxic nature, easily acceptability and affordability. A decoction made from the bark of jamun is given internally for dysentery, enema and diarrhea. The fresh juice of the bark with goat's milk is a good medicine for diarrhoea in children. The bark decoction is also used as a gargle or a mouthwash for gingivitis, spongy gum, mouth ulceration and as an astringent wash for wounds.

Juice of leaves is used for dysentery. A paste made from leaves is applied on skin to treat infections and wounds. An infusion of fresh tender leaves taken with honey or butter milk is an effective remedy for sterility and miscarriage due to ovarian or endometrial functional disorder.

The dried seed powder apart from being a potent medicine for diabetes is also used in diarrhoea, flatulence, constipation, pancreatic and gastric disorders, muscle spasms, fatigue, depression, anxiety and metrorrhagia. The powder is used as a liver tonic, to enrich the blood, strengthen the teeth and gums and as lotion to remove ringworms of the head (*Tinea capitis*) in the Unani medicine system. The fruits are used as astringent in diarrhoea, stomach ache, as a gargle for sore throat and as lotion in *Tinea capitis*. The fruit is also a food remedy for bleeding piles. It is eaten in morning with salt or with honey to treat piles. The natural acids in the fruit play an important role in the secretion of digestive enzymes and stimulate the liver function. Vinegar from juice of ripe fruit is used as stomachic, carminative, and diuretic.

Accumulation of Anthocyanins in *Syzygium cumini*

Most of the immature fruits are green in colour and very a few of these fruits remain green when mature. This can be attributed to the general pattern of chlorophyll loss and accumulation of plant pigments like anthocyanins, carotenoids, and betalains which characterize the ripening phase in most of the fruits (Macheix *et al.*, 1990). Anthocyanin accumulating fruits often display a range of intermediary colours during fruit development which ranges from green to white to the final colour of the fruit with decreasing chlorophyll levels (Wheelwright and Janson, 1985; Wilson and Thompson, 1982). Accumulation of the pigment finally gives the different shades of pink, red, purple or black colour to the fruits. In *S. cumini*, the immature fruits which are green in colour first become pale green and transform to a white fruit. This white fruit stage starts accumulating pink colour which turns to the final stage or the purplish black ripe fruits.

S. cumini is a very rich source of anthocyanin. It has numerous health benefits and is used in traditional medicines to cure a number of diseases. There are many studies to establish the best extraction method for getting the highest amount of nutrients and were applied to animal cell models. Ruan *et al.* (2008), used different solvents like methanol, water, chloroform, n-hexane and ethyl acetate to determine antioxidant activity of leaf extracts against free radicals like 1,1-diphenyl-2-picrylhydrazy (DPPH) and ferric reducing antioxidant power (FRAP). Among these solvents ethyl acetate was the best solvent in terms of antioxidant activity. Another experiment by Kheaw-on *et al.*, 2009 used methanol and acidified methanol to check the total anthocyanin content and antioxidant capacity of the fruit extracts. Acidified methanol was the best solvent in the study.

The anthocyanin biosynthetic pathway, an extension of the phenylpropanoid pathway, is shared among all higher plants, enabling engineering of the pathway for purposeful accumulation of desirable anthocyanins (Tanaka and Ohmiya, 2008). For successful engineering, the first and essential step is to isolate and characterize the pertaining genes. Since delphinidin and malvidin anthocyanin are present in *S. cumini*, which is violet to blue in colour, genes involved in this particular pathway might provide resourceful genetic material for future transformation experiments in top selling cut-flowers like carnation, chrysanthemum and gerbera which lack the bluish hue.

Types of Anthocyanins Present in Fruit

The type of anthocyanins present can determine many important factors regarding stability and other properties of the extract. Three types of anthocyanins were identified in *S. cumini* as glucosides of delphinidin, petunidin and malvidin by HPLC-ESI-MS (Veigas *et al.*, 2007). The study also used extracts to check the lipid peroxidation in biological models like rat brain, liver mitochondria, testes and human erythrocytic ghosts. The extract was very active in the animal cell systems. Another HPLC-MS/MS study of *S. cumini* extracts by Brito *et al.* (2007) identified five out of six types of anthocyanins (delphinidin, cyaniding, petunidin, peonidin and malvidin) in their di glucoside forms.

Two types of tannins in fruits were identified by NMR, MALDI-TOF MS and HPLC in *S. cumini.* Hydrolysable tannin ellagitannins and condensed tannin epiafzelechins were identified and the structure has been predicted with the help of NMR (Zhang and Lin., 2009). The extracts have a good DPPH radical scavenging and FRAP activity which shows that the fruit is a potential source of antioxidants.

Propagation of the Tree through *in vitro* Culture

There is a very high anthocyanin content in *S. cumini* fruits which attributes to its antioxidant and free radical scavenging activity. Anthocyanin pigments can be a good source of natural food colourants for food processing industries. But since the fruit is seasonal there is a limitation in the constant supply and hence an *in vitro* system needs to be established for pigment production throughout the year (Rao and Ravishankar 2002; Lila 2004). Another major limitation with *S. cumini* is low viability of its seeds. The seed germinates when fresh but after a few days at room temperature, the seed is not viable (Krishnamurthy *et al.*, 1997). Seed storage behavior is recalcitrant; seeds germinate well when fresh, but viability is lost within 2 weeks of open storage at room temperature.

Due to very high amounts in *S. cumini* of polyphenols, the culture media turns brown due to exudates from explants. A stable pigmented callus and suspension cultures need to be developed for large scale production of pigments from this tree *in vitro*. The pigment can be further checked for its stability to different physical parameters for its use as a natural colourant.

The tree was cloned by micropropagation technique by using nodal or meristem explants from mature trees (Jain and Babbar, 2000, 2003, Rathore *et al.*, 2004). Use of *in vitro* grown seedlings as explants for micropropagation proves beneficial to control browning of culture media with phenolics.

Processed Food Products from the Fruit

Since all parts of *S. cumini* tree has proved to have medicinal properties against a number of diseases, it has high economic value for application in food processing industries. Different types of fruits are used to prepare jam, jellies, syrups and ice creams since ages. Processing of food adds value to fruits and also makes the seasonal fruits available for consumption throughout the year. According to the Ministry of Food Processing Industries (MoFPI), Government of India, the installed capacity of fruits and vegetables processing industry has increased from 1.1 million tonnes in January 1993 to 2.1 million tonnes in 2006. The processing of fruits and vegetables is estimated to be around 2.2 per cent of the total production in the country.

The seed powder of *S. cumini* is used by diabetic patients to control blood sugar level naturally and is usually prepared at home by drying the seeds and grinding it to a fine powder. Seed powder is also commercially available by a number of herbal product manufacturers in the Asian market. The fruit pulp is used to prepare vinegar beneficial for patients suffering from diabetes, digestion problems and is a diuretic (Kirtikar and Basu, 1987).

The fruits of good size and good quality have a sweet or sub acid flavor with minimum astringency. These are eaten raw and may be made into tarts, sauces and jam. The astringent fruits are soaked in salt water or are pricked and rubbed with a little salt. By letting fruits stand for an hour with the salt, reduces the astringent taste. The fruits can also be utilized for making juice, much like grape juice. When extracting juice from mashed fruit pulp, it is recommended that it should be allowed to drain out on its own without squeezing the fruit. This makes the juice less astringent in taste. The juice can be bottled and stored for a long time by combining it with sugar and water and adding citric acid and sodium benzoate as a preservative The white-fleshed Jamun has adequate pectin and makes a very stiff jelly. The more common purple-fleshed Jamun yields richly coloured jelly but are deficient in pectin and requires the addition of a commercial jelling agent or must be combined with pectin rich fruits like unripe or sour guavas.

In Goa, India and the Philippines, Jamun is an important source of wine, somewhat like Port, and the distilled liquors, brandy and "jambava" are also made from fermented fruit. For the fermentation of the ripe fruits, *Saccharomyces cerevisiae* is used. Chowdhury and Ray(2007) analyzed the Jamun wine and compared the properties with the red grape wine. The Jamun wine is reported to be high in tannins and a strong astringent taste with 6 per cent alcohol. The fruits are also used to make vinegar by various commercial brands throughout India. The Jamun vinegar is attractive, clear purple, with a pleasant aroma and mild flavor. The vinegar is used as a salad dressing.

Since processing of food may affect availability of its nutrients, the stability of the processed food products is a major concern among the researchers. The various types of products made from *S. cumini*

Table 14.3: Processed food from the fruit available in the Asian market*:

Sl.No.	*Products*	*Manufacturers*	*Locations*
1.	Juice or squash	A.I. International	Mumbai, Maharashtra
		Konkan Bazar	Mumbai, Maharashtra
		Disha Consultancy Services	Kolhapur, Maharashtra
		Hahneman Charitable Mission Society	Jaipur, Rajasthan
		Aloecare Arogya Life	Mumbai, Maharashtra
		Kamadhenu Ventures Pvt. Ltd.	Indore, M. P
2.	Ice-cream	Natural Ice-creams	Mumbai, Maharashtra
		Fruity freeze	Kolkata, W.B.
		Gelato Vinto	New Delhi
		Fresh and Naturelle	Kolkata, W.B.
		Shankar	Ahmedabad, Gujarat
3.	Vinegar	Vikash Bharti	Gumla, Jharkhand
		Shanti Kunj Krishi Ghar	Ghaziabad, U.P.
		D.R.V. Marketing Pvt. Ltd.	Ghaziabad, U.P.
		Sushev agricultural farms	Pilibhit, U.P.

**Source*: Information freely available online.

were analyzed to check if the health benefits of the unprocessed fruits are still present in processed products or not (Shahnawaz *et al.*, 2009). There are many commercial brands available in the Asian market especially in India which manufactures jam, jellies, ice cream and vinegar made from the *S. cumini* fruits (Table 14.3). These products are gaining popularity not only among the diabetics but also among the health conscious consumers due to its varied health promoting properties.

Future Prospects

The food processing industries use synthetic colourants mostly which are proved to be carcinogenic, if consumed continuously for a long period of time. This is a big concern among the processed food producers and the consumers who look for all natural ingredients in their food. There is a lot of research on this issue to make the food safe for human consumption and to use natural food colourants instead of the synthetic ones. The best candidate to serve as a natural food colourant would be from the plant itself. Plants have various pigments which help to make the fruits and flowers attractive which primarily attract animals and insects. Anthocyanins are one such naturally occurring plant pigment which apart from having a wide range of shades is also a nutraceutical which can make the food more appealing to the consumers due to its health promoting value added nutrition.

S. cumini fruits are a very rich source of anthocyanins and have numerous health benefits. The fruit pulp is used to prepare different processed food items. Some of these products are used just for the flavor and some used as medicine by the consumers. The astringent flavor is sometimes not preferred by some people but the bright purple colour of the fruit can be used as a natural food colourant. There is still a lot of research that needs to be done to make these fruits suitable candidates for use as colourants in the food industries where the food items go through many types of physical and chemical processing.

Acknowledgement

Authors are highly thankful to Ministry of Food Processing Industries GOI and to BIT, Mesra, Ranchi for providing fellowship.

References

Afify, A. M. R., Fayed, S. A., Shalaby, E. A., and El-Shemy, H. A. 2011. *Syzygium cumini* (pomposia) active principles exhibit potent anticancer and antioxidant activities. *African Journal of Pharmacy and Pharmacology* 5: 948-956.

Ayyanar, M. and Babu, P. S. 2012. *Syzygium cumini* (L.) Skeels: A review of its phytochemical constituents and traditional uses. *Asian Pacific Journal of Tropical Biomedicine* 2: 240-246.

Banerjee, J. and Narendhirakannan, R. T. 2011. Phytochemical analysis, antibacterial, *in vitro* antioxidant and cytotoxic activities of ethanolic extract of *Syzygium cumini* (L.) seed extract. *International Journal of Pharmaceutical Sciences and Research* 2: 1799-1806.

Barh, D. and Viswanathan, G. 2008. *Syzygium cumini* inhibits growth and induces apoptosis in cervical cancer cell lines: a primary study. *ecancermedicalscience* 2: 1-9.

Bhanuprakash, V., Hosamani, M., Balamurugan, V., Gandhale, P., Naresh, R., Swarup, D., and Singh, R. K. 2008. *In vitro* antiviral activity of plant extracts on goatpox virus replication. *Indian Journal of Experimental Biology* 46: 120-127.

Brito E S D, Araujo M C D, Alves R E, Carkeet C, Clevidence B A and Novotny J A 2007. Anthocyanins present in selected tropical fruits: acerola, jambolao, jussara, and guajiru. Journal of Agriculture and Food Chemistry 55: 9389-9394.

Bopp, A., Bona, K. S., Belle, L. P., Moresco, R. N., and Moretto, M. B. 2009. *Syzygium cumini* inhibits adenosine deaminase activity and reduces glucose levels in hyperglycemic patients. *Fundamental and Clinical Pharmacology* 23: 501–507.

Brito, E. S. D., Araujo, M. C. D., Alves, R. E., Carkeet, C., Clevidence, B. A., and Novotny, J. A. 2007. Anthocyanins present in selected tropical fruits: acerola, jambolao, jussara, and guajiru. *Journal of Agriculture and Food Chemistry* 55: 9389-9394.

Chalabi, N., Gallon, D. J. B., Vasson, M. P., and Bignon, Y. J. 2008. Nutrigenomics and antioxidants. *Personalized medicine* 5: 25-36.

Chattopadhyay, P., Chatterjee, S., and Sen, S. K. 2008. Biotechnological potentials of natural food grade colourants. *African Journal of Biotechnology* 7: 2972-2985.

Chaturvedi, A., Kumar, M. M., Bhawani, G., Chaturvedi, H., Kumar, M., and Goel, R. K. 2007. Effect of ethanolic extract of *Eugenia jambolana* seeds on gastri ulceration and secretion in rats. *Indian Journal of Physiology and Pharmacology* 51: 131-140.

Chowdhury, P. and Ray, R. C. 2007. Fermentation of jamun (*Syzgium cumini* L.) fruits to form red wine. *ASEAN Food Journal* 14: 15-23.

Espin, J. C., Garcia-Conesa, M. T., and Tomas-Barberan, F. A. 2007. Nutraceuticals: Facts and fiction. *Phytochemistry* 68: 2986–3008.

Faria, A. F., Marques, M. C., and Mercadante, A. Z. 2011. Identification of bioactive compounds from jambolao (*Syzygium cumini*) and antioxidant capacity evaluation in different pH conditions. *Food Chemistry* 126: 1571-1578.

Farswana, M., Mazumder, P., and Parcha, V. 2009. Modulatory effect of an isolated compound from *Syzygium cumini* seeds on biochemical parameters of diabetes in rats. *International Journal of Green Pharmacy* 3: 128-133.

Gowri, S. S. and Vasantha, K. 2010. Phytochemical screening and antibacterial activity of *Syzygium cumini* (L.) (Myrtaceae) leaves extracts. *International Journal of PharmTech Research* 2: 1569-1573.

Grotewold, E. 2006. The genetics and biochemistry of floral pigments. *Annual Review of Plant Biology* 57: 761-80.

Jagetia, G. C. and Baliga, M. S. 2002. *Syzygium cumini* (Jamun) reduces the radiation-induced DNA damage in the cultured human peripheral blood lymphocytes: a preliminary study. *Toxicology Letters* 132: 19-25.

Jain, N. and Babbar, S. B. 2000. Recurrent production of plants of black plum, *Syzygium cumini* (L.) Skeels, a myrtaceous fruit tree, from *in vitro* cultured seedling explants. *Plant Cell Reports* 19: 519-524.

Jain, N. and Babbar, S. B. 2003. Regeneration of 'juvenile' plants of black plum, *Syzygium cuminii* Skeels, from nodal exp of mature trees. *Plant Cell, Tissue and Organ Culture* 73: 257-263.

Kheaw-on, N., Chaisuksant, R., and Suntornwat, O. 2009. Antioxidant capacity of flesh and seed from *Syzygium cumini* fruits. *Acta Horticulturae* 837: 73-77.

Kirtikar, K. R. and Basu, B. D. 1987. Indian medicinal plants, Second edition, International Book Distributors, Dehradun, pp. 1052-1054.

Krishnamurthy, K. S., Shaanker, R. U., and Ganeshaiah, K. N. 1997. Seed abortion in animal dispersed species, *Syzygium cumini* (L.) Skeels (Myrtaceae): The chemical basis. *Current Science* 73: 869–873.

Kumar, A., Ilavarasan, R., Jayachandran, T., Deecaraman, M., Aravindhan, P., Padmanabhan, N., and Krishan, M. R. V. 2008a. Anti-diabetic activity of *Syzygium cumini* and its isolated compound against streptozotocin-induced diabetic rats. *Journal of Medicinal Plants Research* 2: 246-249.

Kumar, A., Ilavarasan, R., Jayachandran, T., Deecaraman, M., Kumar, R. M., Aravindhan, P., Padmanabhan, N., and Krishan, M. R. V. 2008b. Anti-inflammatory activity of *Syzygium cumini* seed. *African Journal of Biotechnology* 7: 941-943.

Leelavinothan, P. and Saravanan, G. 2006. Effects of *Syzygium Cumini* bark on blood glucose, plasma insulin and C-peptide in streptozotocin-induced diabetic rats. *International Journal of Endocrinology and Metabolism* 4: 96-105.

Li, L., Adams, L. S., Chen, S., Killian, C., Ahmed. A., and Seeram, N. P. 2009. *Eugenia jambolana* Lam. berry extract inhibits growth and induces apoptosis of human breast cancer but not non-tumorigenic breast cells. *Journal of Agriculture and Food Chemistry* 57: 826-831.

Lila, M. A. 2004. Anthocyanins and human health: an *in vitro* investigative approach. *Journal of Biomedicine and Biotechnology* 2004: 306–313.

Loganayak, N. and Manian, S. 2010. *In vitro* antioxidant properties of indigenous underutilized fruits. *Food Science and Biotechnology* 19: 725-734.

Macheix, J. J., Fluriet, A., and Billot, J. 1990. Fruit phenolics. CRC press, Boka Raton: 378.

Meshram, G. A., Yadav. S. S, Shinde, D., Patil, B., and Singh, D. 2011. Antibacterial Study and Effect of Ethanolic Extracts of *Syzygium cumini* Seeds Powder on Glucoamylase *in vitro*. *Journal of Pharmaceutical Sciences and Research* 3: 1060-1063.

Mir, Q. Y., Ali, M., and Alam, P. 2009. Lignan derivatives from the stem bark of *Syzygium cumini* (L.) Skeels. *Natural product research* 23: 422-430.

Modi, D. C., Patel. J. K., Shah, B. N., and Nayak, B. S. 2010. Antiinflammatory activity of seeds of *Syzygium cumini* linn. *Journal of Pharmaceutical Education and Research* 1: 68-70.

Rao, S. R. and Ravishankar, G. A. 2002. Plant cell cultures: chemical factories of secondary metabolites. *Biotechnology Advances* 20: 101-153.

Rathore, V., Shekhawat, N. S., Singh, R. P., and Rathore, J. S., and Dagla, H. R. 2004. Cloning of adult trrees of Jamun (*Syzygium cuminii*). *Indian Journal of Biotechnology* 3: 241-245.

Rekha, N., Balaji, R., and Deecaraman, M. 2010. Antihyperglycemic and antihyperlipidemic effects of extracts of the pulp of *Syzygium cumini* and bark of *Cinnamon zeylanicum* in streptozotocin-induced diabetic rats. *Journal of Applied Bioscience* 28: 1718-1730.

Roy, A., Bhattacharya, S., Pandey, J. N., and Biswas, M. 2011. Anti-inflammatory activity of *Syzygium cumini* leaf against experimentally induced acute and chronic inflammations in rodents. *Alternative Medicine Studies* 1: 23-25.

Ruan, Z. P., Zhang, L. L., and Lin, Y. M. 2008. Evaluation of the antioxidant activity of *Syzygium cumini* leaves. *Molecules* 13: 2545-2556.

Shahnawaz, M., Sheikh, S. A., and Nizamani, S. M. 2009. Determination of Nutritive Values of Jamun Fruit (*Eugenia jambolana*)Products. *Pakistan Journal of Nutrition* 8: 1275-1280.

Tanaka, Y. and Ohmiya, A. 2008. Seeing is believing: engineering anthocyanin and carotenoid biosynthetic pathways. *Current Opinion in Biotechnology* 19: 190-97.

Veigas, J., Narayan, M., Laxman, P., and Neelwarne, B. 2007. Chemical nature, stability and bioefficacies of anthocyanins from fruit peel of *Syzygium cumini* Skeels. *Food Chemistry* 105: 619-627.

Wheelwright, N. T. and Janson, C. H. 1985. Colours of fruit displays of bird-dispersed plants in tropical forests. The *American Naturalist* 126: 777-799.

Willson, M. F. and Thompson, J. N. 1982. Phenology and ecology of colour in bird-dispersed fruits, or why some fruits are red when they are "green". *Canadian Journal of Botany* 60: 701-713.

Zhang, L. L. and Lin, Y. M. 2009. Antioxidant tannins from *Syzygium cumini* fruit. *African Journal of Biotechnology* 8: 2301–2309.

2014, Future Crops Vol. 2 Pages 293–295
Editor: Professor K.V. Peter
Published by: DAYA PUBLISHING HOUSE, NEW DELHI

Chapter 15
Lasoda (*Cordia myxa* Roxb.)

S.K. Sharma, Hare Krishna and A. Parashar
Central Institute for Arid Horticulture,
Beechwal, Bikaner – 334 006, Rajasthan
E-mail: ciah@nic.in

Cordia myxa Roxb. commonly known as "*lasoda*" or "cherry of the desert" is an underutilized, multipurpose, fruit tree spread over the arid and semi-arid regions. The other folk names include *lehsua, goonda/goondi, lasura,* assyrian plum, sebestem, Indian cherry, naruvilli (Hanelt, 2001; Samadia, 2007). This species is widely distributed in India and Sri Lanka, especially in warmer regions. This is favoured for cultivation in marginal eco-system as it has several xerophytic characters like deciduous nature, sunken stomata, waxy coatings, hairyness on leaves and coincidence of maximal growth phase with the period of water availability. Almost all parts of "*lasoda*" are used for various purposes (Yadav and Goel, 2006). The fruits are rich in sugars but are acidless. Mature fruits of *lasoda* are quite wholesome as they are rich in carbohydrates (12.2 g/100 g of edible portion), total ash (2.13 per cent), vitamins and minerals, with high medicinal values. Unripe fruits are pickled and mixed pickle with mango is excellent or used as vegetable. Ripe fruits are eaten raw. Besides, fruits have medicinal value and are considered as antihelmintic, diuretic, demulcent and expectorant. Since, the ripe fruits are full of vitamins and regular use is helpful in good growth of hair. *Lasoda* preparations are, thus, good for people whose constitution might have tendency to go baldy. In addition to fruit, *lasoda* bark and roots are also very effective as a local remedy against cough, cold and various other ailments connected with indigestion and throat problems. Leaves are used as fodder for goat, sheep and cattle and making *pattal* (trays/plates). Wood is used for making ornamental furniture, house posts, beams, scantlings, planks, dugout canoes, boats, tea boxes, cart shafts, axles, yokes, well-curbs, agricultural implements, combs, gun stocks, naves, spokes, etc. It is also used for making quality toys, bowls and wooden utensils for kitchens for handling sour recipes. (Ouedraogo-Kone *et al.,* 2006).

Despite its high values, it remained underexploited economically and hence needs attention. It is a native of India and occurs throughout the warmer regions of the world. It belongs to family Boraginaceae. It is an introduced tree in tropical Asia, Africa, Australia and the Americas (Warfa, 1988) and distributed in the western tropical Africa to Italy, South-East Asia, Australia, Europe and tropical Americas (Hanelt, 2001).

Botany

Lasoda (*Cordia myxa* Roxb.) belongs to family Boraginaceae. It is a fairly fast growing tree and behaves like a deciduous tree during unfavorable environmental conditions. Another related species of *lasoda* or *goonda*, having small fruits is *Cordia gharaf* Forsk. (*gondi*) distributed in arid and Semi arid zones of the country. Other economically important species of family Boraginaceae are *C. rothii, C. macleodii, C. vestita* and *C. wallichii.*

Lasoda is a medium size tree, which attains a height of 15 m with dense foliage and crooked trunk. Bark is grayish brown with longitudinal and vertical fissures. Young shoots are silvery-grey with hairs. Leaves are simple, broadly ovate or sub-orbicular rounded-cordate, alternate, 3–5 nerved, 7–15 cm long. In matter of external appearance these are glabrous above and pubescent below. The young leaves tend to be hairy. New flush comes in spring (March) when plant enters in floriferous phase. It has cyme inflorescence. Flowers are pedicillate, complete, perfect, actinomorphic, hypogymous and pentamerous. Individual florets are nearly 5 mm in diameter. At places these are somewhat hairy and white. The calyx part of an independent flower is about 8 mm long and glabrous, but not pubescent. It splits irregularly at the opening of its bud into flower. The filaments are hairy. Bunches of light yellow coloured hermaphrodite fragrant flowers are borne axillary on current season's growth. Fruits are formed soon after flowering and ripen during May–July, while the immature green fruits are available during April-June. Fruit is a drupe, 1–3 cm long, yellowish brown and pink at ripening. The pulp is sweet, viscid, and transparent surrounding the stone (Yadav and Goel 2006; Barraso *et al.*, 2009).

Germplasm Resources

A large variability exists in plantation of *lasoda* in relation to morphological characters particularly plant height, spread, leaf size, fruit size; fruiting behaviour; quality parameters like fruit colour, pulp content, sweetness, pickling quality, seed and pulp ratio etc. However, there is no improved variety so far. In general, two types of plants *viz.*, Large Fruited and Small Fruited are found. Large fruited cultivars have an average fruit weight of 8.55 g, whereas small-fruited cultivars have fruit weight of 3.0 g. In case of large fruited cultivars, fruits have comparatively more pulp thickness and therefore are suitable for consumption. Variability with respect to fruit size, skin and pulp colour (yellow, pink, dark pink), flesh content and pulp-stone ratio exist as a result of seed propagation. Polyploidy (x=7, 8) is reported in *lasoda* which also widens the genetic base. Some elite types have been identified which have large fruits, small stone and less mucilage content. Large fruited types are recognized as 'Paras Gonda' in Gujarat and 'Naruvili' in Tamil Nadu. Similarly in Rajasthan 'Puskar Local' is of big size with good fruit shape. Some small fruited types have prolific bearing. Small fruited types of *lasoda* locally called as *gundi* of which fruit is of very small size (about 1 cm), orange to light pink in colour at maturity but very much liked by the rural people of western Rajasthan and Gujarat.

A few plus trees have been identified by the CHES, Bhubaneswar, NBPGR, Jodhpur and CIAH, Bikaner. At Bhubaneswar, CHL-1 a promising *lasoda* genotype recorded the high content of TSS (17 °Brix), pulp content and phenols. At CCSHAU, Hisar, different genotypes were tried by Saini *et al.* (2002) and they have reported variability in plant height and spread. At CIAH, Bikaner under *ex- situ* conservation, more than 60 types of *lasoda* were collected and planted under field conditions to identify a few ideal cultivars. At CIAH, Bikaner, Selection 1 is performing well under irrigated hot arid ecosystem with respect to fruit size, pulp content and productivity. The average annual production of tender fruit is more than 100 kg tree (Nath *et al.*, 2008). Kaushik and Dwivedi (2004) reported wide range of biodiversity in morphological and quality characters from 45 accessions of *lasoda* from Haryana.

References

Hanelt, P. 2001. Institute of Plant Genetics and Crop Plant Research (IPK) (eds). Mansfeld's encyclopedia of agricultural and horticultural crops. Springer, Berlin, p. 1874.

Kaushik, R. A. and Dwivedi, N. K. 2004. Genetic diversity in *lasoda*. *Indian Horticulture* 1: 27.

Nath, V., Kumar, D. and Pandey, V. 2008. Fruits for the Future, Vol. 1: Well Versed Arid and Semi-arid Fruits. Satish Serial Publishing House, Delhi. p. 512.

Ouedraogo-Kone, S., Kabore-Zoungrana, C. Y., Ledin, I. 2006. Behaviour of goats, sheep and cattle on natural pasture in the sub-humid zone of West Africa. *Livest Science* 105: 244–252.

Saini, R. S., Kaushik, R. A. and Singh, S. 2002. Research note on the evaluation of *C. myxa* Roxb. Germplasm for vegetative growth character under semi arid conditions. *Haryana Journal of Horticulture Sciences* 31: 62-63.

Samadia, D. K. 2007. Variability and Scope of Improvement in Lasoda (*Cordia myxa*). *Indian Journal of Agro forestry* 9: 111–115.

Warfa, A. M. 1988. *Cordia* (Boraginaceae) in NE tropical Africa and tropical Arabia. Acta Universitatis Upsaliensis 174, Uppsala, Sweden, p. 78.

Yadav, P. K. and Goel, M. 2006. Lasoda (*Cordia dichotoma*). (In) Saroj P. L, Awasthi OP (Eds) Advances in arid horticulture, Vol-II production technology of arid, semiarid fruits. International Book Distributing Co., Lucknow, p. 305–318.

2014, Future Crops Vol. 2 Pages 297–308
Editor: Professor K.V. Peter
Published by: DAYA PUBLISHING HOUSE, NEW DELHI

Chapter 16

Mappia: An Endangered Life Saving Anticancerous Medicinal Plant

S. Shashi Kumar[1], *T. Vasantha Kumar*[2], *Renuka Jain*[3] *and P.E. Rajasekharan*[3]
[1]*Chromus Seeds, Bangalore*
[2]*Indian Institute of Horticulture Research, Bangalore – 560 089*
[3]*Avesthagen Pvt.Ltd., Bangalore*
E-mail: thuppil@iihr.ernet.in

Nothapodytes foetida (Wt.) Sleumer (Syn: *Mappia foetida* Miers), belonging to family Icacinaceae, is a rare and endemic plant found in the Western Ghats. It has gained considerable importance in recent times because of its medicinal value. It is used in treating various types of cancer, HIV and malaria. It is known commonly as Durvasane mara in Kannada; Kalgur, Kalagaura, Ghanera in Marathi; Arali, Perum pulagi, Kal kurinji in Tamil. It is grown in Maharashtra, Goa, Kerala, Karnataka, Assam, Tamil Nadu in India. Crushed vegetative parts have a disagreeable odour, from which the specific epithet Durvasane mara is derived. Camptothecin (CPT) is one of the most important alkaloids of the 21st century because of their clinical applications in treating cancer (Takeuchi *et al.*, 1991; Potsmesil, 1994) and HIV-I (Priel *et al.*, 1991). Recently, camptothecin was found active against parasitic trypanosomes and *Leishmania* (Bodley and Shapiro, 1995) and falciparum malaria (Bodley *et al.*, 1998). It also has antibacterial activity.

The estimated global turnover of CPT derivatives in 2004 was one thousand million US dollars which represents approximately one tonne of CPT in terms of raw material (Watase *et al.*, 2004). Reported trade of *N. nimmoniana* woodchips in the last decade from Maharashtra was around 400 metric tonnes. In the Indian market, the current demand for biomass (woodchips) of *N. nimmoniana* is around 500–700 metric tonnes; the supply being 50 per cent less than the current demand. Ever increasing worldwide demand for CPT from pharmaceutical industries and subsequent pressure on the wild populations of *N. nimmoniana* has endangered the plant (CAMP, 2001). There is need to find alternative sources of CPT to cater to the demands of pharma industry.

Origin and Distribution

Nothapodytes foetida is distributed in the Western Ghats of South India, some parts of Assam in Northeast India, and in the Himalayan foothills, besides being reported in Sri Lanka, Myanmar and Thailand (Hombe Gowda *et al.*, 2002) and also in the Indo-Malaysian region and China. In India, it is distributed in moist and dry deciduous forests and sometimes evergreen forests of the Western Ghats and in the eastern states of West Bengal and Assam. Natural populations of *Nothapodytes* are vulnerable to indiscriminate harvest for trade and an estimated 20 per cent of the population of this species is believed to have declined over the last decade (Ravikumar and Ved, 2000). There was unprecedented pressure on the natural populations in the Western Ghats. In fact, owing to this threat, *Nothapodytes nimmoniana* was classified as a "Vulnerable" species (Ravikumar and Ved, 2000). It is distributed throughout India (Cooke, 1967; CSIR, 1956).

Medicinal Uses

It is a rich source of anticancerous quinoline alkaloids camptothecin (CPT) and 9-methoxy camptothecin (9-mCPT) (Govindachari and Viswanathan, 1972). CPT is the most promising anticancer drug of the 21st century, which breaks single stranded DNA in mammalian systems (Hsiang *et al.*, 1985). CPT was effective in the treatment of colon, head, breast, lungs and uterine cervical cancers (Takeuchi *et al.*, 1991; Potmesil, 1994). In addition to its antitumor property, CPT also possesses effectiveness against HIV, parasitic trypanosomes and *Leishmania* (Priel *et al.*, 1991; Satheesh Kumar and Seeni, 2000). It was reported that *N. nimmoniana* also contains high amounts of alkaloids and essential oils like acetyl-camptothecin, hydroxy benzaldehyde, hydroxy propioguaiacone, scopoletin, sitosterol, sitosteryl-β-D-glucoside, 3-β-hydroxy stigmast-5-en-7-one, 3-keto-octadec-cis-15-enoic acid, palmitic acid, stearic acid, oleic acid, linoleic acid and linolenic acid (Wu *et al.*, 1995; Hosamani and Pattanashettar, 2005).

Camptothecin, approved as an anti-cancer drug, is an expensive chemical. Camptothecin, an active component with anti-cancer properties is extracted from the wood chips (Govindachari and Viswanathan, 1972; Suhas *et al.*, 2007) and the herbal extract carrying the trade name 'ghanera' is traded intensively. The mode of action of camptothecin is not well understood. An interesting issue is that camptothecin was originally known from the Chinese plant species *Camptotheca acuminata*, belonging to family Nyssaceae, which is botanically very distant from the Icacinaceae The content of camptothecin in *Nothapodytes* tree is much higher (0.14±0.24 per cent) as compared to that of *Camptotheca acuminata* (0.005 per cent) (Aiyama *et al.*, 1988). Camptothecin is active against leukemia and carcinosarcoma. Camptothecin has an exceptional anti retroviral activity and is effective in AIDS chemotherapy (Priel *et al.*, 1991). The active constituents are present in variable amounts in stem wood, root, leaves and bark of *N. foetida*. Stem woods of *Nothopodytes foetida* found in Western Ghats of India are an even better source of camptothecin. A close analogue, 9-methoxycamptothecin, is also present in *the Mappia foetida* Miers and is widely used in the treatment of colon and ovarian cancer.

Anti fungal and anti bacterial activity of the plant were studied using various organic solvent extracts. Hexane extracts of leaf and stem showed high inhibitory activity against *Trichothecium roseum*. Extracts prepared in acetone was proved to be most effective against *Trichothecium roseum*. Thus ethanolic, hexane and acetone extracts of stem had potential anti fungal activity. Successive petroleum ether, chloroform and methanol extracts of leaves and stem were tested for their antibacterial activity. The methanol fractions were the most effective against all the tested organisms. The cellular target of camptothecin is DNA topoisomerase-I. Camptothecin inhibits HIV replication *in vitro* and is also effective in the complete remission of lung, breast, uterine and cervical cancer. The detailed pharmacognostical evaluation of leaf including anatomy (light and confocal microscopy), physicochemical characters and phytochemical analysis with HPTLC fingerprinting of extracts revealed presence of camptothecin (Rf value 4.7) (Nandha Kumar *et al.*, 2002).

Species and Cultivars

It belongs to Kingdom: Plantae; Phylum: Magnoliophyta, Class: Magnoliatae, Order: Celastrales, other synonyms are *Mappia foetida* (Wight) Miers, *Nothapodytes foetida* (Wight) Sleumer, *Premna nimmoniana* J. Graham, *Stemonurus foetidus* Wight, *Premna nimonianan* Graham, and *Stemonurus foetidus.* Aruna Planta Medica established plantations of *Mappia foetida* sp (*Mappia tomentosa*, *Mappia weightii*, *Mappia ovata*) in Yercaud hills of Tamil Nadu in Coffee plantations. *Mappia foetida* is being cultivated in Tamil Nadu to make anti cancer products like topotecan and irrinotecan. In Pune district, the cultivation has already started in Mulshi, Maval, Velha and Purandar talukas as an inter-crop with custard apple and sapota. Mappia does not affect the yield of main crop.

Botany

Mappia is a small tree, 3-8 m tall, with smooth, grey, wrinkled bark, about 5 mm thick. Branchlets are slightly angled, corky, with prominent leaf scars. Leaves are simple, alternate, spiral, clustered at twig ends. Leaf base is often unequal, tip is pointed to long-pointed. Leaves are crowded at the ends of branchlets. Leaf stalks are 3-6 cm long. Petiole 1.2-6 cm long, flat above, puberulous; lamina is 9-30 x 5-14 cm, broadly ovate-oblong to elliptic-oblong and sometimes obovate, apex acuminate, base acute to rounded attenuate or asymmetric, margin entire, chartaceous to coriaceous with strong nerves (at high elevations), dark green above, pale beneath, glabrous; midrib flat; secondary nerves 6-11 pairs; tertiary nerves, distantly percurrent. Trunk barks brownish, lenticellate. Young branchlets subterete, lenticellate, puberulous. Inflorescence is borne in terminal cymes. Flowering is in August. Flowers are bisexual, creamy yellow, foul smelling, about 5 mm across, in flat-topped clusters at the end of branches. Petals are hairy inside. Fruit is a drupe, purplish red, and smooth, oblong to ellipsoid 2 x 1 cm, with a single seed. A textual diagnostic description of the species that is not necessarily structured.

Biochemical Constitution

Camptothecin (CPT), a pyrole quinolone alkaloid was isolated from extracts of *Camptotheca acuminata*, a deciduous tree native to China and Tibet, extensively used in traditional Chinese medicine (Wall *et al.*, 1966). CPT has a unique pentacyclic ring structure which gives itself and its derivatives the necessary antitumor activity and such spatial configuration cannot be synthesized artificially (Lorence and Nessler, 2004). As a consequence, drug manufacturers depend on the natural source of CPT. The main sources of this alkaloid are bark and roots of *Camptotheca acuminata* (Vincent *et al.*, 1997) and *Nothapodytes nimmoniana* (Aiyama *et al.*, 1988).

CPT was first discovered in the Chinese deciduous tree, *Camptotheca acuminata* (Nyssaceae) (Lorence and Craig, 2004). Later, the alkaloid was reported from several plant species, which by far the highest concentration (about 0.3 per cent on a dry weight basis) from *Nothapodytes nimmoniana.* (Lilenbaum *et al.*, 1995). The roots of *Ophiorrhiza mungos* and *Ophiorrhiza pumila* produce CPT and 10-methoxycamptothecin. *Ophiorrhiza rugosa* var decumbens is also found to produce CPT (Vineesh *et al.*, 2007). The structure of camptothecin derivative was explained by spectral data as (205)-18,19-de hydro camptothecin (Aiyama *et al.*, 1988). Two new alkaloids, Nothapodytines A and B were discovered from ethanol extracts of stem of *Nothopodytes nimmoniana* (Wu *et al.*, 1996). Camptothecins have unique mechanism of action namely poisoning of eukaryotic DNA topoisomerase-I (Tanizawa *et al.*, 1994). The antitumor drug camptothecin (CPT) and its analogs inhibit the rejoining step of the breakage/

rejoining reaction, which traps the enzyme in covalent linkage with DNA (the cleavable complex) (Desai *et al.*, 1997). Camptothecin-(CPT), {4-(S)-4-ethyl-4-hydroxy-1-H-pyrano},[3,4,6,7]-indolizino-(1,2b) quinoline-3,14 (4H,12H) dione} CPT and minor camptothecinoids are obtained in high percentage from Indian tree *Nothapodytes foetiad* (Govindachari,1972) and Japanese species of *Nothapodytes obscura*, *N. obtusifolia*, *N. piltosporsides*, *N. tomentosa* and *N. collina*.

Table 16.1: Comparative account of CPT content in different species.

Species	*Plant Parts*	*Origin*	*CPT(g/g dry weight)*	*Reference*
N. nimmoniana	Stem bark	Uttara Kannada	0.23 per cent	Padmanabha *et al.* (2006)
(Family Icacinaceae)	Root bark	Okinawa, Japan	0.33 per cent	Aiyama *et al.* (1988)
	Stem bark	Mahabaleshwar, India	1400–2400	(Roja and Heble, 1994)
	Shoot		750	
C. acuminata	Bark	Texas, USA	1800–2000	López-Meyer *et al.* (1994)
(Nyssaceae)	Roots		400	
	Hairy roots		1000	
	Young leaves		4000–5000	
	Seeds		3000	
O. rugosa var. *prostrata*	Stem	North Western Ghats	0.08 per cent	Gharpure *et al.* (2010)
(Rubiaceae)	Root		0.16 per cent	
	Young leaves		0.0062 per cent	
	Old leaves		0.0022 per cent	
	Fruits		0.0165 per cent	

Source: Gharpure *et al.*, 2010.

Fulzele and Satdive (2005) evaluated the different extraction methods and found that maximum per cent of CPT (2.67 per cent w/w) was obtained by microwave assisted extraction (MAE).

Microscopical examination of powder of aerial parts (leaves, stem and root parts) showed fragments of epidermis, glandular trichomes, stone cells, lignified xylem elements, anomocytic stomata and abundant calcium oxalate crystals. Successive extractive value was the highest in aqueous extract (35 per cent on dry weight basis). Mean ash values (per cent) were 2.50 (total), 0.05 (acid insoluble ash) and 1.29 (water soluble ash), the lowest in stem and greater in leaf. Loss on drying was 8.34 per cent and pH values of aqueous extract was 6.50. Foaming index was less than 100. Screening of all extracts indicated presence of all phytoconstituents except resin. TLC fingerprints of extracts of all parts were developed. The proposed parameters presented help to establish the authenticity of drug, differentiate the drug from other camptothecin containing species and drawing the pharmacopoeia standards for this species (Khan Nazeerullah, *et al.*, 2012).

The oil extracted from dry ripe fruits with benzene was reddish yellow with slight fluorescein. Physical and chemical constituents of fatty oil from seeds of ripe fruits of *Mappia foetida* are given in Table 16.2.

Lopez-Meyer *et al.*(1994) reported that all parts of *Camptotheca acuminata* contain CPT and the highest level was in young leaves (4-5 mg/g dry wt) approximately 50 per cent higher than in seeds (3 mg) and in bark (1.8-2 mg/g dry wt). Significant differences in CPT content were noticed across individuals and across sites of collection. Individuals from Ulvi had the highest av. CPT content (0.508±0.16 per cent). Differences in the level of CPT is function of genetic and environmental background (Padmanabha *et al.*, 2006)

Table 16.2: Physical and chemical constituents of fatty oil from seeds of ripe fruits of *Mappia foetida*.

Yield	48 per cent
Specific gravity at 27ºC	0.9319
Refractive index at 27ºC	1.4781
Acid number	3.7
Saponification value	185.4
Iodine value	45.6
Reichert-Meissel value	0.68
Polenske number	0.41
Acetyl value	5.77
Unsaponifiable matter	0.81 per cent

Source: Nadkarni *et al.*, 1944.

Floral Biology and Breeding Behavior

The species is polygamous in nature with a wide array of breeding types including male, female, hermaphrodite, monoecious, andromonoecious, gynomonoecious and trimonoecious trees. The species displays intriguing sexual variation. Apart from the flowers which are comparable to typical thrum and pin morphs of distylous species, some flowers show intermediate features. The functional gender of the different flower types were assessed for the sexuality of individual plants in two populations in the Biligiri Rangana Hills Wildlife Sanctuary, Western Ghats, India. The species showed functional/cryptic dioecy. Thrum flowers are functionally males, pin flowers are females and intermediate flowers are bisexual with functional pollen and stigma. Apart from male and female plants, a small proportion of plants which are essentially males produce a small number of female and bisexual flowers. The sexual system in *Nothapodytes* is described as subdioecy and it appears to be a transitory stage in the evolution of dioecy from heterostyly (Hombegowda *et al.*, 2002). The population of *N. nimmoniana* growing in Dandeli Wildlife Sanctuary, Karnataka, showed a wide array of sexuality with male, female, hermaphrodite, monoecious, andromonoecious and gynomonoecious individuals, indicating that the species is polygamodioecious (Hombegowda *et al.*, 2002). Sexual differentiation in dioecious flowering plants is often incomplete (Ross, 1982).

Floral resource availability, pollinator diversity and activity were carried out in the canopy located in the Protected Area Network in the Western Ghats of Karnataka State. Flowers belong to the fly pollination syndrome and are open to visits by generalists like *Apis dorsata* and *Trigona iridipennis*. Fruiting success did not vary between exploited and unexploited sites, but there was a significant difference in pollinator assemblage. In a lean male flowering year, fruit set was low, suggesting pollen limitation in the population. In common with species with unspecialized flowers, *N. nimmoniana* flowers attract a generalist range of pollinating agents, although flowers display fly-pollination syndrome. Being largely a dioecious species, pollination success and seed set are dependent on the distribution of male trees in the neighborhood of female species. When flowering male trees are lean in the neighborhood of female trees, pollination becomes a major limitation. A lower diversity and abundance of pollinators in the exploited site in comparison to the unexploited site, but there was little pressing evidence. Long-range studies may be needed to conclusively bring out the relationship between pollinators and habitat disturbance in a species with generalist pollination system like *N. nimmoniana*. Hand-pollination of female and bisexual flowers with pollen of male and bisexual flowers of the same tree as well as male flowers of other trees showed good pollen germination and fruits set, suggesting that the female and bisexual flowers produced on polygamous trees are self-compatible (Sharma *et al.*, 2010).

Propagation

Indiscriminate cutting and rapid deforestation caused a severe shortage of source material for many important medicinal plants. Development of propagation technology is highly desirable as no systematic cultivation of this tree species was reported. Sharma *et al.* (2000) found GA (100 ppm) was the best for promoting seed germination, seedling growth, plant establishment and field performance Seeds incubated at 30°C and 20/30°C (16/8 hours) showed maximum germination (89.8 per cent) and 10°C was optimum for seed storage which indicated that seeds are desiccation sensitive and cannot tolerate storage under ambient conditions (Anonymous, 2009-10).

Soft wood cuttings, semi-hardwood cuttings and hardwood cuttings were dipped in powder and liquid forms of 1000,2000,3000 and 4000 ppm concentrations of IAA and IBA by quick dip method and then planted in polythene bags filled with sand, vermiculite and pot mixture. Rooting of cutting was the best achieved on sand medium followed by vermiculite and potting mixture. Treatment with IBA (200ppm) in talc form in softwood cuttings, IBA (3000ppm) talc form in semi-hardwood cuttings and IBA (4000ppm) talc form in hardwood cuttings registered maximum sprouting, rooting, root length, shoot length, roots per cutting and survival percentage. In all, the best results were obtained in semi-hardwood cuttings treated with IBA (3000ppm) in talc form planted in sand medium(Panneerselvam *et al.,* 2004). It is reported that this species is a good coppice and withstand pruning. Due to its shade tolerant nature and without any allelopathic effects, it can fit in any agro forestry systems (Vasudeva *et al.,* 2010).

High yielding trees of *N. immoniana* were identified that could be used in the clonal multiplication of trees or in developing appropriate *in vitro* production systems. One hundred forty seven individuals from 11 populations in Western Ghats, south India, were assessed for their CPT content and found that approximately 23 of these individuals yielded CPT in excess of 1 per cent by dry weight (Suhas *et al.,* 2007). The CPT content in stem and root bark tissues of 17 of these 23 individuals were determined using liquid chromatography mass spectrometry (LC–MS). Occurrences of certain camptothecines, namely 10-hydroxy camptothecin, diacetoxycamptothecin, and acetoxy-camptothecin glycoside were reported. The trees which produce at least 5- to 8-fold more CPT were identified. Production of a few minor camptothecines, including 10-hydroxy camptothecin, in the stem and root bark extracts of the tree were reported, which not only help in harnessing the high-yielding individuals for clonal multiplication but also for exploiting some of the minor camptothecines, which also have been important anti-cancer and anti-viral activity (Ramesha *et al.,* 2008).

Regeneration of adventitious shoots was achieved using different seedling explants. Direct regeneration of shoot buds was observed in Murashige and Skoog's (MS) basal medium supplemented with various concentrations of thidiazuron. The optimum levels of thidiazuron concentrations were 0.91–4.45 M. Leaf explants formed more shoots followed by hypocotyls or cotyledons. The shoot buds elongated and rooted on MS basal medium with N 6 -benzyladenine (0.88–2.22 µM) and indole-3-butyric acid (0.49 µM) (Krishnamurthy *et al.,* 2001).

In vitro Propagation

Development of *in vitro* multiplication and conservation techniques are highly desirable, as there is no reported cultivation or conservation protocols of this endangered species. MS medium containing BA and 2,4-D promoted *in vitro* callus cultures from embryos. Medium containing BA and NAA promoted multiple shoot development and camptothecin and 9-methoxy camptothecin were detected in *in vitro* cultures (Roja and Heble, 1994). A protocol for somatic embryogenesis was developed from callus cultures. The embryos on maturation germinated into plantlets (80 per cent) (Anuradha *et al.*, 2006). A medium comprising of MS medium with picloram 2 mg/l), N6 benzyl adenine (1 mg/ml) and Gibberellic acid (1mg/ml) was developed for establishment of callus cultures from leaves (Sundarvelan *et al.*, 2004). Thengane *et al.* (2003) reported that camptothecin production was high in callus grown in medium with combination of 2,4-D with BA and 2,4,5-T and NAA. The isolated seed embryos showed a positive response on MS with 0.91 µM thidiazuron (TDZ), 3 per cent sucrose, 0.8 per cent agar and pH 5.8 in Standard Culture Conditions (SCC). From regenerating callus, multiple shoots were regenerated and shoot elongation of 2–19 mm was observed on MS basal medium after one month of incubation. One mg/l IBA was ideal for root initiation of the *in vitro* plants. Some the cultures in (SCC) were shifted to low temperature (+10°C) for *in vitro* conservation. Under reduced temperature and light, *in vitro* plants grew well (without any sub culturing) even after six months of incubation. Cultures in MS basal medium showed shoot elongation of 5–10 mm compared to MS with TDZ (3–50 mm). It is possible to conserve the species under low temperature for a longer period without regular sub culturing (Rajasekharan *et al.*, 2010).

A comparative study of accessions from different geographical regions to demonstrate the extent of influence of the genetic and environmental factors on the presence of secondary metabolites was carried out. Callus cultures of different genotypes of *Nothapodytes foetida* for camptothecin content were analyzed. The four different accessions selected for the study were from Bangalore, Sirsi (Karnataka), Kolhapur (Maharashtra), and Munnar (Kerala). Callus initiation was observed only in the embryos. MS medium supplemented with Naphthalene acetic acid (NAA 10.74 mM) and Benzyladenine (BA 2.22 mM) resulted in a better response. Alkaloids were detected in different stages of callus formation and maximum amount of camptothecin was seen in 50-day-old callus (Karadi *et al.*, 2008).

Chemical Characterization

The leaves, stem and root parts of *N. nimmoniana* collected from different geographical sources were subjected to pharmacognostic standardization including phytochemical screening which would be helpful to ascertain the correct identity of *N. nimmoniana* for which there is adulterant available in the drug market for taking more profit (Khan Nazeerullah *et al.*, 2012). A total of 144 trees sampled over 14 populations were analyzed for CPT content. The CPT and 9-MCPT were the highest in young leaves and significant variation was observed across localities for these two components. Two trees from Amboli and Sirsi recorded high CPT content (> 0.71 per cent). These elite trees could be the potential sources for clonal and *in vitro* multiplication (Anonymous,2009-10). Kavitha *et al.* (2010) reported accumulation of CPT and 9-MCPT in various plant parts showing a linear relationship in stem, stem bark and root bark but not in leaves.

Molecular Characterization

Extraction of good quality DNA from *Nothapodytes nimmoniana,* was a challenging task as many of the scondary compounds co-precipitate with DNA and render the DNA unsuitable for downstream applications. DNA extraction protocol was optimized with cetyltrimethylammonium bromide (CTAB) for isolation of genomic DNA. The yield of DNA ranged from 10-250 µg g^{-1} of leaf tissue. The PCR protocol comprising 50 ng template DNA, 0.5 mM dNTPs, 0.5 U Taq DNA polymerase and 0.5 µM primer concentrations were optimum for carrying out ISSR analysis in *N. nimmoniana* (Anuradha Sane *et al.*, 2012).

The genetic fidelity of *in vitro* propagated cultures was assessed using ISSR markers. Twelve samples selected were used as repeated sub-cultured samples at standard culture condition, hardened samples and cultures at reduced culture condition. Repeated sub-culturing in the presence of growth hormones (TDZ) caused somaclonal variation in micropropagated plants (Anonymous, 2008-09).

For molecular characterization, eight polymorphic microsatellite loci were isolated and characterized using enrichment hybridization protocol. Analysis of 36 individuals representing two populations revealed 3 to 12 alleles per locus. Observed heterozygosity ranged from 0.21 to 0.94 for these two populations. None of the loci tested showed linkage disequilibrium. These markers are invaluable for evaluating the genetic structure and assessing the genetic impacts of harvesting of *N. nimmoniana* in the Western Ghats to formulate strategies for conservation of the species (Ramesha *et al.*, 2008).

ISSR markers were employed to characterize the genetic diversity of *Mappia foetida* populations from Western Ghats of India. 57 samples were analyzed from 12 populations of 4 states (Kerala, Tamil Nadu, Karnataka and Maharashtra) using 16 ISSR primers. High percentage of genetic polymorphism (73.7 per cent) was observed among the samples collected from different regions of Western Ghats. Collections from Coorg showed the highest polymorphism (53.3 per cent) followed by those from Chickamagalur (52.4 per cent), whereas samples collected from Kodaikanal recorded the lowest percentage of polymorphism (31.6 per cent). UPGMA dendrogram of 57 samples collected from different regions of Western Ghats, analysed based on Ward's minimum variance algorithm generated two major clusters. First cluster comprised of all 34 samples of Karnataka whereas the second cluster comprised of samples from Kerala, Tamil Nadu and Maharashtra. The first cluster of Karnataka is further subdivided into 3 sub-clusters. The first sub-cluster contained all samples from Sirsi and Jambotti and two samples from Agumbe, while the second sub-cluster composed of 8 samples from Chickamagalur, 7 from Coorg, 2 from Charmadi Ghat and one from Agumbe. The third sub-cluster consisted of the remaining Chickamagalur and Coorg samples. The second major cluster was divided into 2 sub-clusters. The first sub cluster had all Maharashtra samples (Amboli and Kolhapur) where as the second sub-cluster comprised of all samples from Idukki (Kerala) Kalpetta ((Kerala), Ooty and Kodaikanal. The clustering of samples corresponds to the geographical region of collection. The molecular data revealed high genetic diversity in *M. foetida* populations of Western Ghats (Anonymous, 2009-10).

Conclusion

Developing a cultivation strategy combined with optimizing the growing conditions to increase the biological yield as well as CPT would result in consistent supply of high quality raw material to the industries. Domestication and cultivation also help to solve problems which are inherent to herbal medicines such as misidentification and sustainable harvesting. There is an urgent need to take up large scale cultivation for commercial and sustainable harvesting without destroying whole tree and incorporate the Nothapodytes species into silvi-horticultural or other cropping systems like Coffee plantations.

References

Aiyama, R. Nagai, H.; Kokata, K. Y. Shinonava, C. and Sawada, S. A. 1988. Camptothecin derivatives from *Nothapodytes foetida*. *Phytochemistry*, **27**: 3664-3666.

Annual Report, 2007-08. Collection, chemical and molecular profiling, domestication and conservation of *Mappia foetida*, an endangered medicinal plant. NMPB, New Delhi.

Annual Report, 2009-10. Collection, chemical and molecular profiling, domestication and conservation of *Mappia foetida*, an endangered medicinal plant. NMPB, New Delhi.

Anuradha Sane, V. K. Abdul Kareem, P. E. Rajasekharan and T. Vasantha Kumar. 2010. Optimization of DNA Isolation and PCR Protocol for ISSR Analysis of *Nothapodytes nimmoniana*, a Threatened Anti-cancerous Medicinal Plant. *Asian Journal of biotechnology*., **4 (2):** 100-107.

Ashworth L, Aguilar R, Galetto L and Aizen MA. 2004. Why do pollination generalist and specialist plant species show similar reproductive susceptibility to habitat fragmentation. *J. Ecol.*, **92**: 717–719.

Bodley, A. L. and Shapiro, T. A., 1995. Antitrypanosomal activity of camptothecin analogs. Structure-activity correlations. *Proc. Natl. Acad. Sci. USA*, **92:** 3726–3730.

Bodley A. L., Cumming J. N. and Shapiro T. A., 1998. Effects of camptothecin, a topoisomerase I inhibitor, on *Plasmodium falciparum*. *Biochem Pharmacol*. **55**: 709–711.

CAMP (Conservation Assessment and Management Plan) for Medicinal plants in Maharashtra State, FRLHT, Bangalore, 2001.

Desai SD, Liu LF, Vazquez-Abad D and D'Arpa, P. 1997. Ubiquitin-dependent destruction of topoisomerase I is stimulated by the antitumor drug camptothecin. *J. Biol. Chem.* **272(39):** 24159-64.

Fulzele D. P. and Satdive R. K., 2005. Distribution of anticancer drug camptothecin in *Nothapodytes foetida*. *Fitoterapia*. **76(7-8):** 643-48.

Govindachari, T. R. and Viswanathan, N. 1972. Alkaloids of *Mappia foetida*. *Phytochemistry*, **11 (12):** 3529-3531.

Gharpure G., B. Chavan, U. Lele, A. Hastak, A. Bhave, N. Malpure, R. Vasudeva and Patwardhan, A., 2010. Camptothecin accumulation in *Ophiorrhiza rugosa* var. *prostrata* from northern Western Ghats. *Current science*, **98 (3):** 302.

Hombegowda, H. C., Vasudeva, R., Mathachen, G. P., Uma Shaanker, R. and Ganeshaiah, K. N., 2002. Breeding types in *Nothapodytes nimmoniana* Graham.: An important medicinal tree. *Curr. Sci.*, **83 (9)**: 1077–1078.

Hosamani, K. M. and Pattanashettar, R. S. 2005. *Mappia foetida* seed oil: A rich source of oil and moderate source of novel 3-keto-octadec-cis-15-enoic acid and its possible industrial utilization. *Industrial crops and products*. **22(2)**: 135-139.

Hsiang, Y. H., R. Hertzberg, S. Hecht, and L. F. Liu. 1985. Camptothecin induces protein-linked DNA breaks via mammalian DNA topoisomerase I. *J. Biol. Chem.* **260:** 14873-14878.

Kallappa M. Hosamani and Raviraj S. Pattanashettar. 2005. *Mappia foetida* seed oil: A rich source of oil and moderate source of novel 3-keto-octadec-*cis*-15-enoic acid and its possible industrial utilization. *Industrial Crops and Products*, **22(2):** 135–139.

Khan Nazeerullah, Kumar Sunil, Singh Rishi Pa and Dhankhar Neelam. Preliminary Phytochemical Screening, Fingerprinting and Pharmacognostic Evaluation of Nothapodytes nimmoniana Leaves, Stem and Root Collected from Different Geographical Region. *Research Journal of Pharmaceutical, Biological and Chemical Sciences*. **3(1)**: 362.

Karadi, R V, E N Gaviraj, P E Rajasekharan, Resmi Joseph, Anuradha Sane and V K Rao, 2008. Assessment of Callus in Different Genotypes of *Nothapodytes nimmoniana* for Camptothecin Content. *The IUP Journal of Genetics and Evolution*: **1(1):** 57-65

Kavitha, P., Vasantha Kumar, T., Rajasekharan, P. E.; Abdul Kareem, V. K. and Rao, V K. 2010. Camptothecin and 9-methoxy camptothecin, anti cancer alkaloids in *Nothopodytes nimmoniana* from western ghats, *Indian Journal of Medicinal and Aromatic Plant Sciences,* **32 (2):** 129-132.

Krishnamurthy K. V; Thengane S. R.; Kulkarni D. K. and Shrikhande V. A. 2001. Effect of Thidiazuron on Adventitious Shoot Regeneration from Seedling Explants of *Nothapodytes foetida*. *In Vitro Cellular and Developmental Biology: Plant*: **37(2)**: 206-210.

Kumar, R. and Ved D K. 2000. 100 red listed medicinal plants of conservation concern in southern India, foundation for revitalization of local health traditions. *Bangalore*: 261–263.

López-Meyer, M., Nessler, C. L. and McKnight, T. D., 1994. Sites of accumulation of the anti-tumor alkaloid camptothecin in *Camptotheca acuminata*. *Planta Med*., **60**: 558–560.

Lorence A and Nessler C. L., 2004. Camptothecin, over four decades of surprising findings. *Phytochemistry*, **65**: 2735–2749.

Nadkarni, M. D., Airan, J. W. and Shah, S V,1944. Fatty oil from the seeds of *Mappia foetida*. *Current Science*, 13: 233.

Nagaraja, T. G., Pawar, P. P. and Yd. Laxmikant B. Patil. 2009. Anti-microbial activity of *Mappia foetida* miers. *Bioinfolet*, **6(4)**: 340-341.

Nandha Kumar, R.; Vishwanathan H; Suresh T and Mohan P. S. 2002. Anti-bacterial Activity of *Mappia foetida* Leaves and Stem. *Fitoterapia,* **73 (7-8):** 734-736.

Padmanabha BV, Chandrashekar M, Ramesha BT, Hombe GHC, Gunaga RP, Suhas S, Vasudeva R, Ganeshaiah K. N and Uma S. R 2006. Patterns of accumulation of camptothecin, an anti-cancer alkaloid in *Nothapodytes nimmoniana* Graham. in the western ghats, India: Implications for identifying high yielding sources of alkaloid. *Curr. Sci.*, **90(1):** 95-99.

Panneerselvam K.; Bhavanisankar,; M. Jayapragasam, M; Ashok Kumar; Rathakrishnan, P; Vijayaraghavan, A and Adalarasan, R, 2004. Effect of Growth Regulators and Planting Media on Rooting of Cuttings in *Nothopodytes nimmoniana* Mabberly. *Indian Journal of Plant Physiology,* **9(3):** 308-312.

Potsmesil, M. 1994. Camptothecins: From Bench Research to Hospital Wards. *Cancer Research,* **54:** 1431-1439.

Priel E, Showalter S. D and Blair DG 1991. Inhibition of human immunodeficiency virus (HIV-1) replication *in vitro* by noncytotoxic doses of camptothecin, a topoisomerase I inhibitor. *AIDS Res Hum Retroviruses*, **7**: 65–72.

Rajasekharan, P. E.; Abdul Kareem, V. K. and Vasantha Kumar, T 2010. Optimization of Protocols for *in vitro* Multiplication and Conservation of *Nothapodytes nimmoniana,* an Endangered Medicinal Plant. *Acta Horticulturae*. **865**: 37-43.

Ramesha B. T., Amna, T, Ravikanth, G, Rajesh P. Gunaga, R. Vasudeva, K. N. Ganeshaiah, Uma Shaanker R, Khajuria, R. K., Puri, S. C. and Qazi G. N., 2008. Prospecting for Camptothecines from *Nothapodytes nimmoniana* in theWestern Ghats, South India: Identification of High-Yielding Sources of Camptothecin and New Families of Camptothecines. *Journal of Chromatographic Science*, **46**: 362-368.

Ramesha, B. T., Srirama, R., Ravikanth G, Ravishankar, K. V., Dayanandan S., Sreejayan N, Vasudeva, R, Ganeshaiah, K N and Uma Shaanker R. 2008. Development of polymorphic microsatellite loci in *Nothapodytes nimmoniana*, a medicinally important tree from the Western Ghats, India, *Molecular Ecology Resources* **9(1):** 365-367.

Ravikumar, K. and Ved, D. K., 2000. (In) *100 Red Listed Medicinal Plants of Conservation Concern in Southern India: Illustrated Field Guide*, Foundation for Revitalisation of Local Health Traditions, Bangalore, 261–263.

Roja, G. and Heble, M. R., 1994. The quinoline alkaloids camptothecin and 9-methoxycamptothecin from tissue cultures and mature trees of *Nothapodytes foetida Phytochemistry*,. 36: 65–66.

Ross, M. D., 1982. Five evolutionary pathways to subdioecy. *Am. Nat.*, **119**: 297–318.

Satheesh Kumar, K. and Seeni S,. 2000. *In vitro* multiplication of *Nothapodytes foetida* (Wight) Slwumer through seedling explant cultures. *Indian J. Exp. Biol.*, **38**: 273-277.

Sharma, M. V., Uma Shaanker R, Vasudeva R and Shivanna K R, 2010. Functional dioecy in *Nothapodytes nimmoniana*, a distylous species in the Western Ghats. *Current Science*, **99(10):** 1444-1449.

Sharma, M. V., Uma Shaanker R, Simon R. Leather, Vasudeva R and Shivanna, K R. 2010. Floral resources, pollinators and fruiting in a threatened tropical deciduous tree. *Journal of Plant Ecology*: 1–9.

Suhas S., Ramesha B. T., Ravikanth G, Gunaga R P, Vasudeva R, Ganeshaiah K. N. and Uma Shaanker R.2006. Chemical profiling of *Nothapodytes nimmoniana* populations in the Western Ghats, India for anti-cancer compound, camptothecin. *Current Science*, **92:** 1142–1147.

Takeuchi S, Dobashi K, Fujimoto S, Tanaka K, Suzuki M, Terashima Y, Hasumi K, Akiya K, Negishi Y and Tayama T. 1991. A late phase II study of CPT-11 on uterine cervical cancer and ovarian cancer. Research groups of CPT-11 in Gynecologic cancers. *Gan To Kagaku Ryoho*, **18:** 1681–1689.

Tanizawa A, Fujimori A, Fujimori Y and Pommier Y. 1994. Comparison of topoisomerase I inhibition, DNA damage, and cytotoxicity of camptothecin derivatives presently in clinical trials. *J. Natl. Cancer Inst.* **86(11):** 836-42.

Thengane S. R., Kulkarni D K, Shrikhande V A, and Krishnamurthy K V. 2001. Effect of thidiazuron on adventitious shoot regeneration from seedling explants of *Nothapodytes foetida*. *In Vitro Cell. Dev. Biol. Plant*, **37:** 206-210.

Vasudeva, R., Triveni H N and Uma Shaanker R, 2010. Nothopodytes nimmoniana: A potential medicinal tree component for agroforestry systems of north east India. In National Seminar on commercial cultivation of medicinal plants in north east India held at regional institute of science and technology campus, Baridua, Ri-Bhoi, Meghalaya from 9th–10th Feb. 2010.

Ved D. K., 1997. Trade in medicinal plants—the state of our ignorance. *Amruth*, **1(10):** 2–8

Vincent, R. M., Lopez-Meyer, Mc-Knight, T. D. and Nessler, C. L., 1997. Sustained harvest of camptothecin from the leaves of *Camptotheca acuminata*. *J. Nat. Prod.*, **60:** 618–619.

Wall M. E., Wani M. C., Cooke C. E., Palmer K. H., MePhail A. T., and Sim G. A. 1966. Plant antitumor agents I. The isolation and structure of camptothecin, a novel alkaloidal leukaemia and tumor inhibitor from *Camptotheca acuminata*. *Journal of the American Chemical Society* **88:** 3888-3890.

Watase, I., Sudo, H., Yamazaki, M. and Saito, K., 2004. Regeneration of transformed *Ophiorrhiza pumila* plants producing camptothecin. *Plant Biotech.*, **21**: 337– 342.

Wu, T. S., Leu Y L, Hsu H C, Ou L F and Chen C C 1995. Constituents and cytotoxic principles of *Nothapodytes foetida*. *Phytochemistry*, **39**: 383-385.

2014, Future Crops Vol. 2 *Pages 309–319*
Editor: Professor K.V. Peter
Published by: DAYA PUBLISHING HOUSE, NEW DELHI

Chapter 17
Palmyrah

Jiji George[1] and Anitha Karun[2*]
[1]Research Scientist-II, Manipal Centre for Virus Research,
(Regional Reference Laboratory for Influenza Viruses, and ICMR Virology Network Laboratory-Grade-I)
Manipal University, Manipal – 576 104, Karnataka
[2]Principal Scientist, Biotechnology Section Central Plantation Crops Research Institute,
Kasaragod – 671 124, Kerala
**E-mail: anithakarun2008@gmail.com*

The palmyrah palm exists both in wild and in cultivation, ranging from sea level to about 800 m amsl. It grows in seasonal tropical or subtropical climates on sandy soils. It is a very adaptable palm, growing well in dry areas and is quite drought resistant. It also grows in per-humid areas and survives water logging quite well. The worldwide distribution of natural and cultivated population of the palmyrah show that it occupies a band on either side of the equator and extends laterally over a considerable girth. The country wide distribution of the palms are Mauritania, Senegal, Mali, Gambia, Guinea-Bissau, Guinea, Ivory Coast, Upper Volta, Nigeria, Gaban, Congo, Sudan, Tanzania, Madagascar, Saudi Arabia, Iraq, Iran, Pakistan, Bangladesh, Sri Lanka, Myanmar (*Lubeigt,* 1977), Thailand, Cambodia, Malaysia and Indonesia (Kovoor, 1983). In India, the palms are wide spread in Kerala, Tamil Nadu, Karnataka, Andhra Pradesh, Maharashtra, Odissa, Madhya Pradesh, Bihar and West Bengal. Isolated patches are also seen in Assam, Gujarat and Uttar Pradesh (Anonymous, 1948). The palmyrah is a slow grower and the growth rate decreases in regions of higher altitude (Tjitrosoepomo and Pudjoarinto, 1982).

Palmyrah is an allotetraploid (x= 8 or 9, n= 18) dioecious palm with sex ratio of 1: 1 (Sankaralingam *et al.,* 1999). It is a solitary, robust and pleonanthic dioecious palm. The stem attains 10 to 30 m height and about 60 cm diameter, deep grey or black in colour with distinct annulate leaf scars. Under normal conditions, the stem can reach a height of four to five meters in 15 years (Theivendirarajah, 2008). In young trees, the stem is covered with persistent leaf bases. Leaves costapalmate, rigid; crown more or less roundish with evenly projecting leaves; leaf base split at base. The inflorescence is interfoliar and a branched spadix, sheathed by many imbricated, fibrous, coriaceous spathes (Sankaralingam *et al.,* 1999).

The palm commences flowering only after 12 to 15 years of planting. Flowering is seasonal and it appears in inflorescence, which emerges, from the axil of the leaves gradually enlarging in size and branching out (Theivendirarajah, 1983). During the flowering season 3 to 6 inflorescences can be formed in both male and female palms. The male palms begin to flower earlier than the females and even in the palms which are already reproductive the males put out the inflorescence a month or two earlier than the females during the season. The male flowering pattern is such that pollen grains are made available in sufficient quantity during the receptive period of female flowers in inflorescence, which appear one after at regular intervals at axil of leaves.

The flowering season of the palm varies from region to region. In Tamil Nadu the palm flowers in March and April and are matured in July and August (Jeyaraman, 1991). In Jaffna, Sri Lanka inflorescence appears by end of monsoon period (October- December) and continues to produce until the end of July. Most of the female palms begin flowering in March/April and fruits mature in September/October. Male palms begin flowering in January/February and complete anthesis by June/July. In Indonesia, the peak flowering period is between June and August. Though flowers bloom throughout the day, most of them open between 8 and 11 am. Pollination is through insects (bees, wasps, beetles) and wind. It takes 4-5 months for the fertilized female flowers to mature into ripe fruits (Theivendirarajah, 2008).

Economic Uses of the Palm

Among the four sugar yielding palms *viz. Borassus flabellifer* (palmyrah), *Phoenix dactylifera* (Date), *Caryota urens* (Sago) and *Cocos nucifera* (Coconut); palmyrah palm ranks first in yielding sugar (Sankaralingam *et al.*, 1999). It is a multipurpose tree of great utility. Palmyrah is a dioecious palm with the great majority of its economic products such as immature endosperm, mesocarp pulp, fibre from the fruits, tuberous seedlings are obtained only from female palms. But sweet sap from the inflorescence, toddy, palm sugar, brush fibre, cordage weaving and plaiting and wood are obtained irrespective of whether the palms are male or female. However, differences in their yield or quality are reported. Thus female palms are supposed to yield more toddy on tapping from the inflorescence and the female tree gives better and hard timber than the male tree, and is also more expensive.

The palms are slow-growing perennials and have no distinguishing features to identify the sex until flowering. The palm commences flowering only after 12 to 15 years of maturity. On account of the dioecious nature and long juvenile period, farmers have hesitated in planting this multipurpose tree. Breeding and crop improvement would be highly facilitated if gender could be determined at the seedling stage itself. This would help farmers while selecting the seedlings and maintain an optimum sex ratio at plantations. A male-specific random amplified polymorphic DNA (RAPD) marker was identified in palmyrah to determine the sex at the seedling stage itself. This marker is a useful tool for early sex determination in palmyrah.

Nungu (Endosperm)

When the fruit is tender the kernel, inside the hard shell, called Nungu in Tamil and Thati Munja in Telugu, is like a jelly and delicious as well. The tender fruits are to be harvested between 60-70 days to get soft nungu (Sankaralingam *et al.*, 1991). The British named this ice apple as it resembled ice. The fruit is an excellent coolant in the hot coastal summers, rich in minerals. It is nutritious when eaten in its natural form. This fruit is loaded with minerals and absolutely no fat and protein. The sellers cut open each fruit and remove the young kernel with white skin. The white skin is removed and glossy white kernel is served. The germinated seed's hard shell is also cut open to take out the crunchy kernel which tastes like a sweeter Water Chestnut. It is called 'Pooraan' in Tamil.

Tuber (Apocolon) Based Products

The seeds are planted and made to germinate and the fleshy sprouts (below the surface) are boiled and eaten. It is very fibrous and nutritious, known as "Panai Kizhangu" or "Panamkizhangu" in Tamil and "Tagalu" in Telugu. It tastes like any other starchy tubers with a little bitterness. It is believed to have medicinal properties to relieve biliousness, dysentery and gonorrhoea. The root of the tuber is a diuretic, anthelmintic and a decoction is given for certain respiratory illnesses (Morton, 1988). The tuber is a rich source of starch. The optimum stage for harvesting tuber is 135 days after sowing and weight of the tuber ranges 90 -110 g (Sankaralingam *et al.*, 1991).

The tuber is sun dried and milled into flour. The main component of the flour is starch. Palmyrah starch grain is ellipsoidal in shape with concentric layers and the average size of a starch grain is about 43 microns. The large grain size makes it an ideal raw material for manufacture of sago. The yield of pure starch from flour is approximately 60 per cent and this starch is devoid of bitterness and has a low viscosity and low gelation temperature, but good setting properties (Janz *et al.*, 1992). Fresh tubers have a moisture content of about 54.5 per cent, organic matter 43.89 per cent and mineral content of 1.61 per cent (Mohanajayaluxmi, 1986). The oleic acid (18: 1) and linoleic acid (18: 2) form 32.1 per cent and 29.7 per cent of lipids respectively with linolenic acid and arachidonic acids being only 3.6 per cent. Palmitic acid was the main saturated fatty acid (37.7 per cent) (Balasubramaniam *et al.*, 1999).

Arseculeratne *et al.* (1982) studied by feeding rats with palmyrah flour and reported that the flour contained neurotoxic substances. Many of the researcher disagree these findings. This study shows that feeds containing 100 per cent and 70 per cent palmyrah flour result in very little or no feed consumption, and deaths reported could have been interpreted as being due to starvation. Sumudunie *et al.* (2004) conducted another study in that the rats were fed a mixture of 50 per cent palmyrah flour and 50 per cent standard breeding feed resulted in the neurotoxic symptoms like muscle in-coordination, spasms and immobility of hind limbs reported previously. These neurotoxic symptoms can be eliminated by heating palmyrah flour at 80°C for 45 min; that is, detoxification. Attempts were made to reproduce the neurotoxic effect by administering two-fold palmyrah flour extractive compared with that contained in the same flour consumed by experimental rats per day.

Keerthi *et al.* (2009) reviewed the neurotoxicity of palmyrah and made a conclusion that as neurotoxicity has not been reported in humans consuming palmyrah flour, it is hypothesized that this may be due to one or more of the following: a species effect; the mode of processing flour and cooking palmyrah flour recipes containing these water-soluble and dry-heat decomposable saponin primary amines; frequency of consumption of palmyrah flour-based products; and the nutritive value of other dietary components. It is hypothesized that the Hepatotoxic syndrome as reported previously is due to a collective effect of a number of biologically active compounds, most of which are water-soluble saponins, like neurotoxins.

Neera (Padaneer)

The chief product of palmyrah palm is the sweet sap obtained by tapping tip of inflorescence, as is done with other sugar palms and coconut palm. *Neera* can be obtained from young inflorescence of either male or female trees. The palmyrah toddy is sweet when quite fresh, but bitter when fermentation sets in. The toddy season is from January to August. Tapping begins when the spadix emerges out from the hard spathe. Slicing of the spathe is carried out both in morning and evening. The oozing of sap (neera) continues for 30 to 45 days until the entire length of the spike is sliced off. To sweeten the toddy, inside of the pot is lined with little lime before it is tied to the spathe. The lime prevents fermentation of toddy and results in a sweet drink. The sweetened toddy is called Karuppany. This Karuppany can be drunk and is very sweet and delicious and is also used to make jaggery (Panam kaddy). After a period of six months, the palmyrah continues to pour its sap at the rate of four to seven litres a day (Thiagarajan *et al.*, 1991). Female inflorescences are tapped for longer periods of more than six months (Kovoor, 1983). Studies conducted by Gokarn (1930) in India estimated the average yield/ tree ranged from 270-450 litres/season. Martin (1950) obtained 60 kg of brown sugar/tree/season, equivalent to about 400 litres. Many studies indicated that the female palms yielded more than males (Gokarn, 1930; Tjitrosoepoma and Pudjorinta, 1982; Theivendirarajah, 1983). It was observed that 30 per cent defoliation (palms with 14-16 leaves) has yielded maximum sap.

Neera contains sugars, proteins, lipids, calcium magnesium, iron, sodium, potassium, copper, zinc, phosphorus, niacin, thiamine, riboflavin, ascorbic acid and vitamin A. Table 17.1 details the contents in 100 ml of neera collected from female/male before and after fermentation (Barh and Mazumdar, 2008). Naknean *et al.* (2010) characterised the quality of palm sap after harvested in Songkhla province, southern Thailand. All results indicated a large variation quality of palm sap although they harvested in the same production area in Songkhla province. The different quality of palm sap was mainly due to the fermentation of sugars by the activity of microorganisms during palm sap collecting time.

Table 17.1: Ingredients in 100 ml. of neera from female/male before and after fermentation (Barh and Mazumdar, 2008).

Sl.No.	*Parameters*	*Before/After Fermentation*	
		Female	*Male*
1.	pH	7.30/5.90	7.20/6.00
2.	Total sugars(g)	11.60/8.20	13.20/8.80
3.	Total proteins (mg)	15.30/8.20	19.70/8.30
4.	Total lipids(g)	0.04/0.01	0.08/0.03
5.	Calcium (mg)	1.07/0.72	1.12/0.91
6.	Magnesium (mg)	5.12/5.10	5.61/5.0
7.	Iron (mg)	1.03/1.00	1.50/1.15
8.	Sodium (mg)	5.3 7/3.72	5.21/3.97
9.	Potassium (mg)	1.67/1.02	1.91/1.31
10.	Cu (mg)	0.04/0.03	0.06/0.04
11.	Zinc (mg)	0.07/0.06	0.04/0.04
12.	P (mg)	1.71/1.12	1.85/1.20
13.	Niacin(mg)	0.41/0.63	0.48/1.06
14.	Thiamine (mg)	0.04/0.12	0.06/0.32
15.	Riboflavin(mg)	0.01/0.03	0.04/0.12
16.	Ascorbic acid (mg)	3.05/2.65	4.95/3.56
17.	Vitamin A (IU)	62.0/56.0	74.0/68.0
18	Ethanol (per cent) V/V	0.0/2.10	0.0/2.80

Chanthachum and Beuchat, (1997) reported that Kiam wood (*Cotylelobium lanceotatum* Carih.) and Payorm wood (*Shorea rofburthii* G Don) are commonly added to the collection receptacle because it can delay spoilage of palm sap by reducing microbial populations as well as keeping the quality of a product by delaying fermentation.

Sap Based Products

Palm Jaggery

The neera is boiled to 110 °C and it gets transformed into thick viscous fluid. The viscous mass is stirred continuously to avoid charring at the bottom of vessel. It is then poured into empty coconut shells and allowed to cool for complete setting of jaggery. The hemispherical shaped jaggery is then removed from coconut shells and stored in room temperature (Thiagarajan *et al.*, 1991). The jaggery contains moisture 8.61 per cent, sucrose 76.86 per cent, reducing sugars 1.66 per cent fats 0.19 per cent, proteins 1. 04 per cent, minerals 3.15 per cent, calcium 0.86 per cent and phospherous 0.052 per cent (Neelakantan and Banumathi, 1991). The palmyrah palm jaggery is much more nutritious than crude cane sugar. This jaggery can be stored at room temperature and used over a long period. One kilogram of jaggery can be obtained from eight litres of neera (Sankaralingam, *et al.*, 1999). The palm jaggery dissolved in water and filtered finely when fed to typhoid patients is known to add strength to the weak patients and in course of time cures the illness. In small children constipation is prevented by giving water dissolved in palm jaggery. The weak babies gain weight when fed with palm jaggery water (Louis *et al.*, 1991). Palm jaggery is quite popular in the Southern States of India. Chemical analysis on jaggery from palmyrah has shown the presence of vitamins like riboflavin, 402 mg/1008, vitamin B12, 15 mg/100, vitamin C, thiamine and nicotinic acid.

A spicy jaggery called 'puttu panagkatty' is made in certain parts of Tamil Nadu. Here spices like cardamom, ginger, green gram etc. are added to concentrated syrup just before pouring into mould. Spicy jaggery is made as smaller semi spherical pieces packed in small boxes (250 g) made of palmyrah leaf. The 'Kuttan panagkatty' is sold at a higher price than ordinary jaggery.

In Thailand, palm sugar concentrate is a natural product made from sap of either coconut tree (*Cocos nucifera* L.) or Palmyra palm tree (*Borassus flabellifer* Linn.). The product from Palmyra palm sap is more popular and well-known. In Southern Thailand, the palm is widely grown in Petchaburi and Songkhla provinces. Now-a-days, Palmyra palms are the most populated in Songkhla province, approximately 1,262,771 million plants. There are approximately 300 palm sugar concentrate producers in Songkla province (Naknean *et al.*, 2010).

Palm Honey

If neera is boiled the colour gradually changes from light yellow to honey colour. The syrup is more tasty nutritious and medicinal. It is used as a medicine for cough and cold and also for urinary infection due to excessive body heat. It is also used as a medicine in combination with ginger extract to quire soar throat. The neera syrup contains vitamins B and C, in the form of carotene, thiamine, riboflavin, nicotinic acid and ascorbic acid (Raj and Williams, 1991). The palm honey could be used as a dessert. It also serves as a side dish with bread, idlies and dosas. Palm honey can also be used to dress ice creams and salads (Sankaralingam *et al.*, 1999). Balasubramaniam (1985) reported that palmyrah neera can be concentrated to a golden syrup or converted to high fructose syrup by hydrolysis of sucrose to fructose and glucose. The resulting glucose in the mixture of sugars is isomerised to fructose. The high fructose syrup is a better sweetener and can be used more profitably than palm sugar (Theivendirarajah, 2008).

Palm Candy

Palm candy is prepared by boiling the Neera in an alloy vessel and then allowed to cool. The sediments in the liquid are removed by sieving it through a gauda cloth. When neera boils, small quantity of super phosphate is added. The liquid is again heated to 110°C for 1.5 to 2 hours until it reaches a honey like consistency. The fluid is allowed to cool and poured into crystalisers. Sugar crystals start forming after 45- 60 days. Patients suffering from small pox and chicken pox are advised to use dried grapes mixed with palm candy to induce cooling effect to the system. Irritation in the throat and dry cough can be contained by use of palm candy mixed with pepper powder. Pregnant woman suffering from urinary trouble are normally advised to take warm water with palm candy dissolved in it. Palm candy dissolved in cold water is the best curative agent for sour eye or reddish colour in the eye (Louis *et al.*, 1991).

Toddy

If the sap is allowed to ferment through the action of air borne microorganisms, an alcohol content of 5 to 6 per cent may be reached. With pure and selected yeast strains the alcohol content can be further increased. Fermented sap is known as toddy, and is a popular alcoholic beverage in Asian countries, while in African countries it is the only use to which sap is reserved. This is also distilled to produce the alcoholic liquor called palm wine, arrack, or arak. Vinegar can also be made out of fermented toddy. The female palms are supposed to yield more toddy on tapping from the inflorescence (Thiagarajan *et al.*, 1991). In addition to alcohol, the toddy contains sugars, proteins, lipids, minerals, vitamins, etc. (Barh and Mazumdar, 2008). *Borassus flabellifer* of Indonesia and India are more nutritious than cane sugar (Murphy, 1999). Fresh sap of *Borassus flabellifer* is a good source of vitamin-B complex. Balanced administration of fresh and fermented sap can be done to reduce anaemia and to improve vitamin-B12 level (Rangaswami, 1977; Morton, 1988).

Palmyrah Fruit Pulp (PFP)

Palmyrah fruit pulp is used for decades by the people in their traditional food items. The bitter principle of the fruit pulp is due to presence of a steroidal saponin named flabellifern II. This compound is a tetra-glycoside containing two glucose and two rhamnose residues (Janz *et al.*, 1994). The five naturally occurring flabelliferns so far isolated are, Flabellifern (FII), Flabellifern (Fb), Flabellifern (Fc), Flabellifern (Fd) and Flabellifern (FI) (Balasubramaniam *et al.*, 1999). Inspite of bitterness of the palmyrah fruit, it has a number of beneficial characteristics. It is proved that the flabelliferins of PFP inhibit glucose uptake and increase cardiac antioxidant activity and decrease blood cholesterol. Flabelliferins decrease the total cholesterol probably after being converted into β-sitosterol (Pathberiya and Chandrika, 2003). Recent studies by Pathberiya and Chandrika showed that PFP is rich in provitamin A carotenoids and lycopene, a known antioxidant, beneficial in cardiovascular ailments and cancer (Theivendirarajah, 2008).

Ethanol

As growing interest emerges in renewable feed stocks for fermentation, the palmyrah based carbon sources are being considered (Mahendran *et al.*, 1993). The palmyrah sap fermented to mixed alcohols like ethanol, fusel oils (higher alcohols), acetaldehyde, vicinal diketones and esters by air borne microorganisms (Balakumar and Arasaratnam, 2009). De Boer, 2010 reported that planting Sugar palms (*Borassus flabellifer*) as a means to offset the CO_2 load of Apenheul Primate Park, Apeldoorn, Netherlands. The sugar is produced from CO_2 and H_2O and can be used for human consumption or for production of bioethanol. It is predicted that by 2015, Apenheul will be able to offset a substantial proportion of its carbon load from 50 ha of Sugar palms in which it invested.

Timber

The wood of a fairly old palmyrah is hard, strong and durable and is used for constructing the frame work of roofs, house-hold posts and pillars and furniture. (Morton, 1981). In some parts of Indonesia, it is used for making emergency bridges, umbrella holders and walking sticks. In the island of Bali the trunk is being used for gutters and the base of trunk is also used as a mortar in the making of salt. The vacant stem of the tree is the best for making water pipes for traditional irrigation system (Ramachandran *et al.*, 2004). The female tree gives better and harder timber than male tree and is also more expensive (Kalarani and Annathurai, 1991). Tjitrosoespoma and Pudjoarinta (1983) reported that the wood of male palm is heavier than the female palm.

Palmyrah Fibre

The basal sheath of leaf called 'kanku mattai' in Tamil, is a good source of long and stiff fibres. These fibres are actually vascular bundles or groups of sclerenchymatous fibre cells. The size and colour of fibres vary. Matured leaf sheaths have more of dark brown or black strong fibres, while the younger leaf sheaths have relatively light coloured fibres. Fibre is usually obtained from young palms of 5-10 years old (Theivendirarajah, 2008). Palmyrah fibre is hard, stiff and is resistant to both alkali and acids, resistant to water and attack by termites and other insects. It is for these reasons that palmyrah fibre is preferred to other vegetable fibres (Jeganathan and Mohanadas, 1983). They are used primarily for making sanitary, industrial, domestic, scrubbing brushes and road sweeping brushes. In Sri Lanka, the sheaths are soaked in water and beaten to remove the fibre. But in India, leaf sheaths (one day old) are beaten with wooden mallets, when fresh, to separate the fibre from rest of tissue.

The mesocarp of Palmyra nuts is an effective alternative raw material of cellulosic pulps for paper making. They provide pulp and paper sheets with acceptable properties and thus allow an agricultural residue from a major economic activity to be profitably exploited (Sridach, 2010).

Other Uses of the Palm

The effect of ethanolic extract of male flowers (inflorescences) of *Borassus flabellifer* for its anti-inflammatory activity in rodents was reported by Paschapur *et al.*,(2009). The methanolic extracts of leaves and roots of *Borassus flabellifer* show antioxidant activity. This is due to presence of phyto chemicals like flavonoids, saponins, tannins and phenolic compounds. It can be used as an anti oxidant and might have potential as "nutraceuticals" for preparation of functional foods (Kommu *et al.*, 2011). Kumar *et al.* (2012) studied the potential of *Borassus flabellifer* mucilage (BFM) to act as a biodegradable carrier for colon specific drug delivery. The study was made to develop matrix tablet of olsalazine sodium based formulation using BFM which protects the drug in upper GIT and release the major amount of drug in colon due to degradation by bacterial enzymes. The antimicrobial activity of Tender Seed Coat Extract of Borassus Flabellifer was reported by Duddukuri *et al.* (2011).

Bakku *et al.* (2012) reported a protocol for the extraction, purification and optimization of industrially important amylase enzyme from palmyrah fruit pericarp. Kumar *et al.* (2012) studied the possibility of using the mucilage from the endosperm of *Borassus flabellifer* as a suspending agent that can be used as an effective alternative for the formulation of pharmaceutical suspensions. The results suggested that the mucilage was found a superior suspending agent than tragacanth indicating that it may be a good source as pharmaceutical adjuvant.

Role of Palmyrah Development Boards

Palmyrah Development Board was established in different states of India and other palmyrah growing counties like Sri Lanka, for the development and economic well being of palmyrah workers, protection, restoration and development of palmyrah plantations and rehabilitation of the palmyrah based Industries and enhance the production and marketing of products obtained from Palmyrah sap, leaf, fruit, fibre, tuber, timber, etc.

Kerala State Palmyrah Products Development and Worker's Welfare Corporation Limited [KELPALM] was formed in 1985 with an overall objective of developing the palmyrah based industry, its cultivation and thereby improving the economic well being of Palmyrah workers. Kelpalm has set up two facility centres-Kottamam (Neyyattinkara) in Thiruvananthapuram District and Kalleppully (Alathur) in Palakkad District with its Registered and Corporate office at Nandavanam in Thiruvananthapuram. In 1991, the Corporation started manufacturing soft drink from it's Kottamam Facility Centre. Its products are named as palm Lahar, Palm Syrup and Palm Squash. Kottamum centre had also initiated training to Palmyrah workers for manufacturing palm leaf and Palm fiber products. The Kalleppully centre was set up for processing palm fibre made out of fronds, which are available in plenty in Palakkad District (http: / /www. kelpalm. com/).

The Tamil Nadu Palm Products Development Board was established in 1995. The board is promoting research work for development of palm products to achieve social objectives of alleviation of rural poverty by creating marketing forums for palm products like palm jaggery, palm sugar, palm candy, palm fruit jam etc., with the help of cooperative societies. Though the Tamil Nadu Palm Products Development Board is not directly involved in trading activities, it promotes production and trade through Primary Palm Jaggery Cooperative Societies, District Palmgur Marketing Cooperative Federations and Tamil Nadu State Palmgur and Fibre Marketing Co-operative Federation. Palm Fibre is being exported to USA, UK, Belgium, Philippines, Japan, German, Australia and Netherlands by the Tamil Nadu State Palmgur and Fibre Marketing Cooperative Federation and exporters from Thuthukudi in Tamil Nadu. Women artisans are engaged in the production of fine Palm Leaf articles in Manapad in Thuthukudi, Pulicat in Tiruvallur, Adukkamparai in Vellore and Manamadurai in Sivagangai Districts. These products are exported to USA, UK, France, Germany, Switzerland, Canada, Japan, and Netherlands. (http: / /www. tn. gov. in/hhtk/palm/palm-welfare. htm).

The Palmyrah Development Board (PDB) is the institution established on 1978 for restoration and development of palmyrah plantations in Sri Lanka. The board started manufacturing many palmyrah based products. It includes Panam Paanam (Palmta or Soft Drink) main ingradient-fruit pulp, Palmyrah Cordia-fruit pulp, Palmyrah Crush, Palmyrah Fruit Pulp (Concentrated Pulp). This is made by heating the fresh pulp and bringing it to boil, Palmyrah Panattu-the product of fruit pulp in Sri Lanka is the 'panattu' which is really slabs or sheets of dried pulp made on mats about 8 cm x 15 cm in size. During preparation of 'panattu' fresh palmyrah pulp is taken out of fruits through a process of manual squeezing and fibres in the pulp are separated by straining pulp through cloth. The strained pulp is spread daily on mats made out of palmyrah leaf, layer after layer allowing for sun drying. Palmyrah Palmy Toothpaste is a value added product. One bottle of palmyrah jam is prepared using approximately 175 ml of palmyrah pulp (http: / /www. katpahachcholai. com/).

Palmyrah palm is the national tree of Cambodia. The government of Tamil Nadu honored the palm during 1978 by declaring it as a '**state tree**'. The Government of Kerala declared 2007 as the year of palmyrah. The potentials of palmyrah palms remain underexploited and there is urgent need to create awareness to grow, protect the existing palms and to effectively utilise the palm products.

References

Anonymous, 1948. The Wealth of India Vol 1. Council Sci. and Industr. *Res.* New Delhi, *Borassus*, p. 203-207.

Arseculeratne S. N., Gunatilleke A. A. L. and Panabokke R. G. 1982. Studies on the toxicity of palmyrah palm (*Borassus flabellifer* L.) Part II. Milk transfer of toxicity to suckling rats. Journal of the National Science Council of Sri Lanka 10: 277-282

Balasubramaniam, K. 1985. An overview of palmyrah. Seminar and Exhibition on the development of palmyrah. University of Jaffna, Sri Lanka.

Balasubramaniam, K., Janz, E. R. and Ariyasena, D. D. 1999. Palmyrah, Janz, E. R., 29/1, Pietersz Place, Nugegoda, Sri Lanka.

Barh, D. and Mazumdar, B. C. 2008. Comparative nutritive values of palm saps before and after their partial fermentation and effective use of wild date (*Phoenix sylvestris* Roxb.) sap in treatment of anemia. Research Journal of Medicine and Medical Sciences 3(2): 173-176.

Chanthachum, S. and Beuchat, L. R. 1997. Inhibitory effect of Kiam (*Cotylelobium lanceotatum* Craih.) wood extract on gram-positive food-borne pathogens and spoilage micro-organisms. Food Microbiology 14: 603-608.

DE Boer, L. E. M. 2010. Planting Sugar palms *Borassus flabellifer* as a means to offset the CO_2 load of Apenheul Primate Park, Apeldoorn, Netherlands. International Zoo Yearbook, 44, (1): 246–250.

Gokaran, M. R. 1920. Palmyrah palm and its uses in gur manufacture. Poona Agricultural College Magazine 21, 241-252.

Govinda Rao Duddukuri, Nagendra Sastry Yarla and Kaladhar, D. S. V. G. K. 2011 Preliminary Studies on *in vitro* Antimicrobial Activity and Phytochemical Analysis of Tender Seed Coat Aqueous Crude Extract of *Borassus flabellifer* Linn. Asian Journal of Biochemical and Pharmaceutical Research. 1(3) : 517-523.

Janz, E. R., Gooneratne, J., Senanayake, S. P. J. N. and Theivendirarajah, K. 1992. Some studies directed at the integrated utilization of constituents of palmyrah fruit pulp. Proceedings of Sri Lankan Association for the Advancement of Science (SLAAS), 48 (1) : 119.

Jeyaraman, S. 1991. Some production technology of palmyrah palm. (In) Workshop/Seminar on Modernizing the Palmyrah Industry. JBS Haldane Research Centre, Carmelnagar, Nagercoil, Tamil Nadu, p 62– 65.

Kalarani, M. K. and Annathurai, G. 1991. Industrial Uses of Palmyrah. Workshop/Seminar on Modernizing the Palmyrah Industry, JBS Haldane Research Centre, Carmelnagar, Nagercoil, p. 125–127.

Keerthi A. A. P; Ekanayake S. and Premakumara G. A. S.2009. Journal of the national science foundation of Sri Lanka 37 (4): 269-271.

Kovoor, A. 1983. The Palmyrah Palm: potential and perspectives. FAO Plant Production and Protection paper, FAO, Rome.

Kumar, R., Rajarajeshwari, N and Narayana Swamy, V. B 2012. Isolation and evaluation of *Borassus flabellifer* Mucilage as a natural suspending agent. International Journal of PharmTech Research.4(4): 1614-1630.

Lubeigt, G. 1977. Le palmier à sucre, *Borassus flabellifer* L., ses différents produits et la technologie associée en Birmanie. *J. Agric. Trad and Bot. Appl.* 24: 311-340.

Mahendran, S., Balasubramaniam, K. and Sivaganeshan, K., 1993. Mechanical extraction of palmyrah fruit pulp. *Bio. Proc. Eng.* **8**: 301-302.

Morton, J. F. 1988. The Palmyra or toddy palm (*Borassus flabellifer* L.). *Economic Botany* 42(3): 420-441.

Morton, J. F., 1988. Notes on Distribution, Propagation, and Products of Borassus Palms (Arecaceae). Economic Botany, 42(3): 420-441.

Murphy, S. T., and B. R. Briscoe, 1999. The red palm weevil as an alien invasive: biology and the prospects for biological control as a component of IPM. Biocontrol News and Information, 20(1): 35N-46N.

Naknean, P., Meenune, M. and Roudaut, G. 2010. Characterization of palm sap harvested in Songkhla province, Southern Thailand. International Food Research Journal 17: 977-986.

Paschapur, M. S., Patil, M. B., Kumar, R. and Patil, S. R. 2009. Evaluation of anti-inflammatory activity of ethanolic extract of *Borassus flabellifer* L. male flowers inflorescences in experimental animals. Journal of Medicinal Plants Research 3(2): 49-54.

Pathberiya, L. G. and Chandrika, U. G. 2003. Variation of Bioactivity of fruit pulp of different cultivars of palmyrah-research plan. National seminar on Palmyrah research and development, Palmyrah Development Board, Colombo, Sri Lanka.

Ramachandran, V. S., Swarupanandan, K and Renuka. C 2004. Traditional Irrigation System Using Palmyra Palm (*Borassus flabellifer*) in Kerala, India. Palms 48(4): 175–181

Rangaswami, G., 1977. Palm tree crops in India. Outlook-on-Agriculture (UK). 9(4): 167-173.

Ranjith Kumar Bakku, Ramana Terli, Subba Rao Somalanka and Mary Anupama Palukurty.2012. Optimization of *Borassus flabellifer* Amylase Extraction Procedure Using Box-Behnken Design and Development of Simple Affinity Chromatographic Technique for Purification of Amylases. British Biotechnology Journal 2(3): 146-156, 2012

Sandrasegarampillai Balakumar and Vasanthy Arasaratnam 2009. Comparison of Industrial Scale Ethanol Production from a Palmyrah-Based Carbon Source by Commercial Yeast and a Mixed Culture from Palmyrah Toddy. Journal of the institute of brewing. 115(2): 105-110.

Sankaralingam, A., Hemalatha, G. and Mohamed Ali. 1999. A Treatise on Palmyrah (H. Hameed Khan, ed.), All-India Corordinated Research Project on Palms., Agricultural College and Research Institute, T. N. A. U., Killikulam, 628 252, Tamil Nadu, India, p 1-40.

Sridach, W. 2010. Pulping and paper properties of Palmyra palm fruit fibers. Songklanakarin J. Sci. Technol. 32 (2): 201-205.

Sumudunie KA, Jansz ER, Jayasekera S and Wickramasinghe SM. 2004. The neurotoxic effect of palmyrah (*Borassus flabellifer*) flour re-visited.; International Journal of food sciences and nutrition. 55(8) : 607-614

Theivendirarajah, K. 1983. *Palmyrah palm products*. FAO/DANIDA, Palmyrah workshop, Jaffna, Sri Lanka.

Theivendirarajah, K. 2008. Palmyrah palm (*Borassus flabellifer* L.). A Monograph, p 1-100.

Theivendirarajah, K. and Chrystopher, R. K. 1987. Microflora and microbial activity in palmyrah (*Borassus flabellifer* L.) palm wine in Sri Lanka. MIRCEN Journal, 3: 23-31.

Thiagarajan, M., Thiagarajan, V., Sundarsingh, S. D. and Verabadran, V. 1991. Tapping of neera, the equipments used and jaggery preparation. Workshop/Seminar on modernizing the palmyrah industry, p. 67-72.

Tjitrosoepomo, G. and Pudjoarinto, A. 1982. Studies on palmyrah (*Borassus flabellifer* L.) in Indonesia. FAO, Rome, p. 1-70.

2014, Future Crops Vol. 2 Pages 321–327
Editor: Professor K.V. Peter
Published by: DAYA PUBLISHING HOUSE, NEW DELHI

Chapter 18

Salacia oblonga Wall. ex Wight and Arn.

T. Vasantha Kumar, Anuradha Sane, V.K. Rao and P.E. Rajasekharan
Indian Institute of Horticultural Research,
Bangalore – 560 089
E-mail: thuppil@iihr.ernet.in

About eighteen species of *Salacia* occur in India (Anonymous,2003) and a few species from this genus (*S. oblonga, S. reticulata*, and *S. prinoides*) are used for thousands of years in traditional medicines particularly for treatment of diabetes and obesity; as also for gonorrhoea, rheumatism, pruritus and asthma. *Salacia oblonga* Wall. belongs to family Hippocrateaceae, and is a large woody climber commonly known as "ponkoranti" due to its golden colored root bark. *S. oblonga* is used in Ayurvedic medicine for oral treatment of diabetes (Flammang *et al.*, 2007).

Common Names

English: Salacia, Malayalam: Ponkoranti, *Sanskrit*: Ekanayakam, Saptharangi, *Tamil*: Chundan, Ponkoranti, Kadal azhinjil, *Sinhalese*: Chundan.

Parts Used

Root and stem.

Threat Status

The Red List Status of the plant is assessed as **"Vulnerable"** ($A1_c$) globally (Ravikumar and Ved, 2000).

Origin and Distribution

It is distributed in tropical rainy forests of Sri Lanka and Southern India. Global: Southern India and Sri Lanka. National: Western Ghats in Maharashtra, Goa, Karnataka, Kerala and Tamil Nadu, rarely seen in Eastern Ghats of Andhra Pradesh. Regional: In Karnataka, occasional in Kodagu district. In Kerala, fairly common in Thrissur, Idukki and Kollam districts. In Tamil Nadu, reported only from Tirunelveli, Coimbatore and Nilgiri hills.

Habitat

Salacia oblonga is a climbing shrub found in moist deciduous to evergreen forests at an altitude of 700–1500 m amsl.

Binomial Nomenclature

Salacia oblonga Wall. ex Wight and Arn. (Syn: *Salacia oppsitifolia* Bottl).

Botanical Identification

Salacia oblonga belongs to family Hippocrateaceae. It is a large woody climbing shrub; young branches terete, densely sprinkled with lenticels, glabrous. Leaves elliptic-oblong or oblong lanceolate, 4.8-11.2cm x 2.2-5.2 cm., glabrous sub coriaceous, turning yellow when dry, apex obtuse or shortly acuminate, margin distantly serrate, veins prominent beneath; petioles 5-7 mm long, glabrous. Flowers axillary, 1-3 in short pedunculate heads, 4-5 mm across, dull yellow; peduncles 2mm long. Calyx lobes 5, round, entire, glabrous. Petals 5, broadly elliptic, 2-2.5 mm long, obtuse, slightly toothed along the margin, erect. Disk flat, with slight elevation in the middle. Stamens 3; filaments much dilated at base; anthers 2 celled, dehiscing longitudinally. Ovary- elongate-conical, 3 locular, with 4 ovules in each cell; style short. Fruits baccate, globose or sub-globose or somewhat pyriform, 2-4.8 cm in diam. supported on persistent calyx, tuberculate, bright orange red when ripe. Seeds 1-8, angular, 2.7 x 1.5 cm., large, immersed in pulp (Gamble 1915., Kirthikar and Basu 2004). The attractive pear shaped orange red fleshy fruits are striking characters (Ravikumar and Ved, 2000). Phenology: Flowering and fruiting from December to May.

Agrotechnology

Well-drained soils are the most suited to *Salacia oblonga.* The crop can be planted in open conditions or in partial shade.

Seed Propagation

Propagation is by seeds, stem cuttings and by root suckers. Seeds are sown in seedbed, take around 50-60 days for seeds to germinate.

Harvest and Yield

The crop is harvested for its roots and stem in the fourth year after planting in the main field. The whole plant is uprooted. Stem and roots are separated with small branches and leaves and dried well and stored in a cool and dry place.

Pharmacognosy Details

Physicochemical and Organoleptic Parameters (Anonymous, 2011)

Physicochemical Constants		*Organoleptic Characters*	
Parameters	*Values (per cent)*	*Parameters*	*Values*
TA	5.6	**Taste**	Mild
AIA	0.75	**Colour**	Light brown
ASE	7.2	**Odour**	Mild
WSE	4.2	**Texture**	Fibrous

TA: Total Ash; AIA: Acid Insoluble Ash; ASE: Alcohol Soluble Extractive; WSE: Water Soluble Extractive.

TLC Finger Printing Profile

TLC finger printing profile of *Salacia oblonga* root extract showed one faint band after spraying with H_2SO_4 and anisaldehyde at Rf values of 0.34 and 0.28 respectively. No bands were observed when visualized under UV radiation of both short and long wavelengths (Anonymous, 2011).

Anatomical Studies

Salient Features of T.S. of Root of *Salacia oblonga*

Thick periderm outermost layer with lenticels. Cambial cells divides tangentially to produce

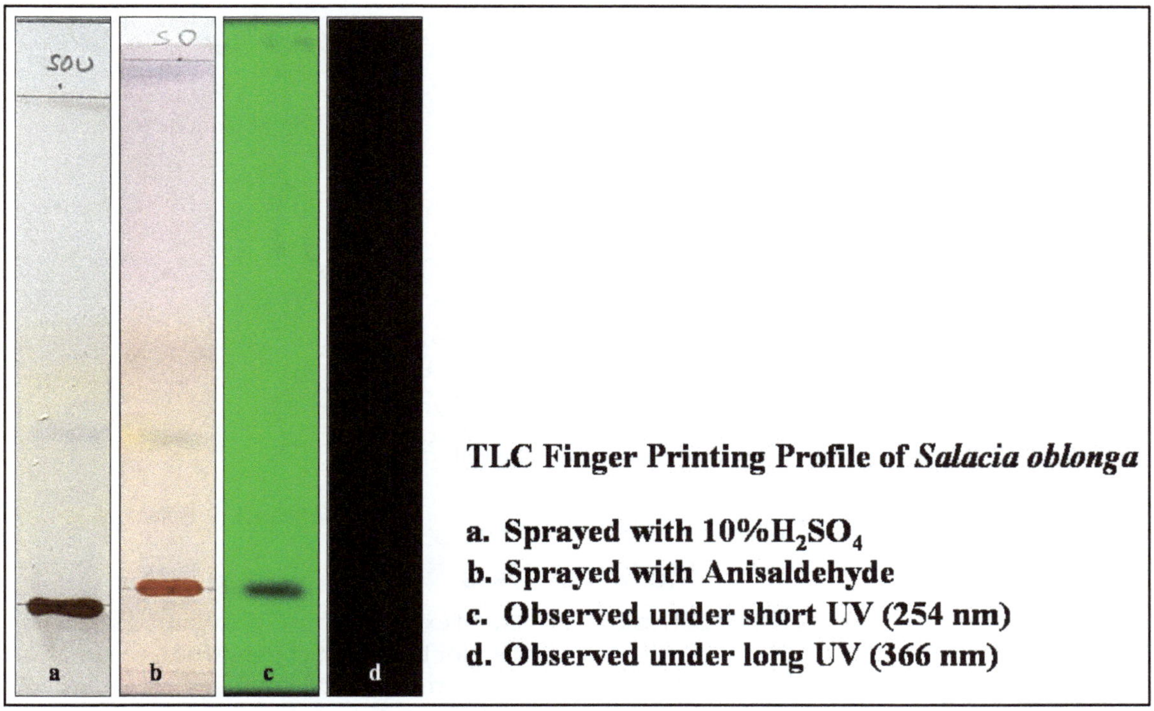

Figure 18.1: TLC finger printing profile of *Salacia oblonga.*

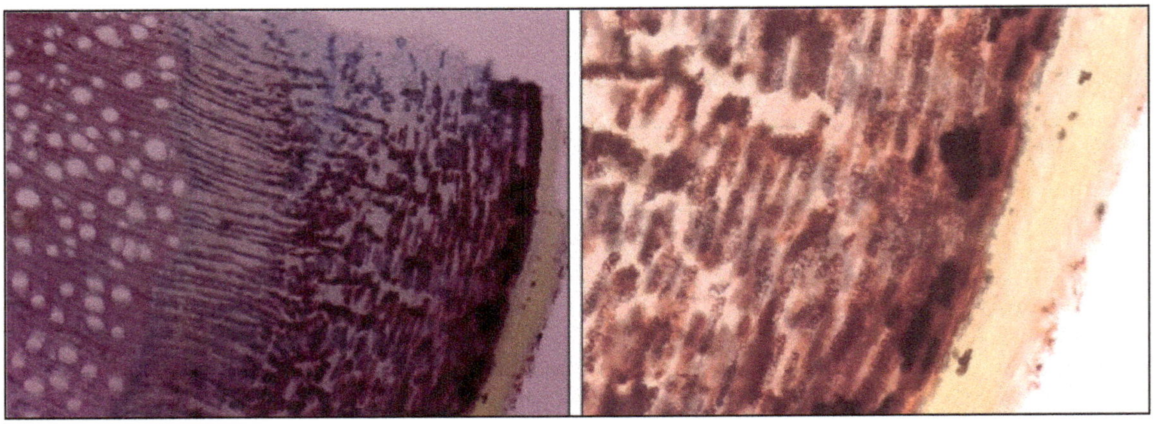

secondary tissues; single layered pericycle becomes meristematic, lying against protoxylem cells and abundant numbers of starch grains are present. First formed cambium produces secondary xylem much earlier and circular tissues. Secondary xylem is present in inner side and secondary phloem towards outer side Well-developed vascular bundles, xylem vessels and xylem trachieds are present. Xylem vessels are many in number. Less number of medullary rays are present in between xylem cells.

Microbial Limit Test

Results of microbial assay of *Salacia oblonga* showed 7200 colonies for aerobic bacteria and 2000 colonies for yeast and moulds which are far below international limits (WHO guidelines 1998 and Indian Herbal pharmacopoeia 2002). Total aerobic bacterial count (TABC): ≤X10^7 and Total yeast and mould count (TYMC): ≤x10^5 (Anonymous, 2011).

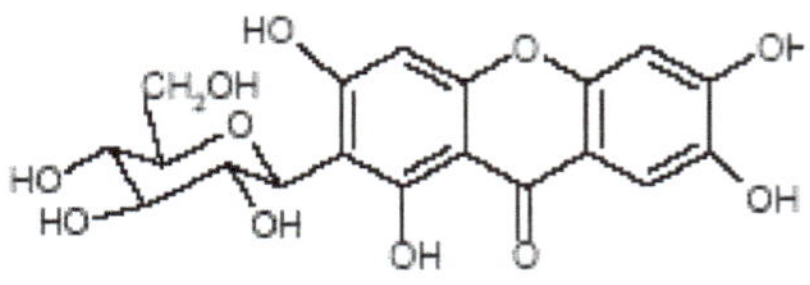

Mangiferin

Phytochemistry

Phytochemical studies conducted on *Salacia oblonga* lead to the isolation of mangiferin, three catechins ((-)-epicatechin, (-)-epigallocatechin, and (-)-4'-Omethylepigallocatechin), two catechin dimers, epi-kokoondio, salacenonal, salaciquinone, and two novel quinonemethide triterpenoids (celastroloids), isoiguesterinol and 30-hydroxypristimerin, from root bark and iguesterin, pristimerin and epi-kokoondiol from stem bark (Li *et al.*, 2008). Mangiferin (a xanthone from roots) and sulfonium ion derivatives, kotalanol (from roots and stems) and salacinol (from roots and stems), were identified as the anti-diabetic principles of *S. oblonga* through pharmacological studies.

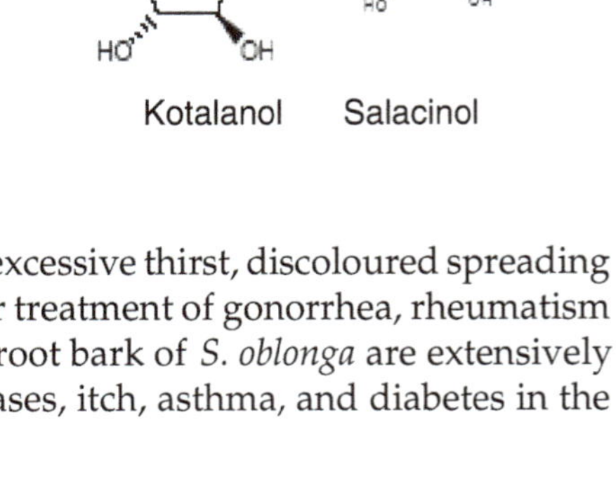

Kotalanol Salacinol

Medicinal Uses

The stem bark is useful in treating polyuria, diabetes and excessive thirst, discoloured spreading patches on the skin, diarrhoea and fever. Root bark is used for treatment of gonorrhea, rheumatism and skin diseases (Ravikumar and Ved 2000). The roots and root bark of *S. oblonga* are extensively used as a remedy in rheumatism, gonorrhoea, and skin diseases, itch, asthma, and diabetes in the

Ayurvedic system of traditional Indian medicine. The root bark powder of this natural drug showed anti-inflammatory effects (Matsuda *et al.*, 1999, Kirthikar and Basu 2004, Nadkarni 2007).

Pharmacological Properties

Hyperlipidaemia - *Salacia oblonga* root is an Ayurvedic medicine with anti-diabetic and anti-obeise properties. The findings from both *in vivo* and *in vitro* suggest that *Salacia oblonga* extract functions as a PPAR-alpha activator, providing potential mechanism for improvement of postprandial hyperlipidemia and hepatic steatosis in diabetes and obesity (Huang *et al.*, 2006).

Cardiovascular Diseases

Salacia oblonga is traditionally used in prevention and treatment of diabetes (Yuhao li *et al.*, 2004). *Salacia oblonga* demonstrated a strong inhibition of α-glucosidase activity *in vitro*, which is suggested to contribute to the improvement of postprandial hyperglycemia. Huang *et al.* (2008) investigated the effect of water extract of *Salacia oblonga* (SOE), an ayurvedic antidiabetic and antiobesity medicine and provided insights into a potential cardioprotective role of traditional herb, which supports further clinical evaluation in obesity and diabetes-associated cardiac hypertrophy.

Diabetes

The extract of *Salacia oblonga*, reduces postprandial glycemia following a standard liquid meal containing mainly maltodextrin. Attenuation of postprandial glycemia is hypothesized to reduce the risk of progression from impaired glucose tolerance to diabetes. It also reduces the number of complications associated with diabetes. Yoshikawa *et al.* (2005) found that the extracts from roots and stems of *Salacia reticulata*, *S. oblonga*, and *S. chinensis* showed substantial inhibitory effects on increase in serum glucose levels in oral sucrose- and maltose-loaded rats and inhibitory effects on intestinal α-glucosidase. In addition, clinical studies and safety evaluations of the extracts from *S. oblonga* are summarized. Fourteen subjects (8 males, 6 females) ate two test meals after an overnight fast following a standardized dinner. In the treatment meal, the tea also contained 480 mg of *Salacia oblonga* extract. In summary, *S. oblonga* extract effectively lowered the postprandial glucose response to a high starch meal, consistent with previous studies. The extract may have value in the management of blood glucose for persons with diabetes. The development of mild to moderate flatulence is a limitation for the use of this herbal extract (Washam Mathew, 2006). The extract of *Salacia oblonga* lowers acute glycemia and insulinemia in persons with type 2 diabetes after a high-carbohydrate meal. The results suggest that *Salacia* may be beneficial to this population for post-prandial glucose control (Williams *et al.*, 2007). In conclusion, *S. oblonga* extract reduces the glycemic response to a high-starch meal. The development of mild flatulence supports alpha-glucosidase inhibition as the main mechanism of action (Steve Hertzler *et al.*, 2007). The other clinical benefits from *Salacia* also show hepatoprotective activity and anti-inflammatory activity

Formulations

'Ilogen-Excel', an Ayurvedic herbal formulation is composed of eight medicinal plants (*Curcuma longa, Strychnos potatorum, Salacia oblonga, Tinospora cordifolia, Vetivelia zizanioides, Coscinium fenestratum, Andrographis paniculata* and *Mimosa pudica*). 'Ilogen-Excel' administration decreased levels of glycosylated haemoglobin, plasma thiobarbituric acid reactive substances, hydroperoxides, ceruloplasmin and vitamin E in diabetic rats. Plasma reduced glutathione and vitamin C were significantly elevated by oral administration of 'Ilogen-Excel'. Administration of insulin normalized all the biochemical parameters studied in diabetic rats. The effect at a dose of 100 mg/kg was more pronounced than 50 mg/kg and brought back all the parameters to near normal levels (Umamaheshwari and Prince, 2007).

References

Anonymous, 2003. The Wealth of Asia. A Compact Disc – Alphabet S. Council of Scientific and Industrial Research CSIR, New Delhi.

Anonymous, 2003. The Wealth of India., Raw Materials, Vol 9, (Rh-So), Council of Scientific and Industrial Research, New Delhi, p. 169.

Flammang, A. M, Erexson GL, Mirwald JM and Henwood S. M., 2007. Toxicological and cytogenetic assessment of a *Salacia oblonga* extract in a rat subchronic study., *Food Chem Toxicol*. 45 (10): 1954-1962.

Gamble, J. S., 1915. Flora of the Presidency of Madras., Vol 1, Bishen Singh Mahendrapal Singh, Dehra Dun, India, p. 214-215.

Huang, T. H, He L, Qin Q, Yang Q, Peng G, Harada M, Qi Y, Yamahara J, Roufogalis B. D and Li Y., 2008. *Salacia oblonga* root decreases cardiac hypertrophy in Zucker diabetic fatty rats: inhibition of cardiac expression of angiotensin II type 1 receptor., Diabetes Obes Metab. 10 (7): 574-585.

Huang, T. H, Peng G, Li GQ, Yamahara J, Roufogalis B. D and Li Y., 2006. *Salacia oblonga* root improves postprandial hyperlipidemia and hepatic steatosis in Zucker diabetic fatty rats: activation of PPAR-alpha., *Toxicol Appl Pharmacol*.210 (3): 225-235.

Huang, T. H, Yang Q, Harada M, Uberai J, Radford J, Li GQ, Yamahara J, Roufogalis B. D and Li Y., 2006. *Salacia oblonga* root improves cardiac lipid metabolism in Zucker diabetic fatty rats: modulation of cardiac PPAR-alpha-mediated transcription of fatty acid metabolic genes., *Toxicol Appl Pharmacol*. 210 (1-2): 78-85.

Kirthikar, K. R. and Basu B. D. 2006. Indian Medicinal Plants, Vol 1. (eds: Blatter E., Caius JF., Mhaskar KS.,) Bishen Singh, Mahendra Pal Singh, Dehra Dun, India, p. 581-583.

Li Y, Huang TH and Yamahara J. 2008. Salacia root, a unique Ayurvedic medicine, meets multiple targets in diabetes and obesity., *Life Sci*. 82 (21-22): 1045-1049.

Matsuda H., Morikawa T., and Yoshikawa M., 2002. Antidiabetogenic constituents from several natural medicines., *Pure Appl. Chem*., 74(7): 1301–1308.

Matsuda H., Murakami T., Yashiro K., Yamahara J., and Yoshikawa M., 1999. Antidiabetic Principles of Natural Medicines. IV.1) Aldose Reductase and a-Glucosidase Inhibitors from the Roots of *Salacia oblonga* WALL. (Celastraceae): Structure of a New Friedelane-Type Triterpene, Kotalagenin 16-Acetate., *Chem. Pharm. Bull*. 47(12) : 1725-1729.

Nadkarni AK. and KM. Nadkarni's, 2007. 'Indian Materia Medica,' Vol. 1, Popular Prakashan Private Limited, Mumbai, p. 1069.

Ravikumar K. and Ved D. K., 2001. 100 Red Listed Medicinal Plants of Conservation Concern In Southern India., *Foundation for Revitalisation of Local Health Traditions (FRLHT)*, Bangalore. p. 327-330.

Steve Hertzler, Matt Washam and Jennifer Williams., 2007. *Salacia oblonga* extract reduces postprandial glycaemia following a solid, high-starch meal., *The FASEB Journal*. 21: 832-839.

Williams JA, Choe YS, Noss MJ, Baumgartner CJ, Mustad VA., 2007. Extract of *Salacia oblonga* lowers acute glycaemia in patients with type 2 diabetes., *Am J Clin Nutr*. 86 (1): 124-130.

Yoshikawa M, Pongpiriyadacha Y, Kishi A, Kageura T, Wang T, Morikawa T and Matsuda H., 2003. [Biological activities of *Salacia chinensis* originating in Thailand: the quality evaluation guided by alpha-glucosidase inhibitory activity] [Article in Japanese]., Yakugaku Zasshi. 123 (10) : 871-880.

Yoshikawa M, Shimoda H, Nishida N, Takada M and Matsuda H., 2002. *Salacia reticulata* and its polyphenolic constituents with lipase inhibitory and lipolytic activities have mild antiobesity effects in rats., *J Nutr*. 132 (7) : 1819-1824.

Yoshikawa M, Zhang Y, Wang T, Nakamura S and Matsuda H. 2008. New triterpene constituents, foliasalacins A (1)-A (4), B (1)-B (3), and C, from the leaves of *Salacia chinensis*., *Chem Pharm Bull*. 56 (7): 915-920.

Yoshikawa M., Kishi A., Morikawa T., Yutana P and Matsuda H.2001 Structures of New Terpenoid Constituents with Aldose Reductase Inhibitory Effect from the Stems of *Salacia chinensis* Originated in Thailand., Koryo, Terupen oyobi Seiyu Kagaku ni kansuru Toronkai Koen Yoshishu., 45: 276-278.

Yoshikawa M., Matsuda H. and Morikawa T. 2002. Exploration of ingredients for obesity and diabetes prevention from plant genus Salacia., Monbu Kagakusho Gakujutsu Furonthia Suishin Jigyo Soyaku Kagaku Furonthia Kenkyu Purojekuto Kenkyu Seika Hokokusho Heisei 13 Nendo, p.124-133

Yoshikawa M., Ninomiya K., Shimoda H., Nishida N. and Matsuda H. 2002. Hepatoprotective and Antioxidative Properties of *Salacia reticulata*: Preventive Effects of Phenolic Constituents on CCl_4-Induced Liver Injury in Mice., Biological and Pharmaceutical Bulletin 25 (1): 72-76.

Yoshikawa M., Xu F., Nakamura S., Wang T., Matsuda H., Tanabe G. and Muraoka O., 2008. Salaprinol and ponkoranol with thiosugar sulfonium sulfate structure from *Salacia prinoides* and α-glucosidase inhibitory activity of ponkoranol and kotalanol desulfate., Heterocycles, 75 (6): 1397-1405.

Yuhao Li, Gang Peng, Qian Li, Suping Wen, Tom Hsun-Wei Huang, Basil D. Roufogalis and Johji Yamahara. 2004. *Salacia oblonga* improves cardiac fibrosis and inhibits postprandial hyperglycemia in obese zucker rats., *Life Sciences* 75, (14): 1735-1746.

2014, Future Crops Vol. 2 Pages 329–335
Editor: Professor K.V. Peter
Published by: DAYA PUBLISHING HOUSE, NEW DELHI

Chapter 19

Saraca asoca (Rose.) De. Willd.: An Epic Tree of Ancient India

T. Vasantha Kumar, Anuradha Sane, V.K. Rao and P.E. Rajasekharan
Indian Institute of Horticultural Research, Bangalore –560 089
Email: thuppil@iihr.ernet.in

Saraca asoca is one of the most sacred trees of the Hindus and Buddhists. Frequently cultivated as an ornamental and avenue tree throughout tropical India, it has been suggested as a shade tree for coffee. Dried stem bark of *Saraca asoca* is collected in spring from mature, wild or cultivated trees, found in Central and Eastern Himalayas, Western Ghats and Deccan plateau.

Origin and Distribution

Tree spreads continental South-East Asia from India to South-West China through Malaysia, Sumatra and eastwards to Celebes. In India, moist zones of Western and Eastern Ghats (Peninsular India), sub-Himalayan tracts (Uttar Pradesh to Arunachal Pradesh), Khasi, Garo and Lushai hills and Eastern India (Bihar, East Bengal, Orissa and North Eastern states), Andhra Pradesh, Kerala, Karnataka and Tamil Nadu up to an altitude of 750 m. sustain Saraca. It is also distributed in Sri Lanka, Myanmar and Andaman Islands. In Karnataka, it is occasional in the wet forests of Chikmagalur, Hassan, Uttara and Dakshina Kannada, Udupi and Shimoga districts. In Kerala, it is sporadic in evergreen forests of Wynaad, Cannanore, Palakkad, Thrissur and Thiruvananthapuram districts. No wild population is recorded in Tamil Nadu.

Botanical Identification

A small evergreen tree, 6-9 m. high, found wild along streams or in shade of evergreen forests. Leaves paripinnate, 15-20 cm. long, leaflets 6-12, oblong or oblong-lanceolate, 7.5-22.5 cm. x 1.25 cm., rigidly sub-coriaceous; flowers orange or orange-yellow, eventually turning vermillion, very fragrant, dense axillary corymbs; pods flat, leathery, 10-25 cm. x 3.5-5 cm., seeds 4-8, ellipsoid-oblong, compressed. Flowering occurs from February to June and fruiting from August to September.

Habitat

Moist deciduous to evergreen forests especially along shady slopes and riversides.

Binomial Nomenclature

Saraca asoca (Rose.) De. Willd

Family

Caesalpiniaceae

Synonyms

Saraca indica Linn., *Jonesia asoca* (Roxb.), *Saraca indica* sensu Baker, auct. non. L.

Common Names

Assamese: Ashoka, Bengali: Ashoka, Asok, Asoka, English: Asok Tree, Gujarati: Ashoka, Asupala, Ashopalava., Hindi: Ashoka, Anganapriya, Asok, Asoka, Kannada: Ashokadamara, Ashokamara, Kankalimara, Seethe mara, Kashmiri: Ashok, Konkani: Assoc, Malayalam: Asokam, Hemapushpam, Vanjulam, Manipuri: Asoka, Marathi: Ashok, Ashoka, Jasundi, Oriya: Ashoka, Punjabi: Asok, Sanskrit: Kankeli, Kankali, Asokah, Vichitrah, Gandhapushpa, Gatasokah, Anganapriya, Apashoka, Chakraguchha, Chitra, Dohali, Doshahari, Gandhapushpa, Hemapushpa, Kantacharandohada, Kantanghridohada, Karnapura, Karnapuraka, Kelika, Krimikaraka, Madhupushpa, Nata, Palladru, Pindapushpa, Pindipushpa, Prapallava, Raktapallava, Rama, Rogitaru, Shhaya, Shokaharta, Shokanasha, Smaradhivasa, Strinirikshanadohada, Subhaga, Tamrapallava, Vamanghrighataka, Vamankayatana, Vanjula, Vanjuldruma, Vichitra, Vishoka, Vitashoka.; Tamil: Asogam, Asogu, Asokam, Asoka maram, Telugu: Ashokapatta, Asokamu.

Plant Parts Used

Stem bark.

Threat Status

Red List status has been assessed *i.e.* 'Endangered' for Karnataka and 'Data Deficient' for Kerala and Tamil Nadu (Ravikumar and Ved, 2000).

Agrotechnology

It can be propagated through seeds. Ashoka grows well in areas with well distributed rainfall and in slightly shady areas. The best season of planting is June-July. Germination takes about 15 to 20 days and varies from 5 to 100 per cent.

In vitro Propagation

Saraca asoca (Roxb.) De Wilde. (Ashoka) is an important medicinal tree species. Woody Plant Medium containing TDZ (0.05 mg/l) was the best for induction of multiple shoots from shoot tip and nodal segment cultures. (Manishankar Babu *et al.*, 2004).

Market Information

Locally, regionally and nationally, pieces of stem bark are sold under the name Ashoka or Ashoka chaal at Rs.15 to 25/kg (Market studies, 1999-2000). Stem bark of *Polyalthia longifolia* and *Shorea robusta* are common adulterants (Ravikumar and Ved, 2000).

Pharmacognosy

Srivastava *et* al. (1988), explained that Ashoka stem bark (*Saraca asoca* De Wilde) is one of the most important ingredients of many Ayurvedic preparations used in the indigenous system of medicine mostly for uterine problems. Due to poor availability of true Ashoka, it is adulterated by other barks including *Polyalthia longifolia* which also bears the same common name. Pharmacognostic study revealed some diagnostic features which facilitate detection of authentic and adulterated materials.

Physicochemical and Organoleptic Parameters

Physicochemical Constants			*Organoleptic Characters*	
Parameters	*Values*	*Limit*	*Parameters*	*Properties*
TA	10.5 per cent	NMT 11 per cent	**Taste**	Slightly bitter
AIA	1.05 per cent	NMT 1 per cent	**Color**	Brownish
ASE	10.2 per cent	NLT 15 per cent	**Odour**	Mild
WSE	12.8 per cent	NLT 11 per cent	**Texture**	Fibrous

TA: Total Ash; AIA: Acid Insoluble Ash; ASE: Alcohol Soluble Extractive; WSE: Water Soluble Extractive.

Limit as per Ayurvedic Pharmacopoeia of India.

TLC Finger Printing Profile

After spraying with derivatization solution Anisaldehyde three bands were observed in *Saraca asoca* bark at Rf Values at 0.14, 0.35 and 0.48 while no bands were observed with sulphuric acid or UV radiation (Anonymous, 2011).

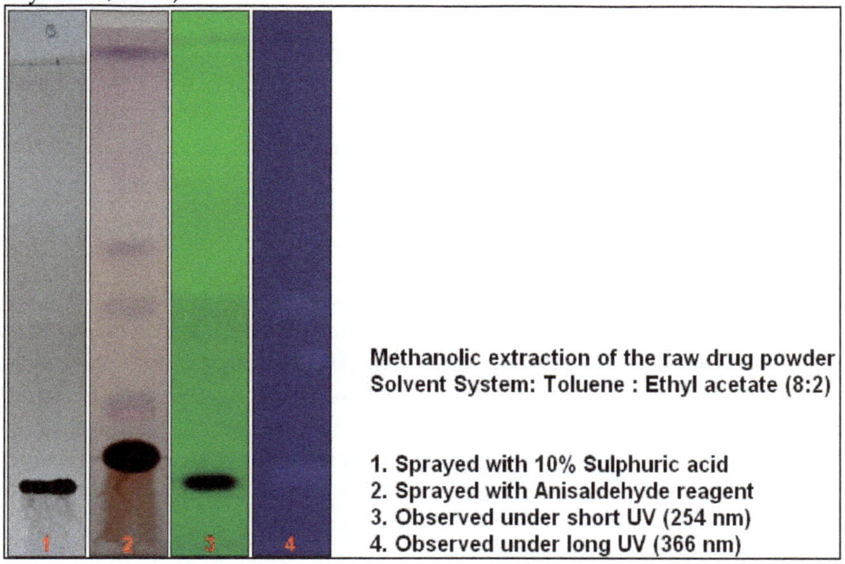

Figure 19.1: TLC finger printing profile of *Saraca asoca* (Bark).

Stem Bark Identification

True bark of *Saraca asoca* is identified by observation of macroscopic and microscopic characteristics of stem bark.

a) Macroscopic Features

Bark channelled, externally dark green to greenish grey, smooth with circular lenticels and transversely ridged, sometimes cracked, internally reddish-brown with fine longitudinal strands and fibers, fracture splintery exposing striated surface, a thin whitish continuous layer is seen beneath the cork layer, taste, astringent.

b) Microscopic Features

Transverse section of stem bark shows periderm consisting of a wide layer of cork, radically flattened narrow cork cambium, secondary cortex wide with one or two continuous layers of stone cells with many patches of sclereids, parenchymatous tissue contains yellow masses and prismatic crystals: secondary phloem consists of phloem parenchyma, sieve tubes with companion cells and phloem fibres occurring in groups, crystal fibres present.

Anatomical Studies

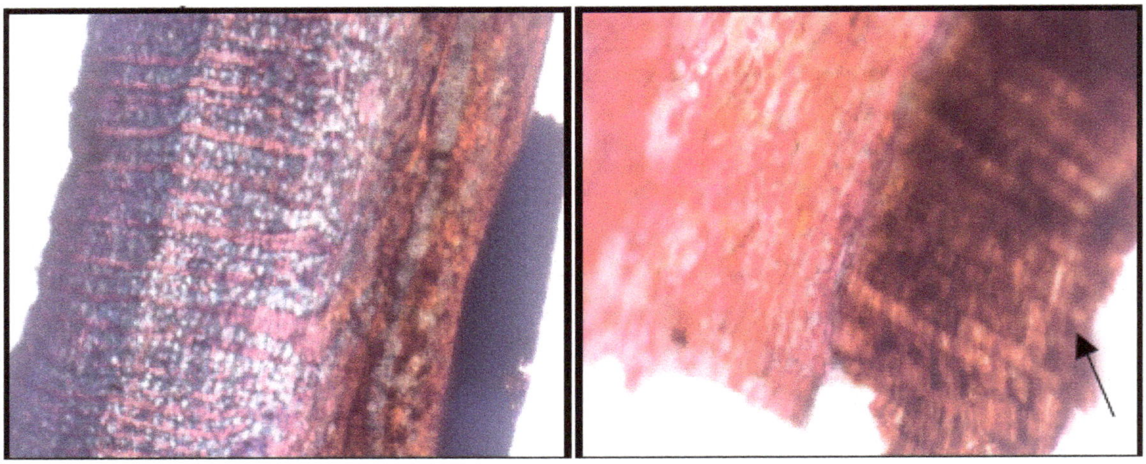

Figure 19.2: T.S. of *Saraca asoca* (Stem bark).

Salient Features: *Saraca asoca* Stem Bark

Outer most layer periderm formed by cork cambium, several layer of periderm, which consists of meristematic cells, parenchymatous in nature with starch content. Lenticels are present, they are round in shape, several layers of cortex region consist of primary phloem and secondary phloem layers. Large cambial region consists of sclerenchymatous thick walled cells. Tannin like brown substance is present and starch grains are present in cortical region. Many calcium oxalate crystals like substance are present. Medullary rays are present in vascular region (Anonymous, 2011).

Microbial Limit Test

Microbial load in *Saraca asoca* (Bark) showed 3000 colonies (CFU/g) for aerobic bacteria and 1800 colonies (CFU/g) for yeast and moulds per gram bark sample which is far below the limits of international guidelines (Total aerobic bacterial count (TABC): ≤X10[7 and] total yeast and mould count (TYMC): ≤X10[5]), (Anonymous, 2011).

Phytochemistry

The bark contains tannin (6 per cent), catechol, an essential oil, haematoxylin, a ketosterol (m. p., 25°), a crystalline glycosidal constituent, a saponin ($C_{10}H_{21}O_{14}$), an organic calcium compound ($C_6H_{10}O_5Ca$), and organic iron compounds. The ketosterol seems to be androgenic in nature. Flowers gave b-sitosterol, flavonoids and flavone glycosides-quercetin, kaempferol-3-O-b-D- glucoside, quercetin-3-O-b-D-glucoside. The anthocyanins present are pelargonidin-3, 5-diglucoside and cyanadin-3, 5-diglucoside. Bark yields catechol and sterols-(24z)-24-methyl cholest-5-en-3b-ol, (22E, 24z)-24-ethylcholesta-5, 22-dien-3 b-ol and (24z)-24-ethyl cholest-5-en-3b-ol, a wax containing n-alkanes, esters and free primary alcohols.

Medicinal Uses

A decoction of the bark is used in treating uterine infections and gynaecological problems. The drug may prove useful in all cases of uterine bleeding where ergot is indicated; the crystalline glycosidal substance is also reported to stimulate uterine contraction. The drug has a stimulating effect on the endometrium and ovarian tissue, and is useful in menorrhagia due to uterine fibroids, in leucorrhoea and in internal bleeding, haemorrhoids and haemorrhagic dysentery. Bark cures biliousness, dyspepsia, dysentery, colic, piles, ulcers and pimples. The (-)-epicatechin, from the bark exhibits antitumor, anticarcinogenic and antiproliferative activities. Leaves possess blood purifying properties and their juice mixed with cumin seeds is used for stomach-ache. Seeds taken with water alleviate suppression of urine and calculus. The flowers are pounded, mixed in water and are used in the treatment of dysentery and the dried ones are used for diabetes. Flowers are an excellent uterine tonic. The alcoholic extract of flowers indicated 50 per cent cytotoxicity in Dalton's lymphoma ascites and Sarcoma-180 tumour cells at the concentration of 38 µg and 54 µg, respectively, with no activity against normal lymphocytes but preferential activity for lymphocytes derived from leukemia patients.

Traditional Uses

The most common traditional formulation used is Ashokarishta (Asokarista); a rishta is a fermented preparation in liquid form (the fermentation yields a small amount of alcohol, about 2 per cent). The popular Ashokarishta products are sold as "female tonics" or "uterine tonics." Traditional system of Indian Medicine shows it to be effective against Uterine fibroids, Dysmenorrhoea, Hemorrhoids and Leucorrhoea. These are taken in doses of 15-30 ml (about 1/2 to 1 fluid ounce), two to three times per day.

Therapeutic Uses

Sotha, Daha, Asrgdara, Apaci, Raktadosa

Dosage

20-30 g of the drug is used for decoction.

Ayurvedic Properties

Rasa: Tikta, Kasaya, Guna: Laghu, Ruksa, Virya: Sita, Vipaka : Katu, Karma : Hrdya, Visaghna, Grahi, Vamya, Sothahara.

Important Formulations

Asokarista, Asokaghrita.

Formulations were made for Menstrual disorders. Suma Venugopal (1998), Mitra *et al.*, (1998). and Mitra *et al.* (1999) evaluated the estrogenic effect of U-3107, it possesses oestrogenic activity only in the presence of functional ovary and is devoid of any progestational activity. The study was conducted and designed to assess the efficacy of a poly-herbal formulation, Menosan-. It was concluded from the clinical trial that Menosan is significantly effective in treating post-menopausal syndrome (Singh and Kala Suhas Kulkarni, 2002). Other poly-herbal formulations for similar disorder include Evecare, Menotab (M-3119) and Speman.

Pharmacological Properties

The bark is bitter and acrid; refrigerant, astringent to the bowels, alexiteric, anthelmintic, demulcent, emollient. It is useful in fractures of the bones; beautifies the complexion.

Antimicrobial Activity

Shahid *et al.* (2007), analyzed *in vitro* cultivation of the explants and used to obtain antibacterial compounds. The extracts of *Saraca asoca* showed antimicrobial activity (Dabur *et al.* (2007) in a range of 75-1200 µg/ml. Studies are carried out on anticancer activity(Ghosh *et al.*, 1999, Varghese *et al.*, 1992) of *Saraca indica* and Ashokarishta for menstrual disorders Middelkoop *et al.* (1983). Antioxidant activity (Samir *et al.*, 2007), oxytocic activity -Teunis *et al.* (1986), larvicidal activity (Mathew *et al.*, 2008) and tolerance to air pollution (Tinku *et al.*, 2003).

References

Anonymous. 2011. Report on Network project on RET Medicinal Plants Conservation and Utilization in South and North East India, NMPB, New Delhi.

Ghosh S, Majumder M, Majumder S, Ganguly NK and Chatterjee BP. 1999. Saracin: A lectin from *Saraca indica* seed integument induces apoptosis in human T-lymphocytes. Arch *Biochem Biophys*. 15; 371(**2**): 163-168.

Kirtikar and Basu. 2006. Indian Medicinal Plants. Volume II (Eds). Blatter E., Caius J. F., and Mhaskar K. S., Bishen Singh Mahendra Pal Singh, Dehradun, 2nd edition. p. 883-884.

Mani Shankar Babu, Narasimhan S, and Nair GM., 2004. *In vitro* culture and biosynthesis of phenolic compounds in *Saraca asoca* (Roxb.) De Wilde., *Int. J. Bioscience Reporter* 2 (**2**): 236-241.

Mathew N, Anitha MG, Bala TS, Sivakumar SM, Narmadha R, and Kalyanasundaram M. 2008. Larvicidal activity of *Saraca indica, Nyctanthes arbor-tristis*, and *Clitoria ternatea* extracts against three mosquito vector species. *Parasitol Res*. Nov 28.

Middelkoop TB and Labadie RP. 1983. Evaluation of Asoka Aristha, an indigenous medicine in Sri Lanka. *J Ethnopharmacol*. 8(**3**): 313-320.

Rajesh Dabur, Amita Gupta, Mandal T K, Desh Deepak Singh, Vivek Bajpai, Gurav A M and Lavekar G S. 2007. Antimicrobial activity of some Indian Medicinal Plants. *African Journal of Traditional, Complimentary and Alternative Medicines*.4(3): 313-318.

Ravikumar K. and Ved D. K. 2000. 100 Red-Listed Medicinal Plants of Conservation Concern in Southern India. Foundation for Revitalisation of Local Health Traditions (FRLHT), Bangalore. December, 1st edition, p. 334-336.

Samir Kumar Sadhu, Amina Khatun, Panadda Phattanawasin, Takashi Ohtsuki and Masami Ishibashi. 2007. Lignan glycosides and flavonoids from *Saraca asoca* with antioxidant activity. *Journal of Natural Medicines*, 61 (**4**): 480-482.

Seetharam V. N, Sujeeth H, Jyothishwaran G, Barad A, Sharanabasappa G and Parveen S. 2003. Antibacterial activity of *Saraca asoca* bark. 65 (**6**): 658-659.

Shahid M, Shahzad A, Malik A, and Anis M. 2007. Antibacterial activity of aerial parts as well as *in vitro* raised calli of the medicinal plant *Saraca asoca* (Roxb.) de Wilde. *Can J Microbiol.* 53(**1**) : 75-81.

Srivastava G. N, Bagchi G. D and Srivastava A. K. 1988. Pharmacognosy of Ashoka Stem Bark and its Adulterants. *Pharmaceutical Biology*, 26 (**2**) : 65-72.

Singh, S. K., and Kala Suhas Kulkarni. 2002. Evaluation of the efficacy and safety of Menosan in Post-menopausal symptoms: A short-term pilot study., *Obs. and Gynae. Today* VII(12): 727-730.

Subhuti Dharmananda. 2004. Excessive uterine bleeding treated with Saraca, an Ayurvedic herb. *Institute for Traditional Medicine*, Portland, Oregon.

Suma Venugopal, 1998. Effect of Eve Care in Oligomenorrhoea., *The Antiseptic* 95 (10): 329-330.

Teunis B. Middelkoop and Rudi P. Labadie. 1986. An *in vitro* Examination of Bark Extracts from *Saraca asoca* (Roxb.) de Wilde and *Rhododendron Arboreum* Sm. for Oxytocic Activity. *Pharmaceutical Biology*. 24, (**1**) : 41-44.

Tinku Ghosh and Ambarish Mukherjee. 2003. Evaluation of some plant species in biomonitoring air pollution. *Environment and Ecology*. 21 (**4**): 747-751.

Varghese CD, Satish, Nair C and Panikkar KR. 1992. Potential anticancer activity of *Saraca asoca* extracts towards transplantable tumours in mice. 54 (**1**) : 37-40.

2014, Future Crops Vol. 2 Pages 337–341
Editor: Professor K.V. Peter
Published by: DAYA PUBLISHING HOUSE, NEW DELHI

Chapter 20

Water Chestnut (*Trapa bispinosa* Roxb.)

S.K. Sharma, Hare Krishna, S.K. Maheshwari and Nitesh Chauhan
Central Institute for Arid Horticulture, Beechwal, Bikaner – 334 006, Rajasthan
E-mail: ciah@nic.in

Water chestnut (*Trapa bispinosa* Roxb.) is commonly known as "*Singhara*" or "*Paniphal*" all over India. The water chestnut is actually not a nut at all, but an aquatic vegetable that grows in marshes. The name "water chestnut" comes from the fact that it resembles a chestnut in shape and coloring. The generic name *Trapa* is derived from the Latin word for "thistle," calcitrappa, is also another common name for the water caltrop. It is an annual aquatic herb of great economic importance and grown for its starch-rich fruits. It is usually grown during rainy season in ponds and lakes, and the edible part is the mature hard kernel.

Origin and Distribution

It is indigenous to Southeast Asia and is in cultivation in China since ancient times. In India it is cultivated mainly in Uttar Pradesh, Bihar, Tamil Nadu, West Bengal, Assam, Odisha and Jammu and Kashmir. Besides India, it is cultivated in Sri Lanka, Africa, South East Asia and U. S. A.; however, the varieties are different.

Composition and Uses

Water chestnut contains approximately 70 per cent moisture, 23.9 per cent carbohydrate, 4.7 per cent protein, 0.2 per cent fat, 1.1 per cent minerals and 20 IU vitamin A. It also has fair quantity of vitamin C (9.8 mg/100 g) with TSS 7° Brix and acidity 0.09 per cent. It is an important food for poor people. Half mature kernel is eaten as such or after frying while, the mature hard Kernel is eaten after boiling. After cooking they retain their crisp texture even after leftovers are re-heated. Processed dry nuts (kernels) seem to be easily digestible and therefore are used to break fasting. Besides, its flour is used for making Abir for holi festival and for preparing bread, Halwa, biscuits and other confectionery items. Certain tribes, use its flour for the treatment of diarrhoea and abdominal pain.

Botanical Description

Water chestnut belongs to family Trapaceae. Two types of leaves *i.e.* finely divided feather-like submerged leaves (borne along the length of stem) and undivided floating leaves (borne in a rosette at the water's surface) are found in water chest nut. The floating leaves have saw-tooth edges and are ovoid or triangular in shape, 2–3 cm long, on inflated petioles 5–9 cm long, which provide added buoyancy for leafy portion.

The inflorescence is solitary, which arises in the axil of leaf. Pedicels are about 15 cm long with woody structure. The flowers are hermaphrodite and are perigynous with a disc above the stamens. Calyx is four in number and gamosepalous. Corolla is also four in number but polypetalous with broad roundish apex. Gynoecium is inferior, bicarpellary, syncarpous, two celled and ovule is solitary in each cell. The floral structures are arranged in such a way that it ensures self-pollination, which results in maintenance of genetic purity of different types in nature. The fruit is a nut with barbed spines. The plant spreads by the rosettes and fruits detaching from the stem and floating to another area on currents or by fruits clinging to objects, and animals.

Climate and Soil

Water chestnut requires warm and humid environment for its optimum growth. Contrarily, a few varieties are reported growing in the temperate zones of the world. The varieties grown in the temperate climate are susceptible to the climatic conditions of tropical and subtropical zones where this crop is cultivated at wider scale. It is largely grown in shallow water of perennial ponds. Water chestnut thrives profusely in pond having heavy soil at the bottom. The yield is noted to be reduced drastically in ponds underlined with sandy soil at the bottom.

In order to augment a successful crop, depth of standing water in the pond should never be allowed to reach below 50 cm and above 2 to 2.5 metres. Further, the pond should be free from water weeds and basin should be covered with a layer of at least 15 cm thick mud. This creates conducive environment for normal development of both root and shoot of water chestnut. It is suggested that a site or pond which could hold water for at least four months after transplanting should, preferably, be selected for water chestnut cultivation.

Varieties

So far no standard varieties are identified for commercial cultivation. A few genotypes are available. They are known by different names in different regions. Four distinct varieties of *Singhara* (or *Trapa*) namely Kanpuri, Patna, Desi Big and Desi Small are reported from Bihar. Similarly, four important varieties namely Green Spinous, Green Spineless, Rosy Spinous and Rosy Spineless are reported from Meerut region of Uttar Pradesh. Of these, the first two are very high yielders. Further, from Lucknow, eleven local cultivars were identified and divided into three groups according to their maturity like (i) Early - Bari Gajvee, Singree and Barkatia; (ii) Mid - Bara Chikanwa, Kanpuria Lal, Murwa and Surkha and; (iii) Late - Gulaba, Desi Lal, Saharanpuria and Kadarna.

Nursery Raising and Transplanting

Before sowing of seeds, nursery bed should be well puddled. Seeds should be sown in the pond having about 40 cm deep water. Seeds may be broadcasted in the pond in December, which take about one month to germinate. Plants should be allowed to grow in the nursery pond till the onset of rainy season *i.e.* end of June or early July.

The crop should be transplanted in July, when there is about 120 to 150cm water in the pond. About 70-100 kg fully mature nuts would be sufficient to cover one hectare of pond. The plants should be pulled out along with the nuts. Two to three plants should be tied up together and planted in the mud at the bottom of the pond at a distance of two metres both ways. Alternatively, seedlings ready for transplanting, may be cut to longitudinal segments with blade in such a way that each shoot should have a well developed root system. Such pieces of shoots should be packed in moist gunny bags and transported to the site of transplantation. Young shoots are better than old ones owing to development of more lateral shoots.

It is highly desirable that bottom surface of the pond must be even to ensure transplanting of seedlings at a suitable and matching depth of standing water in the pond. For instance, if the length of a particular shoot is 90 cm long, it should never be planted with more than 70 to 80 cm depth of water. While transplanting, roots of the nursery seedlings should be thoroughly pressed in the mud. If water depth is up to 50 cm or more then the transplanting should be done either by plunging water or by wrapping some mud around the roots.

Ten-15 lateral shoots may develop from transplanted seedlings after about 20 to 30 days. Transplanting may be done in May and June in the perennial ponds, natural or artificial, or with onset of the monsoon. Care should be taken to maintain water level of pond, especially, if transplanting are done before monsoon to avoid submergence of plants.

Self Seeding or Natural Seeding Method

Self seedling is not desirable and advantageous for commercial cultivation. After the last picking of the fruits (nuts) from main crop, some of the lately developed fruits and weakly developed flowers which set in under most uncared conditions, settle down in the mud on surface of pond and thereby serve as seed material, but this generally have a poor stand.

Maintenance of Water Level in the Pond

Utmost care should be taken to maintain optimum water level in pond for commercial cultivation of water chestnut. Water depth in the pond should be between 0.5-2.5 metres during different growth stages because under excess water level crop starts rotting whereas, under shallow conditions, plants will not develop properly. The optimum water level for pond should be 150 to 180 cm, 90 to 120 and 80 to 120 cm, respectively for branching stage, flowering stage and for maturity stage respectively.

Planting Distance

Water chest nut is usually planted at a distance of 2 metres; however, three metres of planting distance can be followed if soil is rich. Some researchers reported that nursery seedling of normal health should be transplanted at a spacing of about 1 to 1.5 metres randomly in the pond.

Intercultural Operations

Due to the fact that the crop remains in standing water, hoeing is not feasible; however, weeding of the pond during rainy seasons, in July-August, is highly advantageous. During weeding, effort should be made to spread out the vines at suitable spacing to allow proper aeration among them. Sometimes dense crop of water chestnut requires judicious thinning so as to ensure improved production.

Manures and Fertilizers

In general, water chestnut does not require additional application of manures and fertilizers as the pond water itself is a good source of nutrients which remain in soluble form. An application of nitrogen @ 30-35 kg/ha. in the form of urea before or at flowering is recommended to obtain a good yield. Ponds wherein water chestnut is being grown are usually located near the villages and therefore as such most of the village wastes and other waste matters flow into the ponds with rain water. The organic wastes rapidly decompose in the pond and serve as source of nutrients for the crop. In addition, the water chestnut suckers have the capacity to utilize carbon dioxide available in the pond water.

Flowering and Fruiting

Water chestnut takes about one month for the plant to come into flowering after transplanting. After the onset of rainy season, generally in July, luxurious growth of the crop starts and towards the end of July, flower bud initiation takes place. Once the plants start flowering, they continue to do so, till they become very old and rot away. Opening of flowers and dehiscence of the anthers occur almost simultaneously thereby ensuring self-pollination. Each and every shoot contains about 30 to 35 leaves and from the axil of each leaf single bud emerges which develops within a period of four to six days. Flower opening and fertilization take place outside the water but fruit development takes place under the water surface. A flower, after pollination, may very well develop into a fruit ready for picking in 10-15 days. Fruits are ready for harvest after 21 days of anthesis. The observations point out that there is a variation in duration between the anthesis and fruit maturity which may be due to varietal or climate differences.

Harvesting

Water chestnut is harvested from September –December depending upon the prevailing climatic conditions. Generally, two pickings are recommended. The first picking may be taken at 100 days of planting and the other over a month thereafter. Six to eight pickings at 10-15 days interval from healthy and normal plants may also be taken. Usually, the fruits are harvested by perching on a floating board, an inflated tube or a huge pitcher, etc. as the float moves from one place to another. After final harvesting, the plants are allowed to remain in the pond, which eventually rot and render organic matter for the next crop.

Yield

A good crop yields about 250 to 300 quintals of fresh fruits/ha. depending on variety, which after drying are reduced to about 80 to 100 quintals. After removal of the testa, the dry weight of the fruits further reduces to about 50 to 60 quintals. According to a study, farmers in Odisha earn Rs. 25,344/ha as net profit by cultivating water chestnut.

Plant Protection

The most important insect-pests of water chestnut are water chestnut beetle and blue beetles. They remain active during rainy seasons *i.e.* from August to September. High temperature and high humidity are congenial for multiplication of the insects.

Among diseases, seed rot, root rot and stem rot (caused by *Fusarisum*, *Pythium*, *Phytophthora* and *Rhizoctonia* spp. are important diseases of this crop. The plant residues should be collected and burnt and water should be cleaned regularly. Seed and seed coat are cleaned and treated with Captan @0.25

per cent. In addition, incidence of leaf spots (caused by *Cercospora* and *Phytophthora* spp.) results in development of small circular to large irregular shaped spots on the leaves. Affected leaves turn light yellow and droop and ultimately the plants may die. The pond should be cleaned thoroughly, while adopting proper manuring practices. As soon as the symptoms of the disease appear on the plant, Blitox-50 @ 2.5 per cent should be sprayed. Care should always be taken that no fish should remain in pond while applying pesticides, if needed.

Previous Volume Contents

Volume 1

2011, xxiii+286p., col. plts., figs., tabls., ind., 25 cm Rs. 2200/US$ 100
ISBN 978-81-7035-754-4

Index

B

C

D

E

G

H

M

O

P

Q

R

T

U

V

www.ingramcontent.com/pod-product-compliance
Ingram Content Group UK Ltd.
Pitfield, Milton Keynes, MK11 3LW, UK
UKHW050845101026
15624UKWH00018B/593

9 789351 301